DISCRETE MATHEMATICS

AND

ITS APPLICATIONS

Series Editor

Kenneth H. Rosen, Ph.D.

AT&T Labs

Continued Titles

Network Reliability: Experiments with a Symbolic Algebra Environment,
Daryl D. Harms, Miroslav Kraetzl, Charles J. Colbourn, and John S. Devitt

RSA and Public-Key Cryptography
Richard A. Mollin

Quadratics, *Richard A. Mollin*

Verificaton of Computer Codes in Computational Science and Engineering,
Patrick Knupp and Kambiz Salari

DISCRETE MATHEMATICS AND ITS APPLICATIONS

Series Editor KENNETH H. ROSEN

RSA and PUBLIC-KEY CRYPTOGRAPHY

Richard A. Mollin

CHAPMAN & HALL/CRC

A CRC Press Company

Boca Raton   London   New York   Washington, D.C.

Library of Congress Cataloging-in-Publication Data

Mollin, Richard A., 1947-
    RSA and public-key cryptography / Richard A. Mollin.
        p. cm. — (Discrete mathematics and its applicatoins)
    Includes bibliographical references and index.
    ISBN 1-58488-338-3
    1. Coding theory. 2. Public key cryptography. I. Title. II. Series.

QA268 .M655 2002
652'.8—dc21   2002031096

Visit the CRC Press Web site at www.crcpress.com

© 2003 by CRC Press LLC

No claim to original U.S. Government works
International Standard Book Number 1-58488-338-3
Library of Congress Card Number 2002031096
Printed in the United States of America 1 2 3 4 5 6 7 8 9 0
Printed on acid-free paper

Preface

This book is intended for a second course in cryptography at the undergraduate level, where the student is assumed to have had a course in introductory number theory. Also, the book is intended as a source book for those in the cryptography business, who will find collected together herein numerous facts that are currently scattered throughout the literature on public-key cryptography and related issues described in the Table of Contents. The impetus for the writing of this text arose from the author's involvement in the establishment of an iCORE (Informatics Circle of Research Excellence) Chair in cryptography at the University of Calgary in September of 2001, and the launching of an associated *Centre for Information Security and Cryptography* (CISAC). In addition to the education of numerous graduate students and postdoctoral fellows in cryptology, we have an ongoing commitment to the development of a new stream of cryptography courses for the Mathematics Department. This text will serve as the text for one of them at the senior undergraduate level. No suitable text for that course was on the market, nor is there one at the time of this writing, hence the appearance of this one.

◆ **Features of This Text**

• The book is ideal for the student since it offers a wealth of exercises with 350 problems. The more challenging exercises are marked with a ☆. Also, complete and detailed solutions to all of the *odd-numbered exercises* are provided at the back of the text. Complete and detailed solutions of the *even-numbered exercises* are included in a *Solutions Manual*, which is available from the publisher for the instructor who adopts the text for a course. Moreover, the exercises are presented at the end of each section, rather than at the end of each chapter.

• The text is *accessible* to anyone from the senior undergraduate to the research scientist, and all levels of readers will find challenging and inspirational data. To ensure that the book is as self-contained as possible, we have three appendices of relevant background information. Appendix A has a brief, but highly informative, overview of letter frequency analysis (of the English language) to assist in our cryptanalytic travels. Appendix B has a solid background review and analysis of elementary complexity theory to provide us with the necessary tools for algorithmic analysis and related phenomena. Lastly, Appendix C contains the fundamentals of the number-theoretic results used in the text together with any other relevant information that we will need such as vector space basics, matrix theory fundamentals, and some facts on continued fractions.

• There are *over 100 footnotes* containing *nearly 40 biographies* of the individuals who helped develop cryptologic concepts, together with historical data of interest, as well as other information which the discerning reader may want to explore at leisure. These are woven throughout the text, to give a human face to the cryptology being presented. A knowledge of the lives of these individuals can only deepen our appreciation of the development of PKC and related concepts. The footnote presentation of this material allows the reader to have immediate information at will, or to treat them as digressions, and access them later without significantly interfering with the main discussion.

• There are *optional topics*, denoted by ☞, which add additional material for the more advanced reader or the reader requiring more challenging material which goes beyond the basics presented in the core data.

• There are more than *60 examples, diagrams, figures*, and *tables* throughout the text to illustrate the concepts presented.

• For ease of search, the reader will find consecutive numbering, namely object $N.m$ is the mth object in Chapter N (or Appendix N), exclusive of footnotes and exercises, which are numbered separately and consecutively unto themselves. Thus, for instance, Diagram 3.5 is the 5th numbered object in Chapter Three; exclusive of footnotes and exercises; Exercise 4.37 is the 37th exercise in Chapter Four; and Footnote 9.2 is the second footnote in Chapter Nine.

• The *bibliography* contains nearly 250 references for further reading.

• The *index* has more than 2,000 entries, and has been devised in such a way to ensure that there is maximum ease in getting information from the text.

• The webpage cited below will contain a file for comments, and any typos/errors that are found. Furthermore, comments via the e-mail address below are also welcome. Lastly, if the instructor for a given course, using this text, has any difficulty in getting the solutions manual for the even-numbered exercises, contacting the author will expedite matters.

◆ **Acknowledgments** First of all, thanks go to Hugh Williams for discussions prior to the writing of this book that inspired my creating the original template for a table of contents. As iCORE Chair in cryptography at U. of C., he has brought a new dimension and exciting prospects for the future. The author is grateful for the proofreading done by the following people, each of whom lent his own (largely non-intersecting) expertise and valuable time: John Brillhart (U.S.A.), John Burke (U.S.A.), Jacek Fabrykowski (U.S.A.), Bart Goddard (U.S.A.), Franz Lemmermeyer (U.S.A.), John Robertson (U.S.A.), Kjell Wooding (Canada), graduate student in cryptography at U. of C., Thomas Zaplachinski (Canada), a former student, now cryptographer, and Robert Zuccherato (U.S.A.).

Richard Mollin, Calgary, September 1, 2002

website: http://www.math.ucalgary.ca/~ramollin/

e-mail: ramollin@math.ucalgary.ca

About the Author

Richard Anthony Mollin received his Ph.D. in mathematics (1975) from Queen's University, Kingston, Ontario, Canada, where he was born. He is now a full professor in the Mathematics Department at the University of Calgary, Alberta, Canada. He has over 160 publications in algebra, number theory, computational mathematics, and cryptology to his credit. This book is his seventh, with [160]–[165] being the other six. He resides in Calgary with his wife Bridget and two cats. When not engaged in mathematics or entertaining friends and mathematical visitors from all over the world, he and Bridget enjoy hiking in the Canadian Rockies.

To Bridget — as always

Contents

Chapter 1

History and Basic Cryptographic Concepts

For secrets are edged tools, and must be kept from children and from fools.
John Dryden (1631–1700) English poet, critic, and playwright

1.1 Terminology

In *The Lives of the Twelve Caesars* [226, p. 45], Suetonius writes of Julius Caesar: "... if there was occasion for secrecy, he wrote in cyphers; that is, he used the alphabet in such a manner, that not a single word could be made out. The way to decipher those epistles was to substitute the fourth for the first letter, as *d* for *a*, and so for the other letters respectively."

What is being described in the above is a means of sending messages in disguised form — an example of *enciphering* or *encrypting*, which is the process of disguising messages. The original (undisguised) message is called the *plaintext* or (less often) *cleartext*, and the disguised message is called the *ciphertext*, whereas the final message, encapsulated and sent, is called a *cryptogram*. The reverse process of turning ciphertext into plaintext is called *decryption* or *deciphering*, or (less frequently used) *exploitation*. The science of *cryptography* is the study of sending messages in *secret*, namely, in enciphered form, so that only the *intended* recipient can decipher and read it. The study of mathematical techniques for attempting to defeat cryptographic methods is called *cryptanalysis*. Those engaged in cryptography are called *cryptographers*, whereas those engaged in cryptanalysis are called *cryptanalysts* — the "enemy" or "opponent". Although cryptography is sometimes called an *art* in the contemporary literature, we reserve the term *science* for modern cryptography, which is concerned with the mathematical techniques that are embraced by the study and the computational tools used to implement them. Certainly from a historical perspective

1

(as we will see shortly) cryptography began as an art in the course of human development, but is now a very serious science in our information-based society.

The study of *both* cryptography and cryptanalysis is called *cryptology* and the practitioners of cryptology are called *cryptologists*. The etymology of cryptology is the Greek *kryptos* meaning *hidden* and *logos* meaning *word*. The term *cryptology* was coined by James Howell in 1645. However, John Wilkins (1614–1672) in his book, *Mercury, or the Secret and Swift Messenger*, introduced into the English language the terms *cryptologia* or *secrecy in speech*, and *cryptographia* or *secrecy in writing*. He also used *cryptomeneses* as a general term for any secret communications. Wilkins, who was a cofounder of the Royal Society along with John Wallis (1616–1703), later married Oliver Cromwell's sister, and became Bishop of Chester. The appearance of the use of the word *cryptology* is probably due to the advent of Kahn's encyclopedic book [117], *The Codebreakers*, published in 1967, after which the term became accepted and established as that area of study embracing both cryptography and cryptanalysis.

Cryptography may be viewed as *overt secret writing* in the sense that the writing is clearly seen to be disguised. This is different from *steganography* (from the Greek *steganos* meaning *impenetrable*) which conceals the very *existence* of the message, namely, *covert secret writing*. For instance, *invisible ink* would be called a *technical* steganographic method, whereas using tiny dissimilarities between handwritten symbols is a *linguistic* use of steganography. One of the most famous of the technical steganographic methods, employed by the Germans during World War II, was the use of the microdot (invented by Emanuel Goldberg in the 1920s) as a period in typewritten documents. Today, hiding messages in graphic images would also qualify. Generally speaking, steganography involves the hiding of secret messages in other messages or devices. The modern convention is to break cryptography into two parts: *cryptography proper* or *overt secret writing*, and *steganography* or *covert secret writing*. The term *steganography* first appeared in the work *Steganographia*, by Johannes Trithemius (1462–1516). We will not be concerned in this text with steganography, but rather with cryptography proper. (See [16] for more data on Steganography.)

The first recorded instance of a cryptographic technique was literally written in stone almost four millennia ago by an Egyptian scribe who used hieroglyphic symbol substitution in his writing on a rock wall in the tomb of a nobleman of the time, *Khnumhotep*. The intention was not to disguise the inscription, but rather to add some measure of majesty to his inscription of the nobleman's deeds, which included the erection of several monuments for the reigning Pharaoh *Amenemhet* II. Although the scribe's intent was not secrecy (the primary goal of modern cryptography), his method of *symbol substitution* was one of the elements of cryptography that we recognize today. Today the use of substitutions *without* the element of secrecy is called *protocryptography*. Subsequent scribes actually added the essential element of secrecy to their hieroglyphic substitutions. However, the end-goal here seems to have been to provide a *riddle* or *puzzle*. As these inscriptions became more complex, this practice abated and suffered the same fate as the inhabitants of the tombs. Hence, the seeds of cryptology were planted in ancient Egypt but did not mature rapidly or continuously. Instead cryptography had several incarnations in various cultures, with probably fewer methods extant than the number lost in antiquity.

We will now continue with the description of cryptographic terms and refer the reader to [117] for an encyclopedic history of cryptography, or to this author's previous book [165] for a very brief overview.

At the outset, we had a glimpse of a cryptographic means invented by Julius Caesar, which we can illustrate as follows. For now we may think of a *cipher* as a means of transforming plaintext into ciphertext. This is often visualized as a table consisting of plaintext symbols together with their ciphertext equivalents, where the convention is that plaintext letters are written in lower case, and ciphertext letters are written in upper case.

Table 1.1 The Caesar Cipher

Plain	a	b	c	d	e	f	g	h	i	j	k	l	m
Cipher	D	E	F	G	H	I	J	K	L	M	N	O	P
Plain	n	o	p	q	r	s	t	u	v	w	x	y	z
Cipher	Q	R	S	T	U	V	W	X	Y	Z	A	B	C

We see that each plaintext letter is substituted by the ciphertext letter three places to the right in the alphabet with X, Y, and Z "looping back" to A, B, and C. A mathematically more satisfying way to display this is to give the letters numerical values as follows.

Table 1.2

a	b	c	d	e	f	g	h	i	j	k	l	m
0	1	2	3	4	5	6	7	8	9	10	11	12
n	o	p	q	r	s	t	u	v	w	x	y	z
13	14	15	16	17	18	19	20	21	22	23	24	25

We can transform plaintext into numerical values and visualize the Caesar Cipher in terms of modular arithmetic. If α is the numerical equivalent of a plaintext letter, then $\beta \equiv \alpha + 3 \pmod{26}$ is the ciphertext letter substituted for it. (See Appendix C for a reminder of the arithmetic of congruences.) To decipher, take each ciphertext numerical equivalent β, perform $\alpha \equiv \beta - 3 \pmod{26}$, and convert back to the letter equivalents via Table 1.2 to recover the plaintext. For instance, suppose that the following was encrypted using the Caesar Cipher:

IOHH DW GDZQ.

Then the numerical equivalents of these ciphertext letters are:

$$8, 14, 7, 7, 3, 22, 6, 3, 25, 16.$$

By calculating $\alpha \equiv \beta - 3 \pmod{26}$ on each one, we get the numerical equivalents of the plaintext letters:

$$5, 11, 4, 4, 0, 19, 3, 0, 22, 13.$$

Then translating them back to letters via Table 1.2, we get the plaintext:

flee at dawn.

The reader may now solve the related Exercises 1.1–1.8.

With the Caesar Cipher, a given plaintext letter is enciphered as the same ciphertext letter each time the cipher is used. This is an example of a *monoalphabetic substitution cipher*, which we will formally define in Section 1.2. This predictability makes this type of cipher very vulnerable to attack. We will discuss various attacks in Section 1.3. For now, we may think of a *substitution* as a permutation of the plaintext letters, and a *monoalphabetic cipher* as one in which only *one* cipher alphabet is used, where the term "cipher alphabet" means a list of equivalents used to transform plaintext into secret form, as in Table 1.1. (See Exercises 1.9–1.14.)

The above illustration is a motivator for some definitions of cryptographic terminology. However, before we can do this, we need to provide a basis of rigorous mathematical language from which we can launch a discussion of cryptography in depth. Both the plaintext and the ciphertext for any given cipher are written in terms of elements from a finite set $\mathcal{A}$, called an *alphabet of definition*, which may consist of letters from an alphabet such as English, Greek, Hebrew, or Russian, and may include symbols such as:

$$\text{⤙,✈,♣,✳,♥,✓,♫,⚕,▢,☞,✪,✿,▲,✎,❦,⤳,○,◆, ☙, ⤚, ⛟, ☎, ◐, ◑,✦,}$$

or any other symbols that we choose to use when sending messages. Although the alphabets of definition for the plaintext and the ciphertext may differ, the usual practice is to use the same for both. A commonly used one is $\mathcal{A} = \{0, 1\}$, the *binary* alphabet of definition. There is also the alphabet of definition consisting of $\mathcal{A} = \{0, 1, 2, \ldots, 25\}$ — the alphabet of definition for the Caesar Cipher.

Once we have an alphabet of definition, we choose a *message space* $\mathcal{M}$, which is defined to be a finite set consisting of strings of symbols from the alphabet of definition, and elements of $\mathcal{M}$ are called *plaintext message units*. A finite set $\mathcal{C}$, consisting of strings of symbols from the alphabet of definition for the ciphertext, is called the *ciphertext message space* and elements from $\mathcal{C}$ are called *ciphertext message units*. For instance, for the Caesar Cipher, $\mathcal{M} = \mathcal{C} = \mathbb{Z}/26\mathbb{Z}$, the ring of integers modulo 26. (See Appendix C for general number theoretic facts.)

Next, we need to foil cryptanalysts, so we need a set of parameters $\mathcal{K}$, called the *keyspace*, the elements of which are called *keys*. For example, with the Caesar Cipher, any $m \in \mathcal{M}$ is enciphered as $c \in \mathcal{C}$ where

$$c = m + 3 \in \mathbb{Z}/26\mathbb{Z}.$$

Thus, the *enciphering key* is $k = 3 \in \mathcal{K}$ since we are using the parameter 3 as the shift from $m \in \mathcal{M}$ to $c \in \mathcal{C}$. The *deciphering key* is also the parameter $k = 3$ since we achieve $m \in \mathcal{M}$ from $c \in \mathcal{C}$ by

$$m = c - 3 \in \mathbb{Z}/26\mathbb{Z}.$$

This motivates the following, which will also give rigour to the preceding discussion.

Definition 1.3 (Enciphering and Deciphering Functions)

An enciphering transformation *or* enciphering function *is a one-to-one function*

$$E_e : \mathcal{M} \mapsto \mathcal{C},$$

where the enciphering key $e \in \mathcal{K}$ *uniquely determines* E_e *acting upon the plaintext message units* $m \in \mathcal{M}$ *to get ciphertext units* $E_e(m) = c \in \mathcal{C}$. *A* deciphering transformation *or* deciphering function *is a one-to-one function*

$$D_d : \mathcal{C} \mapsto \mathcal{M},$$

which is uniquely determined by a given decryption key $d \in \mathcal{K}$ *acting upon ciphertext message units* $c \in \mathcal{C}$ *to get plaintext message units* $D_d(c) = m$.

The operation of applying E_e *to* m *is called* enciphering, *or* encrypting, *whereas the operation of applying* D_d *to* c *is called* deciphering, *or* decrypting.

Remark 1.4 *This remark is devoted to a discussion of distinctions between* encoding/decoding *and* enciphering/deciphering. *We may think of "codes" as transformations that replace the words and phrases in the plaintext with alphabetic or numeric* code groups. *(For instance, a code for* Launch *the* attack! *might be* $L - 343$.) *Historically, the first code vocabularies were called* nomenclatures, *which consisted of plaintext corresponding to (order-preserving) literal or numerical codegroups. This required only one codebook, a* one-part code. *The notion of a two-part nomenclature was introduced by A. Rossignal (1600–1682). For the first part, he used what was called a* tables à chiffer, *consisting of plaintext letters in alphabetical order, and ciphertext letters in random order. The second part, called a* tables à déchiffer, *consisting of the plaintext letters jumbled, while the ciphertext letters were in alphabetical order, what we now call a* two-part code *(see Footnote 1.5 on page 14). Later these nomenclatures were expanded into what were called* repertories, *then ultimately* codes. *We use the term* cryptographic codes *in the above context in order to avoid confusion with* error-correcting codes *that detect or correct errors to ensure messages transmitted over a noisy channel are recovered intact, which involves no secrecy.*

The distinction between encoding and enciphering, for example, is that enciphering transformations applied to plaintext units result in ciphertext units, done independent of the meaning of the plaintext. On the other hand, cryptographic encoding means that words, for instance, are replaced by codegroups. The key is the codebook, which lists the plaintext units and their corresponding codegroups. Cryptographic codes are not used for modern communications since they are, by their very nature, open to what is called a known-plaintext attack *— see Section 1.3. This means that the key — the codebook — would have to be replaced often. In the modern world, the time and money this would cost renders this means of securing transmissions unacceptable for most applications.*

Definition 1.3 has given us another building block in our quest for the necessary mathematical precision to carry out our exploration of cryptographic concepts in depth, and has given rigour to the somewhat informal discussion that has gone before. We now can illustrate the Caesar Cipher in light of our new notions.

The key e, which is addition of 3 modulo 26, uniquely determines the enciphering transformation $E_e(m) \equiv c \equiv m + 3 \pmod{26}$, namely, $E_e(m) = c = m + 3 \in \mathcal{C} = \mathbb{Z}/26\mathbb{Z}$. Also, the key d, subtraction of 3 modulo 26, uniquely determines the deciphering transformation $D_d(c) \equiv m \equiv c - 3 \pmod{26}$, which is the same as saying $D_d(c) = m = c - 3 \in \mathbb{Z}/26\mathbb{Z}$. We also make the very important observation that $D_d(E_e(m)) = m$, and $E_e(D_d(c)) = c$. In other words, D_d is the the inverse function of E_e, which we denote by $D_d = E_e^{-1}$. This is the crucial relationship between enciphering and deciphering functions that allows us to define the term *cipher* in more precise mathematical terms.

Definition 1.5 (Cryptosystems)

A cryptosystem *or* cipher *is comprised of a set* $\{E_e : e \in \mathcal{K}\}$ *of enciphering functions and a set* $\{D_d = E_e^{-1} : d \in \mathcal{K}\}$ *of deciphering functions, which corresponds to the former set in such a way that for each* $e \in \mathcal{K}$, *there exists a unique* $d \in \mathcal{K}$ *such that* $D_d = E_e^{-1}$, *that is, so that* $D_d(E_e(m)) = m$ *for all* $m \in \mathcal{M}$. *Individually e and d are called* keys, *and* (e, d) *is called a* key pair. *The set of pairs* $\{(m, E_e(m)) : m \in \mathcal{M}\}$ *is called a* cipher table.

The following illustrates a generic cryptosystem.

Diagram 1.6 An Illustrated Cryptosystem

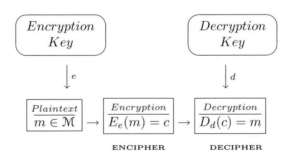

Remark 1.7 *The reader is cautioned that there is not a general consensus in the literature about the use of the term* cipher. *Some sources use the term as we have in Definition 1.5, whereas others use it to mean what we have called a cipher table. Moreover, at least in practice, the term "cipher" and "cipher table" are used synonymously. We will use the terms "cipher" and "cryptosystem" interchangeably to mean the notion given in Definition 1.5, whereas we*

reserve the term "cipher table" as defined therein. Thus, the cipher table *for the Caesar Cipher is Table 1.1, for instance, whereas the* Caesar Cipher *is the system of enciphering and deciphering transformations discussed above. Moreover, Definition 1.5 provides a formal mathematical formulation of the terms* cipher *and* key *used informally in the previous discussion.*

Now a natural question (and one that is central to the main theme of this book) is to ask about the distinctions (or lack thereof) between the enciphering and deciphering keys. The following is the ancestor of the type of cryptosystem that will dominate the discussions in the balance of the chapters.

Definition 1.8 (Symmetric-Key Cryptosystems)

A cryptosystem is called symmetric-key (*also* single-key, one-key, *and* conventional) *if for each key pair* (e, d), *the key* d *is "computationally easy" to determine knowing only* e, *and similarly* e *is easy to determine knowing only* d.

In practical symmetric-key cryptosystems, we usually have that $e = d$, thus justifying the term "symmetric-key".

Remark 1.9 *We will use the term "computationally easy problem" to mean one that can be solved in expected polynomial time and can be attacked using available resources (the reader unfamiliar with these terms may find them in Appendix B on complexity issues). The reason for adding the caveat on attacks is to preclude problems that are of polynomial time complexity but for which the degree is "large". The antithesis of this would be a "computationally infeasible problem", which means that, given the enormous amount of computer time that would be required to solve the problem, this task cannot be carried out in* realistic *computational time. Therefore, "computationally infeasible" means that, although there (at least theoretically) exists a unique answer to the problem, we could not find it even if we devoted every scintilla of the time and resources available. Note that this is distinct from a problem that is* unsolvable *with any amount of time or resources. For example, an unsolvable problem would be to cryptanalyze ABC, assuming that it was encrypted using a monoalphabetic substitution. There is simply no unique verifiable answer without more information. However, the bottom line is that there is no* proven *example of a computationally infeasible problem (see Footnote B.9 on page 211).*

Symmetric-key ciphers come in two flavours. The first is described as follows.

Definition 1.10 (Block Ciphers)

A block cipher *is a cryptosystem which separates the plaintext message into strings of plaintext message units, called* blocks, *of fixed length* $k \in \mathbb{N}$, *called the* blocklength *and enciphers one block at a time.*

The Caesar Cipher is a block cipher where the enciphering key e (addition of 3 modulo 26) is $-d$ where d is the deciphering key (subtraction of 3 modulo 26) and the blocklength is $k = 1$.

More generally, we have the following.

Definition 1.11 (Shift Ciphers)

A shift cipher consists of the following. The enciphering function is given by

$$E_e(m) \equiv c \equiv m + b \,(\mathrm{mod}\ n)$$

for any $b, n \in \mathbb{N}$ where $m \in \mathcal{M} = \mathbb{Z}/n\mathbb{Z}$ and $c \in \mathcal{C}$, which is the same as saying

$$E_e(m) = c = m + b \in \mathbb{Z}/n\mathbb{Z} = \mathcal{C}.$$

The deciphering transformation is given by

$$D_d(c) \equiv m \equiv c - b \,(\mathrm{mod}\ n),$$

or simply

$$D_d(c) = m = c - b \in \mathbb{Z}/n\mathbb{Z}.$$

Hence, the shift cipher is a symmetric-key cipher with $d = -e$ since e is the addition of b modulo n, and $d = -e$ is the subtraction of b modulo n, the additive inverse of e. This is an example of a block cipher of length $k = 1$. The Caesar Cipher is the special case obtained by taking $b = 3$ and $n = 26$. Also, for fans of the Stanley Kubrick film 2001: *A Space Odyssey*, take $b = -1$ and $n = 26$, from which *HAL* is deciphered as *IBM*. Note that shift ciphers are vulnerable to letter frequency analysis (see Appendix B on page 205).

The above provides a warm-up for the discussion of classical ciphers in Section 1.2, where we will discuss generalizations of the shift cipher. Before we can look at the second type of symmetric-key cipher, we need the following notion.

Definition 1.12 (Keystreams, Seeds, and Generators)

If $\mathcal{K}$ is the keyspace for a set of enciphering transformations, then a sequence $k_1 k_2 \cdots \in \mathcal{K}$ is called a keystream. *A keystream is either randomly chosen, or is generated by an algorithm, called a* keystream generator, *which generates the keystream from an initial small input keystream called a* seed. *Keystream generators that eventually repeat their output are called* periodic.

Remark 1.13 *The term "randomness" is not easy to define precisely, and could consume numerous pages to discuss (see Knuth [123, 149–189]). However, it suffices to say that since computers are finite state devices, then any random-number generator on a computer must be periodic, which means it is predictable, so not truly random. Hence, the most one can expect from computer, generating sequences of bits (bitstrings) say, is* pseudorandomness. *Thus, by a* pseudorandom generator *we will mean a deterministic algorithm that takes a short random* *seed and expands it into a bitstring for which it is computationally infeasible (see Remark 1.9) to tell the difference from a truly random bitstring. However, keystreams so generated must be* cryptographically secure, *which means that they must satisfy the additional property that, for a given output bit, the next output bit must be computationally infeasible to predict, even given knowledge of all previous bits, knowledge of the algorithm being used, and knowledge of the hardware. Since these issues are not of concern to us herein, we will assume that we have a cryptographically secure pseudorandom number generator for our keystream, or a truly randomly chosen keystream, and proceed with our discussion of cryptographic techniques.*

Definition 1.14 (Stream Ciphers)
   Let $\mathcal{K}$ be a keyspace for a cryptosystem and let $k_1 k_2 \cdots \in \mathcal{K}$ be a keystream. This cryptosystem is called a Stream Cipher *if encryption upon plaintext strings $m_1 m_2 \cdots$ is achieved by repeated application of the enciphering transformation on plaintext message units,*

$$E_{k_j}(m_j) = c_j,$$

and deciphering occurs as

$$D_{k_j}(c_j) = m_j$$

for $j \geq 1$. If there exists an $\ell \in \mathbb{N}$ such that $k_{j+\ell} = k_j$ for all $j \in \mathbb{N}$, then we say that the Stream Cipher is periodic *with* period ℓ.

   Perhaps the most famous example of a stream cipher is the following, wherein $\oplus$ denotes addition modulo 2, also called *XORing*.

◆ **The Vernam**[1.1] **Cipher**
   The Vernam Cipher is a Stream Cipher with alphabet of definition $\mathcal{A} = \{0, 1\}$ that enciphers in the following fashion. Given a bitstring $m_1 m_2 \cdots m_n \in \mathcal{M}$, and a keystream $k_1 k_2 \cdots k_n \in \mathcal{K}$, the enciphering transformation is given by

$$E_{k_j}(m_j) = m_j \oplus k_j = c_j \in \mathcal{C},$$

and the deciphering transformation is given by

$$D_{k_j}(c_j) = c_j \oplus k_j = m_j.$$

[1.1]Gilbert S. Vernam was born in Brooklyn, New York, and was a graduate of Massachusetts College. Within a year of graduating, he joined the American Telegraph and Telephone Company (AT&T). After a year of working for AT&T, he married Alline Eno, and they had one child together. On December 17, 1917, he wrote down the cipher for which he has earned the title of the *Father of Automated Cryptography*. This was the first Polyalphabetic Cipher automated using electrical impulses. Throughout his life, he was granted sixty-five patents, including the fully automated telegraph switching system, and he even invented one of the first versions of a binary digital enciphering of pictures. However, for all his accomplishments, he died in relative obscurity on February 7, 1960 in Hackensack, New Jersey after years of battling Parkinson's disease.

The keystream (pad) is randomly chosen and never used again. For this reason, the Vernam Cipher is also called the *one-time pad*.

We could have used a different alphabet of definition such as $\mathbb{Z}/n\mathbb{Z}$, but since we convert to binary on a computer, we may as well stick with the binary alphabet. See Exercise 1.15, for instance, to see that we can essentially use $\mathbb{Z}/26\mathbb{Z}$ as with the Caesar Cipher, even though our alphabet of definition is binary. The big drawback to the Vernam Cipher is that the keylength (the number of characters in the key) and plaintext length are the same. This creates key management problems so the Vernam Cipher usually is employed only for the most sensitive transmission of data such as missile launch "codes".

Example 1.15 *Assume that the following bitstring is a randomly chosen key for the one-time pad:*[1.2]

$$k = (1100101000110011110001010111000101111110101010001).$$

Also, assume that the following was enciphered using k:

$$c = (1011100101111010111100010100000111110000010101010010).$$

Then to find the the plaintext string, we perform addition modulo 2 on $k + c$:

$$(1100101000110011110001010111000101111110101010001) +$$
$$(1011100101111010111100010100000111110000010101010010) =$$
$$(0111001101001001001101000011000010001110000000011) = m$$

A one-time pad can be shown to be theoretically unbreakable. In 1949 with Shannon's development [205] of *perfect secrecy* the one-time pad was proved to be unbreakable, something assumed to be true for a long time prior to the proof.

It is known that the revolutionary, Che Guevara, used a version of the Vernam Cipher in base ten with additions and subtractions carried out by hand. In 1967 the Bolivian army captured and executed him. On his body they found some handwritten pages detailing how he created cryptograms for transmission to Fidel Castro. Guevara used numerical equivalents of his plaintext, written in Spanish, as decimal numbers of one or two digits according to a fixed cipher table. The key was a random sequence of digits known only to Guevara and Castro. The plaintext message was then added to the key (without carries) and this produced the cryptogram, a process now called the *Guevara Cipher*.

Also, inspired by the Cuban missile crisis of the 1960s, a *hot-line* between Washington and Moscow was established that used what they called the *one-time tape*, which was a physical manifestation of the Vernam Cipher. In the United States, this took the form of the ETCRRM II or *Electronic Teleprinter*

[1.2]Recall that the base $b \in \mathbb{N}$ representation of an integer is given by $(a_{t-1}a_{t-2} \ldots a_1 a_0)_b = \sum_{j=0}^{t-1} a_j b^j$ where $t \in \mathbb{N}$, $a_{t-1} \neq 0$, and $0 \leq a_j \leq b-1$ for $0 \leq j \leq t-1$. We often suppress the binary subscript for convenience as in our example above. See Exercises 1.17–1.22.

Cryptographic Regenerative Repeater Mixer II. The one-time tape worked via the existence of two magnetic tapes, one at the enciphering source, and one at the deciphering end, both having the same keystream on them. To encipher, they performed addition modulo 2 with the plaintext and the bits on the tape. To decipher, the receiver performed addition modulo 2 with the ciphertext and the bits on the (identical) tape at the other end. Thus, they had instant deciphering and perfect secrecy as long as they used truly random keystreams, each was used only once, and the tapes were destroyed after each use. Today, one-time pads are in use for military and diplomatic purposes when unconditional security is of the utmost importance.

We now have sufficient framework with which to work, so we turn to the next section for a description of some classical ciphers.

Exercises

In Exercises 1.1–1.8, determine the plaintext from the given ciphertext using the Caesar Cipher 1.1.

1.1. **L WKLQN WKHUHIRUH L DP**

1.2. **IURP RQH OHDUQ WR NQRZ DOO**

1.3. **EHKROG WKH VLJQ**

1.4. **WKLV LV KDUG ZRUN**

1.5. **QRQ VHTXLWXU**

1.6. **WUXWK LV PLJKWB DQG ZLOO SUHYDLO**

1.7. **WUXWK FRQTXHUV DOO WKLQJV**

1.8. **WKH HQG FURZQV WKH ZRUN**

1.9. The following is a cipher for a monoalphabetic substitution.

Plain	a	b	c	d	e	f	g	h	i	j	k	l	m
Cipher	M	L	O	P	R	Q	T	S	N	W	U	V	Z
Plain	n	o	p	q	r	s	t	u	v	w	x	y	z
Cipher	Y	X	C	B	A	F	E	D	G	K	J	I	H

Using this cipher, decrypt the following:

KMA NF NZZNYRYE

In Exercises 1.10–1.14, use the cipher given in Exercise 1.9 to decipher each of the ciphertext messages as follows.

1.10. **YX FDOS MTRYOI**

1.11. **YRGRA FMI MYIESNYT**

1.12. **QNAR MYP NOR**

1.13. **GMYNEI**

1.14. **KR MAR PXYR**

1.15. Interpret the plaintext solution m to Example 1.15 as a bitstring that is the concatenation of bitstrings of length 5, each corresponding to an English letter equivalent given in decimal form by Table 1.2. Find the English text equivalent of the bitstring. For instance, the first such bitstring, 01110, is 14 in decimal, which is **O** from Table 1.2. Proceed from there to the full English plaintext.

1.16. Given the key

$$k = (11010111101111010101111011101011010)$$

for the one-time pad, and ciphertext

$$c = (01001011111011010000101101111001000),$$

determine the plaintext m and interpret it as the concatenation of bitstrings of length 5, each corresponding to an English letter equivalent given in decimal form by Table 1.2. Find the English text equivalent of the bitstring.

(For instance, the sum modulo 2 of the first five bits of k and c is

$$11010 + 01001 = 10011,$$

which is binary for the decimal 19. In Table 1.2, 19 corresponds to the letter **T**. Continue in this fashion for the modulo 2 additions of each remaining five-bit integer in k and c to decipher the balance of the cryptogram.)

In Exercises 1.17–1.22, use the key k given in Exercise 1.16 and the ciphertext given in each of these following exercises to determine the English equivalents of the plaintext by the method outlined in Exercise 1.16.

1.17. $c = (10010100111110111011010011101101001)$.

1.18. $c = (01000110101110000100110010100101000)$.

1.19. $c = (11010100111110111110101011111001000)$.

1.20. $c = (11001110101110010001101010111111110)$.

1.21. $c = (11111100000111110001010001111001011)$.

1.22. $c = (11010101011010111011110000111001000)$.

1.2 Classical Ciphers

We shall see that cryptography is more than a subject permitting mathematical formulation, for indeed it would not be an exaggeration to state that abstract cryptography is identical with abstract mathematics.

A.A. Albert[1.3]

Perhaps the best known classical cipher is the Caesar Cipher described in Section 1.1. We demonstrated how this cipher is a special case of a shift cipher. As it turns out, shift ciphers are special cases of the following concept.

◆ **Affine Ciphers**
An *Affine Cipher* is defined as follows. Let $\mathcal{M} = \mathcal{C} = \mathbb{Z}/n\mathbb{Z}$, $n \in \mathbb{N}$,

$$\mathcal{K} = \{(a,b) : a, b \in \mathbb{Z}/n\mathbb{Z} \text{ and } \gcd(a,n) = 1\},$$

and for $e, d \in \mathcal{K}$, and $m, c \in \mathbb{Z}/n\mathbb{Z}$, set

$$E_e(m) \equiv am + b \pmod{n}, \text{ and } D_d(c) \equiv a^{-1}(c - b) \pmod{n}.$$

Example 1.16 *Let $n = 26$ and $\mathcal{M} = \mathcal{C} = \mathbb{Z}/26\mathbb{Z}$. Define an Affine Cipher as follows.*

$$E_e(m) = 5m + 7 = c \in \mathcal{M}.$$

Since $5^{-1} \equiv 21 \pmod{26}$, then

$$D_d(c) = 21(c - 7) = 21c + 9 \in \mathcal{M}.$$

We want to encipher and send the following message:

launch the missles.

To do this, we get the numerical equivalents for the letters from Table 1.2 on page 3:

$$11, 0, 20, 13, 2, 7, 19, 7, 4, 12, 8, 18, 18, 8, 11, 4, 18.$$

Then apply E_e to each value of m to get:

$$10, 7, 3, 20, 17, 16, 24, 16, 1, 15, 21, 19, 19, 21, 10, 1, 19.$$

[1.3]Abraham Adrian Albert (1905–1972) was born in Chicago, Illinois on November 9, 1905, and remained in Chicago most of his life. He studied under L. E. Dickson at the University of Chicago receiving his Ph.D. in 1928 for his dissertation entitled *Algebras and Their Radicals and Division Algebras*. His impressive work in classifying division algebras earned him a National Research Council Fellowship, which allowed him to obtain a postdoctoral position at Princeton. After that he spent a couple of years at Columbia University, then returned to the University of Chicago in 1931. Albert's book *Structure of Algebras*, published in 1939, remains a classic to this day. His work on algebras earned him the Cole Prize in that same year. World War II induced Albert to take an interest in cryptography. In fact, the quote given above is taken from his lecture on mathematical aspects of cryptography at the American Mathematical Society meeting held in Manhattan, Kansas on November 22, 1941. His honours and awards are too numerous to list here, but it is clear that he had a lasting influence. He died on June 6, 1972 in Chicago.

Then we use Table 1.2 to get the ciphertext English equivalents:

KHDURQ YQB PVTTVKBT,

which is sent as the cryptogram. The reader may now verify that $D_d(c)$ yields the original plaintext. (See Exercises 1.23–1.28.)

In turn, Affine Ciphers are special cases of the following notion. Recall first that a *permutation* σ on a finite set S is a bijection on S; namely, σ is a one-to-one and onto function on S.

Definition 1.17 (Substitution Ciphers)

Let A be an alphabet of definition consisting of n symbols, and let M be the set of all blocks of length r over A. The keyspace K will consist of all ordered r-tuples $e = (\sigma_1, \sigma_2, \ldots, \sigma_r)$ of permutations σ_j on A. For each $e \in K$, and $m = (m_1 m_2 \ldots m_r) \in M$, let

$$E_e(m) = (\sigma_1(m_1), \sigma_2(m_2), \ldots, \sigma_r(m_r)) = (c_1, c_2, \ldots, c_r) = c \in C,$$

and for $d = (d_1, d_2, \ldots, d_r) = (\sigma_1^{-1}, \sigma_2^{-1}, \ldots, \sigma_r^{-1}) = e^{-1}$,

$$D_d(c) = (d_1(c_1), d_2(c_2), \ldots, d_r(c_r)) = (\sigma_1^{-1}(c_1), \sigma_2^{-1}(c_2), \ldots, \sigma_r^{-1}(c_r)) = m.$$

This type of cryptosystem is called a substitution cipher. *If all keys are the same, namely, $\sigma_1 = \sigma_2 = \cdots = \sigma_r$, then this cryptosystem is called a* simple substitution cipher *or* monoalphabetic substitution cipher. *If the keys differ, then it is called a* polyalphabetic substitution cipher.

Definition 1.17 gives the promised formal interpretation of the notions that we discussed in Section 1.1 wherein we gave examples of monoalphabetic block ciphers such as the Caesar Cipher, and polyalphabetic stream ciphers such as the Vernam Cipher — one-time pad.[1.4] Another classical polyalphabetic block cipher is given as follows.

◆ Vigenère[1.5] Cipher

Fix $r, n \in \mathbb{N}$, and let $M = C = (\mathbb{Z}/n\mathbb{Z})^s$, the elements of which are ordered s-tuples from $\mathbb{Z}/n\mathbb{Z}$, and $K = \mathbb{Z}^r$ where $s \geq r$. For

$$e = (e_1, e_2, \ldots, e_r) \in K, \text{ and } m = (m_1, m_2, \ldots, m_s) \in M,$$

[1.4]The Vernam Cipher was, in fact, the *first* polyalphabetic substitution cipher that was automated using electrical impulses, wherein each 5-bit key value determined one of thirty-two fixed alphabetic substitutions.

[1.5]Blaise de Vigenère (1523–1596), while on a diplomatic mission to Rome in his mid-twenties, first came into contact with cryptology at the Papal Cevria. This inspired him to read the books by the pioneers Alberti, Belaso, Cardano, Porta, and Trithemius. One of the books that Vigenère wrote, *Traicté des Chiffres* published in 1586, contains the first discussion of plaintext and ciphertext autokey systems. However, it was not until the seventeenth century that two-part "codes" were used (see Remark 1.4 on page 5). The first to actually put them into practice were Kings Louis XIII and Louis XIV of France. Vigenère died of throat cancer in 1596.

let $E_e(m) = (m_1 + e_1, m_2 + e_2, \ldots, m_r + e_r, m_{r+1} + e_1, \ldots, m_s + e_{s-kr})$, for some integer k, and let $D_d(c) = (c_1 - e_1, c_2 - e_2, \ldots, c_r - e_r, c_{r+1} - e_1, \ldots, c_s - e_{s-kr})$, for $c = (c_1, c_2, \ldots, c_s) \in \mathcal{C}$, where $+$ is addition, and $-$ is subtraction, modulo n. This cryptosystem is called the Vigenère Cipher with period s. If $r = s$, then this cipher is often called a Running-key Cipher. For the case $r \neq s$ see Example 1.24.

The Vigenère Cipher is symmetric-key, given that knowing e is tantamount to knowing d. It is a Block Cipher with blocklength r, and it is polyalphabetic if we ensure that not all the keys e_j for $j = 1, 2, \ldots, r$ are the same.

There is an easy means of visualizing the Vigenère Cipher when we use $n = 26$ and the English alphabet via Table 1.2, as follows.

THE VIGENÈRE TABLEAU

	a	b	c	d	e	f	g	h	i	j	k	l	m	n	o	p	q	r	s	t	u	v	w	x	y	z
a	A	B	C	D	E	F	G	H	I	J	K	L	M	N	O	P	Q	R	S	T	U	V	W	X	Y	Z
b	B	C	D	E	F	G	H	I	J	K	L	M	N	O	P	Q	R	S	T	U	V	W	X	Y	Z	A
c	C	D	E	F	G	H	I	J	K	L	M	N	O	P	Q	R	S	T	U	V	W	X	Y	Z	A	B
d	D	E	F	G	H	I	J	K	L	M	N	O	P	Q	R	S	T	U	V	W	X	Y	Z	A	B	C
e	E	F	G	H	I	J	K	L	M	N	O	P	Q	R	S	T	U	V	W	X	Y	Z	A	B	C	D
f	F	G	H	I	J	K	L	M	N	O	P	Q	R	S	T	U	V	W	X	Y	Z	A	B	C	D	E
g	G	H	I	J	K	L	M	N	O	P	Q	R	S	T	U	V	W	X	Y	Z	A	B	C	D	E	F
h	H	I	J	K	L	M	N	O	P	Q	R	S	T	U	V	W	X	Y	Z	A	B	C	D	E	F	G
i	I	J	K	L	M	N	O	P	Q	R	S	T	U	V	W	X	Y	Z	A	B	C	D	E	F	G	H
j	J	K	L	M	N	O	P	Q	R	S	T	U	V	W	X	Y	Z	A	B	C	D	E	F	G	H	I
k	K	L	M	N	O	P	Q	R	S	T	U	V	W	X	Y	Z	A	B	C	D	E	F	G	H	I	J
l	L	M	N	O	P	Q	R	S	T	U	V	W	X	Y	Z	A	B	C	D	E	F	G	H	I	J	K
m	M	N	O	P	Q	R	S	T	U	V	W	X	Y	Z	A	B	C	D	E	F	G	H	I	J	K	L
n	N	O	P	Q	R	S	T	U	V	W	X	Y	Z	A	B	C	D	E	F	G	H	I	J	K	L	M
o	O	P	Q	R	S	T	U	V	W	X	Y	Z	A	B	C	D	E	F	G	H	I	J	K	L	M	N
p	P	Q	R	S	T	U	V	W	X	Y	Z	A	B	C	D	E	F	G	H	I	J	K	L	M	N	O
q	Q	R	S	T	U	V	W	X	Y	Z	A	B	C	D	E	F	G	H	I	J	K	L	M	N	O	P
r	R	S	T	U	V	W	X	Y	Z	A	B	C	D	E	F	G	H	I	J	K	L	M	N	O	P	Q
s	S	T	U	V	W	X	Y	Z	A	B	C	D	E	F	G	H	I	J	K	L	M	N	O	P	Q	R
t	T	U	V	W	X	Y	Z	A	B	C	D	E	F	G	H	I	J	K	L	M	N	O	P	Q	R	S
u	U	V	W	X	Y	Z	A	B	C	D	E	F	G	H	I	J	K	L	M	N	O	P	Q	R	S	T
v	V	W	X	Y	Z	A	B	C	D	E	F	G	H	I	J	K	L	M	N	O	P	Q	R	S	T	U
w	W	X	Y	Z	A	B	C	D	E	F	G	H	I	J	K	L	M	N	O	P	Q	R	S	T	U	V
x	X	Y	Z	A	B	C	D	E	F	G	H	I	J	K	L	M	N	O	P	Q	R	S	T	U	V	W
y	Y	Z	A	B	C	D	E	F	G	H	I	J	K	L	M	N	O	P	Q	R	S	T	U	V	W	X
z	Z	A	B	C	D	E	F	G	H	I	J	K	L	M	N	O	P	Q	R	S	T	U	V	W	X	Y

To use the above table, we do three things to encipher: (1) Put the plaintext letters in a row; (2) Above each plaintext letter put the keyword letters, repeated as often as necessary to cover all the plaintext; (3) Replace each letter of the plaintext with the letter at the intersection of the row and column of the table containing the keyword letter and plaintext letter, respectively. For example, suppose that our keyword is **watchers**, and our plaintext is **immortals never die**. Then we write the keyword as many times as necessary above the plaintext, so the third row is the ciphertext determined by the intersections in the table.

w	a	t	c	h	e	r	s	w	a	t	c	h	e	r	s	w
i	m	m	o	r	t	a	l	s	n	e	v	e	r	d	i	e
E	M	F	Q	Y	X	R	D	O	N	X	X	L	V	U	A	A

For instance, the letter w is in a row that intersects the column containing the letter i at the ciphertext letter E, and so on.

To decipher using the Vigenère Tableau, we do three things: (1) Put the ciphertext letters in a row; (2) Put the keyword letters above the ciphertext letters, repeating them as required to cover all ciphertext; (3) For each column in which a keyword letter sits, locate the row in which the ciphertext letter below it sits. Then the letter in the first column of that row is the corresponding plaintext. For instance, let's decipher our previous example. The third row is the plaintext.

w	a	t	c	h	e	r	s	w	a	t	c	h	e	r	s	w
E	M	F	Q	Y	X	R	D	O	N	X	X	L	V	U	A	A
i	m	m	o	r	t	a	l	s	n	e	v	e	r	d	i	e

For instance, since the letter w sits over the ciphertext letter E, and the row in which E sits in w's column has the letter i in its first column, this is the first letter of plaintext, and so on. (See Exercises 1.29–1.34.)

The following is an example of how the Vigenère Cipher may be viewed as a periodic stream cipher.

Example 1.18 *The Vigenère Cipher with key e of length r may be considered to be a periodic stream cipher with period r. The key $e = (e_1, e_2, \ldots, e_r)$ provides the first r elements of the keystream $k_j = e_j$ for $1 \leq j \leq r$, after which the keystream repeats itself.*

Example 1.18 illustrates that there are circumstances when a block cipher may be considered to be a stream cipher. For instance, even the shift cipher may be considered to be a Vigenère Cipher with period length one. This should not lead to confusion since all interpretations are consistent with the definitions. That there may be a blurring at the simplest levels between the two types that is of no concern. Example 1.18 is actually an illustration of the following concept.

Definition 1.19 (Synchronous Stream Ciphers)

A Stream Cipher is said to be synchronous *if the keystream is generated without use of either the plaintext or the ciphertext, called keystream generation where that generation is independent of the plaintext and ciphertext.*

Basically, in a synchronous stream cipher both the sender and the receiver must be synchronized in the sense that they must be at exactly the same position in their shared key. Any loss or insertion of bits means that they must re-synchronize. Definition 1.19 is one of the two categories into which stream ciphers are classified. The other is given as follows.

Definition 1.20 (Self-Synchronizing Ciphers)
  A Stream Cipher is called self-synchronizing (*or* asynchronous) *if the keystream is generated as a function of the key and a fixed number of previous ciphertext units. If the Stream Cipher utilizes plaintext in the keystream generation, then it is called* nonsynchronous.

Vigenère had another idea that provides us with an example of a nonsynchronous stream cipher as follows.

Example 1.21 This is a description of a variant of the Vigenère Cipher. Let $n = |\mathcal{A}|$ where $\mathcal{A}$ is the alphabet of definition. We call $k_1 k_2 \cdots k_r$ for $1 \leq r \leq n$ a *priming key*. Then given a plaintext message unit $m = (m_1, m_2, \ldots, m_s)$ where $s \geq r$, we generate a keystream as follows: $k = k_1 k_2 \cdots k_r m_1 m_2 \cdots m_{s-r}$. Then we encipher via:

$$E_k(m) = (m_1 + k_1, \ldots, m_r + k_r, m_{r+1} + m_1, m_{r+2} + m_2, \cdots, m_s + m_{s-r}) = c,$$

where $+$ is addition modulo n, so $m_j + k_j, m_{r+j} + m_j \in \mathbb{Z}/n\mathbb{Z}$. We decipher via: $D_k(c) = (c_1 - k_1, \ldots, c_r - k_r, c_{r+1} - m_1, \ldots, c_s - m_{s-r}) = m$. This cryptosystem is nonsynchronous since the plaintext serves as the key, from the $(r+1)^{st}$ position onwards, with the simplest case being $r = 1$.

Example 1.21 is an instance of the following concept.

Definition 1.22 (Autokey Ciphers)
  An Autokey Cipher *is a cryptosystem wherein the plaintext itself (in whole or in part) serves as the key (usually after the use of an initial priming key).*

For instance, the cipher given Example 1.21 is called the *Autokey Vigenère Cipher*, wherein the plaintext is introduced into the key generation after the priming key has been exhausted.
  There is also a means of interpreting the Vernam Cipher as a Vignère Cipher.

Example 1.23 The definition of the Vignère Cipher is a Running-key Cipher. In other words, the keystream is as long as the plaintext. The Vigenère Cipher becomes a Vernam Cipher if we assume that the keystream is truly random and never repeats.

A variant of the Vigenère Cipher is the following.

◆ **Beaufort**[1.6] **Cipher**

Fix $r, n \in \mathbb{N}$. Both the encryption and decryption functions are given by $x \mapsto (e_1 - x_1, e_2 - x_2, \ldots, e_r - x_r)$, for $e = (e_1, \ldots, e_r) \in \mathcal{K}$ and $x = (x_1, \ldots, x_r) \in (\mathbb{Z}/n\mathbb{Z})^r = \mathcal{M} = \mathcal{C}$. In other words, the encryption and decryption functions are the same; namely, they are their own inverses.

The self-decrypting Beaufort Cipher was used in a rotor-based cipher machine called the Hagelin[1.7] M-209.

Example 1.24 *In the Beaufort Cipher, let $r = 4$, $n = 26$, and choose the key HARD. We wish to decipher the following cryptogram.*

HXFVQAGCDAXYTJYLFAGS.

First convert the key and the cryptogram to its numerical equivalents via Table 1.2: $e = (7, 0, 17, 3)$, and ciphertext

$$(7, 23, 5, 21, 16, 0, 6, 2, 3, 0, 23, 24, 19, 9, 24, 11, 5, 0, 6, 18).$$

Then apply the key to each $r = 4$-block modulo $26 = n$:

$$e(7, 23, 5, 21) = (7 - 7, 0 - 23, 17 - 5, 3 - 21) = (0, 3, 12, 8) = (ADMI),$$

[1.6]This cryptosystem was invented by Francis Beaufort (1774–1857). He was born in County Meath, Ireland. He began his nautical career at age thirteen as a cabin boy in the British Navy. By age twenty-two, he attained the rank of lieutenant, and by 1805, he was given his first command on the H.M.S. Woolwich. His assignment was to do a hydrographic survey of the Rio de la Plata region of South America. It was during this time that he began to develop what later became known as the *Beaufort Wind Force Scale*, which is an instrument that meteorologists use to indicate wind velocities on a scale from 0 to 12, where, for instance, 0 is calm, 6 is a strong breeze, and 12 is a hurricane. In 1838, his scale was put into use by the British fleet. By 1846, Beaufort was promoted to Rear Admiral, and by 1855 Sir Francis Beaufort retired from the Admiralty. He died two years later but left an admirable legacy. The cipher that bears his name originated with him, but was not published until a few months after his death by his brother.

[1.7]Boris Caesar Wilhelm Hagelin, who was born on July 2, 1892, invented this device in the early 1940s. In 1922, Emanuel Nobel, nephew of the famed Alfred Nobel, put Hagelin to work in the firm *Aktiebolaget Cryptograph* or *Cryptograph Incorporated*, a company owned by Avid Gerhard Damn, who invented cipher machines of his own. Hagelin simplified and improved one of Damn's machines. After Damn's death in 1927, Hagelin ran the firm. Later he developed the M-209, which became so successful that in the early 1940s more than 140000 were manufactured. The royalties from this alone made Hagelin the first to become a millionaire from cryptography. Due to Swedish law that allowed the government to confiscate inventions required for national defence, Hagelin moved the company to Zug, Switzerland in 1948, where it was incorporated as *CRYPTO AG* in 1959. The firm is still in operation although it got embroiled in a controversy over sales of a cipher product to Iran in the early 1990s. It went so far that a senior salesman for CRYPTO AG, Hans Bueler, was arrested in Tehran on March 18, 1992, where he spent nine and a half months in prison being questioned about leaking cryptographic data to the Western powers. He knew nothing as it turns out, and CRYPTO AG ultimately paid a million dollars for Bueler's release in January, 1993. The company fired him shortly thereafter. The controversy continued into whether or not CRYPTO AG's products were compromised by Western intelligence services. The company denies all allegations, but rumours abound.

$$e(16, 0, 6, 2) = (7 - 16, 0 - 0, 17 - 6, 3 - 2) = (17, 0, 11, 1) = (RALB),$$

$$e(3, 0, 23, 24) = (7 - 3, 0 - 0, 17 - 23, 3 - 24) = (4, 0, 20, 5) = (EAUF),$$

$$e(19, 9, 24, 11) = (7 - 19, 0 - 9, 17 - 24, 3 - 11) = (14, 17, 19, 18) = (ORTS),$$

$$and \quad e(5, 0, 6, 18) = (7 - 5, 0 - 0, 17 - 6, 3 - 18) = (2, 0, 11, 11) = (CALL).$$

The end result is: **ADMIRAL BEAUFORT'S CALL** (*with the apostrophe understood tacitly*). (*See Exercises 1.35–1.36.*)

Definition 1.17 tells us that a simple substitution cipher encrypts single plaintext symbols as single ciphertext symbols, such as the Caesar Cipher described in Section 1.1. When groups of one *or more* symbols are replaced by other groups of ciphertext symbols, then this cryptosystem is called a *polygram substitution cipher*. For instance, we have the following classical polygram substitution cipher, which is a *digraph cipher* — where two adjacent symbols, called a *digram*, are enciphered/deciphered at a time.

◆ **The Playfair[1.8] Cipher[1.9]**

In this cipher the letters I and J are considered as a single entity.

Table 1.25

A	Z	W	IJ	D
E	U	T	G	Y
O	N	K	Q	M
H	F	X	L	S
V	R	P	B	C

Pairs of letters are enciphered according to the following rules.

(a) If two letters are in the same row, then their ciphertext equivalents are immediately to their right. For instance, VC in plaintext is RV in ciphertext. (This means that if one is at the right or bottom edge of the table, then one "wraps around" as indicated in the example.)

(b) If two letters are in the same column, then their cipher equivalents are the letters immediately below them. For example, ZF in plaintext is UR in ciphertext, and JB in plaintext is GI in ciphertext.

[1.8]This cipher was conceived by Sir Charles Wheatstone, and it was the first literal digraphic cipher in cryptographic history. Although his friend, Lord Lyon Playfair, never claimed that the cipher was his idea, he was zealous in his support for the invention. Playfair had even discussed it with Prince Albert, suggesting its use in the Crimean War. It turns out to have been used in the Boer War and possibly elsewhere. In any case, Britain's War Office ostensibly kept the cipher a secret due to its being used as the British Army's field cipher. Thus, it ultimately came to be known as the *Playfair Cipher*.

[1.9]John F. Kennedy's PT-109 was sunk by a Japanese cruiser in the Soloman Islands during World War II. Kennedy swam to shore on the Japanese controlled Plum Island where he was able to send a message using the Playfair Cipher to arrange rescue for the survivors of his crew. In May of 2002, it was announced that Dr. Robert Ballard found the wreckage of PT-109, nearly sixty years after it went down.

(c) If two letters are on the corners of a diagonal of a rectangle, then their cipher equivalents are on the other corners, and the cipher equivalent of each plaintext letter is on the same row as the plaintext letter. For instance, UL in plaintext becomes GF in ciphertext and SZ in plaintext is FD in ciphertext.

(d) If the same letter occurs as a pair in plaintext, then we agree by convention to put a Z between them and encipher. Also, if a single letter remains at the end of the plaintext, then a Z is added to it to complete the digraph.

Example 1.26 *Suppose that we wish to decipher:* **BP DV GW VY FD OE HQ YF SG RT CF TU WC DH LD KU HV IV WG FD**, *assuming that it was encrypted using the Playfair Cipher. One merely reverses the rules to decipher. For instance, the first pair BP of ciphertext letters occurs on the same row. So we choose the letters to their left, PR. The second set DV occurs on a diagonal with AC as the opposite ends (respectively) of the other diagonal. Then GW occurs in diagonal with TI, which is chosen as plaintext, and so on to get:* **practices zealously pursued pass into habits**, *where the last letter Z is ignored as the filler of the digraph. Such fillers are called* nulls. (*See Exercises 1.37–1.40.*)

Another classical polygram substitution cipher is the following, which requires some elementary matrix theory. In what follows, the symbol $M_{r \times r}(\mathbb{Z}/n\mathbb{Z})$ denotes the ring of $r \times r$ matrices with entries from $\mathbb{Z}/n\mathbb{Z}$ (see Appendix C).

◆**The Hill**[1.10] **Cipher**

Fix $r, n \in \mathbb{N}$, let $\mathcal{K} = \{e \in M_{r \times r}(\mathbb{Z}/n\mathbb{Z}) : e \text{ is invertible}\}$, and set $\mathcal{M} = \mathcal{C} = (\mathbb{Z}/n\mathbb{Z})^r$. Then for $m \in \mathcal{M}$, $e \in \mathcal{K}$, $c \in \mathcal{C}$, $E_e(m) = em$ and $D_d(c) = e^{-1}c$. (Note that e is invertible if and only if $\gcd(\det(e), n) = 1$.)

Example 1.27 *Assume that* **SVJYYCDOIMWPCVDXCH** *has been encrypted with with $n = 26$, and $r = 3$ using the Hill Cipher with key*

$$e = \begin{pmatrix} 1 & 2 & 3 \\ 0 & 5 & 1 \\ 2 & 0 & 1 \end{pmatrix}.$$

[1.10]Lester S. Hill invented this cipher. He achieved his Ph.D. in mathematics from Yale in 1926, and taught mathematics at Hunter College in New York from 1927 until his retirement in 1960. Hill was the first to successfully employ general algebraic concepts for cryptography. A. A. Albert (see Footnote 1.3 on page 13) liked Hill's ideas so much that he tailored them to work on some simple cryptosystems. Hill's rigorous mathematical approach may be said to be one of the factors which has helped foster today's solid grounding of cryptography in mathematics. He died in Lawrence Hospital in Bronxville, New York after suffering through a lengthy illness.

In order to find the plaintext, we first calculate

$$e^{-1} = \begin{pmatrix} 1 & 10 & 13 \\ 16 & 25 & 5 \\ 24 & 6 & 1 \end{pmatrix}.$$

Then we decipher each triple of ciphertext numerical equivalents as follows.

$$e^{-1} \begin{pmatrix} 18 \\ 21 \\ 9 \end{pmatrix} = \begin{pmatrix} 7 \\ 0 \\ 21 \end{pmatrix}, e^{-1} \begin{pmatrix} 24 \\ 24 \\ 2 \end{pmatrix} = \begin{pmatrix} 4 \\ 6 \\ 20 \end{pmatrix}, e^{-1} \begin{pmatrix} 3 \\ 14 \\ 8 \end{pmatrix} = \begin{pmatrix} 13 \\ 22 \\ 8 \end{pmatrix}$$

$$e^{-1} \begin{pmatrix} 12 \\ 22 \\ 15 \end{pmatrix} = \begin{pmatrix} 11 \\ 11 \\ 19 \end{pmatrix}, e^{-1} \begin{pmatrix} 2 \\ 21 \\ 3 \end{pmatrix} = \begin{pmatrix} 17 \\ 0 \\ 21 \end{pmatrix}, e^{-1} \begin{pmatrix} 23 \\ 2 \\ 7 \end{pmatrix} = \begin{pmatrix} 4 \\ 11 \\ 25 \end{pmatrix}.$$

Now, using Table 1.2, we get the letter equivalents of each plaintext triple and decipher the plaintext as **have gun will travel**, *where there is an extra Z on the end, enciphered to ensure a final triple in plaintext, so we discarded it. (See Exercises 1.41–1.44.)*

Block ciphers come in two categories. The first of these is described in Definition 1.17 and the second is given as follows.

Definition 1.28 (Transposition/Permutation Ciphers)

A simple transposition cipher, *also known as a* simple permutation cipher, *is a symmetric-key block cryptosystem having blocklength $r \in \mathbb{N}$, with keyspace $\mathcal{K}$ being the set of permutations on $\{1, 2, \ldots, r\}$. The enciphering transformation is given, for each $m = (m_1, m_2, \ldots, m_r) \in \mathcal{M}$, and given $e \in \mathcal{K}$, by*

$$E_e(m) = (m_{e(1)}, m_{e(2)}, \ldots, m_{e(r)}),$$

and for each $c = (c_1, c_2, \ldots, c_r) \in \mathcal{C}$,

$$D_d(c) = D_{e^{-1}}(c) = (c_{d(1)}, c_{d(2)}, \ldots, c_{d(r)}).$$

The cryptosystems in Definition 1.28 have keyspace of cardinality $|\mathcal{K}| = r!$. Permutation encryption involves grouping plaintext into blocks of r symbols and applying to each block the permutation e on the numbers $1, 2, \ldots, r$. In other words, the places where the plaintext symbols sit are permuted. We will use the notation $(\sigma_1, \sigma_2, \ldots, \sigma_r)$ to mean that $j \to \sigma_j$ for $j = 1, 2, \ldots, r$. For instance, take $e = (1, 2, 3, 4, 10, 7, 8, 9, 5, 6, 11, 12, 13)$ and apply it to *They flung hags*, and we get *They hung flags*. To see this, let m_j be the jth letter in *They flung hags* for $j = 1, 2, \ldots, 13$. Then, for instance, $m_{e(5)} = m_{10} = h$; $m_{e(6)} = m_7 = u$; $m_{e(7)} = m_8 = n$; and so on. This is a permutation of the *places* where the plaintext letters sit. Therefore, plaintext letters get moved according to the given place permutation. To decipher, one merely reverses the permutation. In contrast, Definition 1.17 provides for a permutation of the actual plaintext symbols themselves.

Example 1.29 Let $r = 6$, $\mathcal{M} = \mathcal{C} = \mathbb{Z}/26\mathbb{Z}$, with the English letter equivalents given by Table 1.2. Then if $e = (2, 3, 6, 1, 4, 5)$ is applied to

agency,

we get

GEYANC

since $m_1 = A, m_2 = G, m_3 = E, m_4 = N, m_5 = C$, and $m_6 = Y$, so

$$e(m) = (m_{e(1)}, m_{e(2)}, m_{e(3)}, m_{e(4)}, m_{e(5)}, m_{e(6)}) = (m_2, m_3, m_6, m_1, m_4, m_5).$$

Since the inverse transformation is

$$d = e^{-1} = (4, 1, 2, 5, 6, 3),$$

then another way to visualize encryption is to write $4, 1, 2, 5, 6, 3$ in the first row, and the plaintext letter equivalents in the second row, then read the letters off in *numerical order*. For instance,

$$\begin{pmatrix} 4 & 1 & 2 & 5 & 6 & 3 \\ A & G & E & N & C & Y \end{pmatrix}.$$

Thus, the first in numerical order is G, the second is E and so on. (See Exercises 1.45–1.58.) To decrypt, write the message under the encryption key, and read off the text in numerical order.

An easy means for finding the inverse of a given key e such as in Example 1.29 is given as follows. The key in that example can be written as

$$e = \begin{pmatrix} 1 & 2 & 3 & 4 & 5 & 6 \\ 2 & 3 & 6 & 1 & 4 & 5 \end{pmatrix},$$

since $1 \mapsto 2$, $2 \mapsto 3$; and so on. To find the inverse, just read off in numeric order (determined by the second row), the terms in the first row. For instance, the term in the first row sitting above the 1 is 4, showing that 4 is the first term in e^{-1}. The term in the first row sitting above the 2 is 1, showing that 1 is the second term in e^{-1}, and so on.

To date, the best known symmetric-key block cipher is the *Data Encryption Standard* (DES) which was the first commercially available algorithm and was put into use in the 1970s (see [165] for a complete description and background). However, by the end of the century DES had reached the end of its usefulness, largely because of a keylength (56-bit) that was too small for modern security. It was replaced in August 2001 by the *Advanced Encryption Standard — Rijndael*, which is also described in detail in [165].

We have described several classical ciphers, and historical background, as a precursor to the study of public-key encryption in this text.

Exercises

In Exercises 1.23–1.28, use the Affine Cipher given in Example 1.16 to decrypt each of the ciphertexts given as follows.

1.23. **TBHORQYQBRHIBT**

1.24. **EHYOZKYQBMZOWBO**

1.25. **MZPMYQBRHPET**

1.26. **TBRDOBYQBMHTB**

1.27. **TDOOZDUWYQBRVYX**

1.28. **EOZYBRYYQBRVIVKVHUT**

In Exercises 1.29–1.34, assume that the keyword, given in the example on page 16, for the Vigenère Cipher was used to encrypt each of the ciphertexts. Use the deciphering technique with the Vigenère Tableau in that example to find each plaintext.

1.29. **OHXEYIRLADTUAEKW**

1.30. **OHXFLPZNARXFALVE**

1.31. **DEWGZXIGUEWVVAEK**

1.32. **PHXAYISMELMCJMKQ**

1.33. **XOMJZXIMCGEGZIEV**

1.34. **BODTDEIDKRWUYYCW**

In Exercises 1.35–1.36, use the Beaufort Cipher key in Example 1.24 to decipher the following cryptograms.

1.35. **CSEAOTNLDYAZONDH**

1.36. **OWARZNRKDTJR**

In Exercises 1.37–1.40, decipher the following cryptograms assuming they were encrypted using the Playfair Cipher.

1.37. **VE AO ZO KE VO VN OU**

1.38. **HN BU UG IJV NE GW UW**

1.39. **UP EG NV YO YO VG ZU**

1.40. **EX MH YA DE LD MF TV GU QC UV**

In Exercises 1.41–1.44, decipher the following cryptograms using the key and setup for the Hill Cipher given in Example 1.27.

1.41. **WNWXCQZYRGSWOOV**

1.42. **TNQPDXVGYONTZXTXYKSBRRRRNLYPIKIATP**

1.43. **QIWTLPTAYZUH**

1.44. **JQDTNQMJRATWEQUEKY**

In Exercises 1.45–1.57, decipher the following cryptograms using the key
$e = (4, 3, 1, 6, 7, 5, 8, 2)$ *in a permutation cipher. (See Definition 1.28.)*

1.45. **STLOLRLE**

1.46. **IFWHTGTEOREISRMR**

1.47. **ELAILVDLSROAIFLE**

1.48. **TCVRYOCILPOTEEDM**

1.49. **SUTHITMR**

1.50. **AWBEORFEEVNBEMRO**

1.51. **ELSDINDP**

1.52. **IMTGINTI**

1.53. **GRDSEIOTAGREUSN**

1.54. **MICNAILRCNSIOTNA**

1.55. **DOGEEDDOERSAIVLP**

1.56. **ATTITLAONARANLDI**

1.57. **LLFDWEIIERTREGTH**

1.58. Apply the key $e = (5, 9, 10, 7, 12, 8, 6, 13, 1, 3, 11, 2, 4)$ as a permutation cipher to: **BRITNEY SPEARS**

1.59. A *superincreasing sequence* is a sequence of natural numbers $b_1, b_2, \ldots, b_n$ such that $b_i > \sum_{j=1}^{i-1} b_j$ for all $i = 2, 3, \ldots, n$. Prove that a set of natural numbers $\{b_1, b_2, \ldots, b_n\}$, satisfying the property that $b_{j+1} > 2b_j$ for all $j = 1, 2, \ldots, n-1$, is a superincreasing sequence.

1.60. The *subset sum problem* is defined as follows. Given $m, n \in \mathbb{N}$ and a set $\mathcal{S} = \{b_j : b_j \in \mathbb{N}, \text{ for } j = 1, 2, \ldots, n\}$, called a knapsack set, determine whether or not there exists a subset $\mathcal{S}_0$ of $\mathcal{S}$ such that the sum of the elements in $\mathcal{S}_0$ equals m.[1.11] Find all subsets of the knapsack set $\mathcal{S} = \{1, 2, 3, 5, 9, 10, 11\}$ that have 13 as their sum. Is $\mathcal{S}$ a superincreasing sequence?

[1.11]Exercises 1.59–1.60 are related to another classical cipher called the *knapsack cipher*. We will not describe this type of cryptosystem herein since the last of the remaining subset sum based knapsack ciphers was broken in 2001. For details and additional comments, see Footnote 3.9 on page 67.

1.3 Classification of Attacks

Attack is the best form of defence

Late eighteenth century proverb

Although our primary goal is to study cryptography, there is an essential need to understand some basics about cryptanalysis, if for no other reason than to understand what the *strength* of a cryptosystem (usually tantamount to its *weakest* point) means from knowledge of the various ways of breaking it. To *break* a cryptosystem means to decrypt the ciphertext without knowledge of the key, and in practical terms this means reconstructing the key via observations of the cryptosystem being employed.

By an *attack*, we mean the use of any methodology that begins with some information about the plaintext or ciphertext, encrypted using a key, which is yet unknown to the cryptanalyst whose end-goal is to break the system. The kinds of observations and the manipulations of the cryptosystem made by the cryptanalyst determine the type of attack. The following provides a description of the basic attacks on cryptosystems, under the assumption that the cryptanalyst has complete knowledge of the enciphering transformation being employed, but has no knowledge of the key. (This is the traditional assumption made for academic cryptanalysis, although this is not always the case in "real-life" cryptanalysis. Nevertheless, this is a reasonable assumption to make if one wants to ensure maximum strength of a cryptosystem.)

● Classification of Cryptanalytic Attacks

There are two basic types of attacks, *passive* and *active*. A passive attack involves the cryptanalyst's monitoring (also called "eavesdropping") of the communication channels, thereby threatening only the confidentiality of the data. An active attack is one where the cryptanalyst attempts to add, delete, or otherwise alter the message, so that not only confidentiality but also the integrity and authentication of the data are threatened. We begin with a classification of passive attacks, which are given in order of the degree of difficulty, on the part of the cryptanalyst, to mount a successful attack.

❶ Passive Attacks

◆ Ciphertext-Only

The cryptanalyst has access only to the ciphertext, obtained through interception of some cryptograms, from which to deduce the plaintext, without any knowledge whatsoever of the plaintext. For instance, with a relatively small amount of ciphertext enciphered using the Caesar Cipher, the cryptanalyst needs only try the twenty-five different deciphering shifts to get the key.

◆ Known-Plaintext

The cryptanalyst has both ciphertext and corresponding plaintext from intercepted cryptograms as data from which to deduce the plaintext in general, or the key. In the case of a shift cipher, for instance, only *one* plaintext-ciphertext pair needs to be known to determine the key, which is instantly known to be how far the enciphered symbol is shifted from the plaintext symbol.

One of the most prominent known-plaintext attacks against block ciphers is *linear cryptanalysis* (LC), using linear approximations to describe the behavior of the block cipher. LC was developed by Matsui [147] in 1994, when he successfully used it (under experimental conditions) against DES (see page 22) to obtain a key with 2^{43} known plaintexts (see [148]).

◆ **Chosen-Plaintext**

The cryptanalyst chooses plaintext, is then given corresponding ciphertext, and analyzes the data to determine the encryption key. It turns out that RSA is extremely vulnerable to this type of attack, as we shall discuss in Chapter 6.

One of the best-known chosen-plaintext attacks against block ciphers is *differential cryptanalysis* (DC) developed by Biham and Shamir [26] in 1993. DC involves the comparisons of pairs of plaintext with pairs of ciphertext, the task being to concentrate on ciphertext pairs whose plaintext pairs have certain "differences". Some of these differences have a high probability of reappearing in the ciphertext pairs. Those that do are called "characteristics", which DC uses to assign probabilities to the possible keys, with an end-goal being the location of the most probable key. When used against DES, the DC attack proved less successful than the LC attack.

◆ **Chosen-Ciphertext**

The cryptanalyst chooses the ciphertext and is given the corresponding plaintext. This attack is most effective against public-key cryptosystems (see Chapter 6), but sometimes is effective against symmetric-key types as well.

One method of mounting a chosen-ciphertext attack is to gain access to the equipment used to encipher. This was done prior to World War II when the Americans were able to reconstruct the Japanese cipher machine that was used for diplomatic communication. This allowed the American cryptanalysts to decipher Japanese cryptograms during the war, but since the Japanese were unable to cryptanalyze American ciphers, they assumed that their cryptograms were also unbreakable — a fatal assumption. Cryptanalysts helped to ensure that Japan's lifeline was rapidly cut, and that German U-boats were defeated.

Another incident involved the downing of the plane carrying the commander-in-chief of the Combined Fleet of the Japanese Navy, Admiral Isoruko Yamamoto. The Americans had been able to decipher a highly secret cryptogram, giving the itinerary of Yamamoto's plane on a tour of the Solomon Islands.

Of vital importance was the Battle of Midway, which was a stunning victory by American cryptanalysts since they were able to give complete information on the size and location of the Japanese forces advancing on Midway. This enabled the Navy to concentrate a numerically inferior force in exactly the right place at the right time, and prepare an ambush that turned the tide of the Pacific War.

◆ Adaptive Chosen-Plaintext

This is a variant of the chosen-plaintext attack where a cryptanalyst can not only choose the plaintext that is enciphered, but also can modify the choice based upon the results of previous encryption. Thus, a cryptanalyst can choose a block of plaintext, then choose another based on the results of the first and continue this iterative procedure. This is, in general terms, the means by which DC attacks *product ciphers* (a notion due to Shannon [205]) which are block ciphers that iterate several operations such as substitution, transposition, modular addition or multiplication, and linear transformation.

◆ Adaptive Chosen-Ciphertext

This is a variant of the chosen-ciphertext attack where a cryptanalyst chooses ciphertext depending upon previously received ciphertexts. This has been used successfully against public-key cryptosystems as well (see Chapter 6).

Remark 1.30 *It should be pointed out that some sources break up the above categories and consider, for instance, chosen-plaintext as an active attack, since the cryptanalysts can choose to have their desired plaintext enciphered and see the ciphertext which results. However, this is still eavesdropping, so we maintain the above as the subcategorization of passive attacks since there is no proactive involvement of the cryptanalyst under our definition.*

There are also passive attacks that do not involve any cryptanalysis. Loss of a key through noncryptanalytic techniques is called a compromise. *For instance, a passive attack on* pretty good privacy (PGP), *which we will study in Chapter 8, called* keypress snooping *involves an attacker installing a keylogger, which can capture the password of the target, and involves no cryptanalysis; the system is completely compromised, and may go completely undetected.*

The following provides some important active attacks, but is by no means exhaustive. Moreover, our primary focus will be on passive attacks on cryptosystems. Active attacks, due to their very nature, are more difficult to carry out than passive attacks.

❷ Active Attacks

◆ Man-in-the-Middle Attack

This is an attack that is particularly relevant for key exchange protocols, which we will study in Section 2.1.

The central idea is perhaps best described if we take this opportunity to introduce our first in a list of a cast of characters, who are well-known in the cryptographic community. There is *Alice* and her friend *Bob* who are participants in all communications, and *Mallory, the malicious active attacker*. The central idea of the man-in-the-middle attack is that Mallory assumes a position between Alice and Bob. Mallory can stop all or parts of the data being sent by Alice and Bob and substitute them with his own messages. In this fashion, he impersonates Alice and Bob who believe they are communicating with each other directly, while they are really talking to Mallory.

Diagram 1.31 (Man-in-the-middle Attack)

$$\boxed{ALICE} \xleftarrow{\quad\quad\longrightarrow\quad\quad} \boxed{MALLORY} \xleftarrow{\quad\quad\longrightarrow\quad\quad} \boxed{BOB}$$

In Chapter 7, we will see how to prevent this type of attack against public-key exchange and the providing of digital signatures.

◆ Rubber-hose Attack

The cryptanalyst threatens, blackmails, or tortures someone until they give up the key. Sometimes bribery comes into play, in which case, the attack is a subcategory called a *purchase-key* attack.

◆ Timing Attack

This is the most recent of the active attacks described here. This involves the repeated measuring of the exact execution times of modular exponentiation operations, the applications of which we will study in Section 2.3. This allows the cryptanalyst to get information on the decryption exponent in RSA; for instance (see Chapter 6). Another interpretation of a timing attack would be the interception of a message, which is re-sent at a later time (called a *replay attack* — see page 133 for this and other types of attacks). As a precaution against this type of attack, this active interference could be detected by a *timestamp* in messages.

We conclude this section with an illustration of a ciphertext-only attack on polyalphabetic cryptosystems, developed in 1863.

■ Kasiski's[1.12] Attack on Polyalphabetic Ciphers

The central idea behind this attack is the observation that repeated portions of plaintext encrypted with the same segment of the key result in identical ciphertext portions. Therefore, if repeated occurrences in ciphertext are not accidental, one would expect the same plaintext portion was enciphered, starting from the same position in the key. Consequently, the number of symbols between the beginnings of repeated ciphertext portions should be a multiple of the keylength. For instance, if the repeated ciphertext is the adjacent triple of letters XYZ, called a *trigram*, and the number of characters between the Z and

---

[1.12]Friedrich W. Kasiski (1805–1881) was born on November 29, 1805 in a western Prussian town. He enlisted in East Prussia's thirty-third Infantry Regiment at the age of seventeen and retired in 1852 as a major. Although he became interested in cryptography during his military career, it was not until the 1860's that he published his ideas. In 1863, he published *Die GeheimschRiften und die Dechiffrir-kunst* which dealt largely with a general solution, sought for centuries, for polyalphabetic ciphers with repeating keywords. His revolutionary work went unrecognized in his time, and he himself turned from cryptography. He took an active interest in anthropology, publishing his findings in archeological journals. He died on May 22, 1881, without knowing the impact his work in cryptanalysis would ultimately have.

the occurrence X in the next XYZ is seventeen, and this is not an accident, then 20 is a multiple of the keylength.

Since some of these repeated segments are coincidental, a means of analyzing these segments, called a *Kasiski examination* is to compute the greatest common divisor (gcd) of the collection of all the distances between repeated segments. Then choosing the largest factor occurring most often among these *gcd*s is probably the keylength. Once a probable keylength ℓ, say, is determined, a frequency analysis can be done on a breakdown of the ciphertext into ℓ classes (with an individual class containing every ℓ-th character (beginning at the n-th character) for a fixed $n \in \mathbb{N}$) to determine the suspected key. Then one can apply knowledge of frequency analysis such as that given in Appendix A on page 203 to actually determine the key, assuming the plaintext is English, which we will do throughout the text. If the plaintext is another language, comparing the frequency of ciphertext letters to that of a suspected language actually may determine that language.

Let's apply this to one of our classical polyalphabetic block ciphers discussed in Section 1.2. We need to introduce another in our cryptographic cast of characters, *Eve, the eavesdropper.*

Example 1.32 (Kasiski's Attack on the Vigenère Cipher)

Suppose that the following ciphertext was encrypted with the Vigenère Cipher and Eve knows this. She wants to use Kasiski's method to decipher it.

EMFQYXRDJEOGYHZW

OHXEYIRLADTUAEKW

OHXFLPZNARXFALVE

DEWGZXIGUEWVVBEK

PHXAYISMELMCJMKQ

XOMJZXIMCGEGZIEV

BODTDEIDKRWUYYCW

First, Eve looks for multiple occurrences in the ciphertext. She discovers that the trigram **OHX** *has a spacing of 16 between the first* **O** *and the next occurrence of an* **O** *in the trigram. Similarly, she finds the following spacings: 16 for* **WOHX***, 8 for* **EW***, and 8 for* **XF***. She quickly guesses that since 8 divides all of these, then it is a good assumption that the keylength is 8. She is now ready to do a frequency analysis, so she breaks the ciphertext down into 8 classes as follows.*

Initial Position	Position in Ciphertext	Ciphertext Class
n	$n + 8k \ (k = 0, 1, \ldots, 13)$	Letters in positions $n + 8k$
1	$1, 9, \ldots, 105$	EJOAOADUPEXCBK
2	$2, 10, \ldots, 106$	MEHDHREEHLOGOR
3	$3, 11, \ldots, 107$	FOXTXXWWXMMEDW
4	$4, 12, \ldots, 108$	QGEUFFGVACJGTU
5	$5, 13, \ldots, 109$	YYYALAZVYJZZDY
6	$6, 14, \ldots, 110$	XHIEPLXBIMXIEY
7	$7, 15, \ldots, 111$	RZRKZVIESKIEIC
8	$8, 16, \ldots, 112$	DWLWNEGKMQMVDW

Eve now must analyze every class to determine each individual letter in the sought-after key. She does this by a series of frequency analysis techniques, depending on information in Appendix A, as follows.

(1) *In the first class, Eve calculates the following frequency of letters and the percentages of the highest total occurrences in the class:* **A**, 2, 14%; **B, C, D**, 1 occurrence each; **E**, 2, 14%; **J,K**, 1 occurrence each; **O**, 2, 14%; **P,U,X**, 1 occurrence each.

Since Appendix A tells us that the most frequently occurring letter in the English language is **E**, *then she has three choices for it. If* **O** *is the ciphertext for it in this class, then this means a shift of* 10 *places from plaintext into ciphertext (the distance from* **E** *to* **O**). *However, by data in Appendix A, we know that the most commonly occurring letters in the English language are the letters in the family* **E,T,A,I,N,O,S,H,R**, *and within this first class of ciphertext, the only consecutively occurring letters are* **RST**, *so if* **O** *were* **E**, *then the letters* **RST** *would encipher to* **BCD** *having the low frequency* 1, 1, 1 *in this class, so Eve discards this as unlikely. She notes that it is also unlikely to have* **E** *as itself in the first letter of a key, so she discards this as well and chooses* **A** *as the enciphering of* **E**. *This means a shift of* 22 *places (the distance from* **E** *to* **A** *modulo* 26). *Hence, the first letter of our key is* **W**, *the distance from* **A** *to* **W** *being* 22.

Now Eve calculates frequencies for each of the other classes to determine the remaining seven letters of the key.

(2) **D**, 1; **E**, 3, 21.4%; **G**, 1; **H**, 3, 21.4%; **L,M**, 1 occurrence each; **O,R**. 2 occurrences each, 14% each.

Since no letter other than **E** *is likely to have the high occurrence of* 21.4%, *Eve concentrates upon* **E** *and* **H** *as the two possibilities. If* **H** *were the candidate, then this would mean a shift of* 3, *so* **D** (3 *places to the right of* **A** *in this class) would be the second key letter. However,* **WD** *is a very unlikely beginning of a meaningful English word, so she discards it and takes* **E** *itself as the enciphering of* **E** *for a shift of zero places. This means that* **A** *itself is the second key letter. (Result:* **WA** _ _ _ _ _ _ .)

(3) D, E, F, 1 *occurrence each;* **M**, 2, 14%; **O,T**, 1 *occurrence each;* **W**, 3, 21.4%; **X**, 4, 29%.

Here Eve sees that there is no contest for **E** *since* **X** *has such a commanding frequency, so the shift is* 19. *Hence,* **T** *is the third letter in the key,* 19 *places from* **A**. (*Result:* **WAT** _ _ _ _ _.)

(4) A,C,E, 1 *occurrence each;* **F**, 2, 14%; **G**, 3, 21.4%; **J**, 1; **Q,T**, 1 *occurrence each;* **U**, 2, 14%; **V**, 1.

Again, Eve sees that **G** *is the most likely candidate for* **E**, *so she chooses this which means a shift of* 2. *Thus, the third letter in the key is* **C**, 2 *places from* **A**. (*Result:* **WATC** _ _ _ _.)

(5) A, 2, 14%; **D,J,L,V**, 1 *occurrence each;* **Y**, 5, 35.7%; **Z**, 3, 21.4%.

Here Eve employs a different strategy since there are some quite high frequencies in this class. She notes that since **RST** *is the only consecutively occurring trigram in the most often used English letters, as we observed for the first class above, and since* **YZA** *is the only possible trigram of consecutive letters in this class, then* **YZA** *must be the enciphering of* **RST**. *This is a shift of* 7, *so* **H** *must be the fourth key letter, since* **H** *is* 7 *places to the right of* **A**. (*Result:* **WATCH** _ _ _.)

(6) B, 1; **E**, 2, 14%; **H**, 1; **I**, 3, 21.4%; **L,M,P**, 1 *occurrence each;* **X**, 3, 21.4%, **Y**, 1.

Here Eve has no consecutive trigrams, so she looks at the most likely candidates, **I** *and* **X**, *for* **E**. *Since* **X** *would mean a shift of* 19, *then the sixth key letter would be* **T**, *but adding a* **T** *at this stage would render it meaningless, so Eve chooses* **I**, *for a shift of* 4, *and* **E** *is the sixth key letter.* (*Result:* **WATCHE** _ _.)

(7) C, 1; **E**, 2, 14%; **I**, 3, 21.4%; **K,R**, 2, 21.4% *each;* **S,V**, 1 *occurrence each;* **Z**, 2, 14%.

Here the method breaks down since there are no trigrams and the most likely candidates for **E** *yield a key letter that does not fit with what we already have in step six. Thus, Eve decides to go to the final family and make an educated guess at the seventh letter.*

(8) D, 2, 14%; **E, G, K, L**, 1 *occurrence each;* **M**, 2, 14%; **N,Q,V**, 1 *occurrence each;* **W**, 3, 21.4%.

Eve sees that **W** *is the most likely for* **E** *so the shift is* 18 *giving us* **S** *a the eighth key letter. Hence, Eve, sees that we have* **WATCHE_S** *and clearly sees the seventh letter should be* **R**, *so we have the key*

WATCHERS.

The reader may now go to the example on page 16 and Exercises 1.29–1.34 for a complete deciphering of the cryptogram.

Exercises

In Exercises 1.61–1.66, use Kasiski's method to find the key assuming the ciphertext was obtained via the Vigenère Cipher. Then use the key to decrypt the message.

1.61.

ZEOLDSTFZEOLDMGTPZUGKQRTU
KCBYOUFZFAGJQSPOMTLVQVXY

1.62.

VFKILRXKFBJYGUYENYR
JAAVADPMNIYGJUXIQ

1.63.

HZVUBLLZXUPBWAJFSBA
MODPFLPPFQXWQBXHW

1.64.

VISHMTEUCRLKMKMXR
FKXVIYLLYFLEHGVSRS

1.65.

OOPETAAYBAARUIWE
JEZEKGIDJTPVTRGQ

1.66.

EALPBNSSVTQVSTZYHGZAVMHYFEK
HYIEPGOFHCEJZBNKJBNULCTGM
UIKYRLSAVOFABTZLPNKTBSRG

1.67.

GZWWNOEJRWEWKFDLYBZWMCWG
HIESROELTIXLYOWZGMDSXSNLUIVC

You see it's like a portmanteau — there are two meanings packed up into one word.   from *Through the Looking-Glass* (1872)

Lewis Carrol (Charles Lutwidge Dodgson) (1832–1898)

English writer and logician

Chapter 2

Protocols, Discrete Log, and Diffie-Hellman

The past, at least, is secure.

Daniel Webster (1782–1852) American politician

2.1 Cryptographic Protocols

In this chapter, we will require increasingly more number theoretic results. Appendix C on page 212 is provided as a quick reminder and finger-tip reference for the reader. We begin this section with a formalization of certain terms; some of which we have been using in Chapter 1. First of all, an *entity* is any person, such as Bob, Alice, Eve, or Mallory introduced in Chapter 1, or a thing, such as a computer terminal, which sends, receives, or manipulates information. We have used the term *communication channel*, which we now define to be any means of communicating information from one entity to another. If two entities, such as Alice and Bob, are communicating and a third entity, such as Eve or Mallory, tries to interfere, passively or actively, with the data over the communication channel, then that third entity is called an *adversary* or *opponent*. A *secure channel* is one that is not physically accessible to an adversary. An *unsecured channel* is one from which entities other than those for whom the information was intended, can delete, insert, read, delay, or reorder data. A cryptosystem is said to be secure (against eavesdropping) if an adversary, such as Eve, who eavesdrops on a channel which is sending cryptograms, gains nothing over an entity which does not listen to the communication channel.

A *protocol*, in general human terms, may be regarded as *prearranged etiquette*, such as understood behaviour at a formal dinner party. On the other hand, a *cryptographic protocol* is an algorithm, involving two or more entities, using cryptography, designed to achieve a security goal. A security goal will involve issues of authentication (see Chapter 7), privacy (see Chapter 9), and secrecy. A protocol, by definition, can be no more secure than the underlying cryptosystem.

Of course there are attacks against protocols, which may be any of the types that we studied in Section 1.3. For instance, Eve could eavesdrop and learn the key in a symmetric-key protocol, and if Mallory knows the key, he can substitute or alter messages. We may also use public-key encryption for protocols, about which we will learn in Chapter 3. We will leave this issue for now and turn to the details of protocols in general.

Before setting out to classify protocols, we need to discuss why protocols are useful and what is necessary to make them work. One of the most important reasons for using cryptographic protocols is to allow us to examine means by which dishonest entities can try to cheat, thereby giving us methods for developing protocols to thwart them. We need to do this since computer protocols are not "face-to-face" so the lack of personal presence means that it is difficult to thwart cheaters.

In order for protocols to operate properly, we assume that all entities in the protocol are aware in advance of the (unequivocal) steps necessary for the protocol to function. Moreover, every entity involved must be in agreement to follow these steps, and to complete the protocol to the finish.

With these conventions in mind, we proceed with a detailed classification.

● Classification of Cryptanalytic Protocols

There are three kinds of protocols that we now outline. In order to describe the first type of protocol, we introduce another entity in our cryptographic cast of characters: *Trent, the trusted arbitrator.*

◆ Arbitrated Protocols

This type of protocol relies on Trent who is trusted not to render preferential treatment to Alice, Bob, or any other participants. Trent has no allegiances to any of the entities involved in the protocol and has no particular reason to complete the protocol. (Think of Trent as a disinterested lawyer.) Thus, a characteristic of this type of protocol is that all entities can accept, not only that what is done is correct, but also that their portion of the protocol is complete. An example of this type of protocol is given as follows.

The following type of protocol is ideally suited for identification of an owner, Alice (who has a PIN or password S) of a credit card, ID card, or computer account. It does so by allowing Alice to convince a merchant, Bob, of knowledge of S without revealing even a single bit of S. This was introduced in [88] by Fiat and Shamir in 1987 as an authentication and digital signature scheme (see Chapter 7), and it was modified in [85]–[86] to an identification protocol. (For a very intriguing story involving patents, the military, and this protocol, see [132].)

■ Feige-Fiat-Shamir Identification Protocol — Simplified Version

In the following protocol, Alice has to *prove* her identity, via demonstration of knowledge of a secret, to Bob.

Setup Stage

(1) Trent chooses a modulus $n = pq$, where p and q are large primes of roughly the same size to be kept secret. (We will formally study these types, called *RSA moduli* in Chapter 3.) Also, a parameter $a \in \mathbb{N}$ is chosen.

(2) Next, Alice and Bob, respectively, randomly select secret natural numbers $s_A, s_B \le n - 1$, with $\gcd(s_A s_B, n) = 1$. Then they compute, respectively, the smallest natural numbers t_A and t_B such that $t_A \equiv s_A^2 \pmod{n}$ and $t_B \equiv s_B^2 \pmod{n}$. They register their secrets s_A and s_B with Trent, whereas t_A and t_B do not need to be kept secret.

Protocol

Execute the following steps.

(1) Alice selects an $m \in \mathbb{N}$, called a *commitment*, such that $m \le n - 1$ and sends $w \equiv m^2 \pmod{n}$, called the *witness*, to Bob.

(2) Bob chooses $c \in \{0, 1\}$, called a *challenge*, and sends it to Alice.

(3) Alice computes $r \equiv m s_A^c \pmod{n}$, called the *response*, and sends it to Bob.

(4) Bob computes r^2 modulo n. If $r^2 \equiv w t_A^c \pmod{n}$, then reset a's value to $a - 1$ and go to step (1) if $a > 0$. If $a = 0$, then terminate the protocol with Bob accepting the proof. If $r^2 \not\equiv w t_A^c \pmod{n}$, then terminate the protocol with Bob rejecting the proof.

Remark 2.1 *This protocol is an example of what is known as a* zero-knowledge *proof of identity, which we will not cover in this text (see [165] for details on the subject of* zero-knowledge *in general). The Feige-Fiat-Shamir Protocol is also what is called a* zero-knowledge proof of knowledge *of a modular square root of* t_A, *namely, the secret* s_A. *(See Exercise 2.1.)*

Suppose that Mallory tries to impersonate Alice. Some methods can have a 50% chance of fooling Bob in any one round as follows. Suppose that Mallory picks an m so that $0 < m < n - 1$. If Mallory decides he will defeat Bob's selection of $c = 0$, he sends $w = m^2$. If he decides he will defeat Bob's selection of $c = 1$, he sends $w = m^2 t_A^{-1}$. Now, if Mallory sends $w = m^2$, having guessed that Bob will send $c = 0$, and Bob does challenge with $c = 0$, then Mallory sends $r = m$, and Bob computes $r^2 \equiv m^2 \equiv w t_A^0 \pmod{n}$. This passes step (4). If Mallory sends $w = m^2 t_A^{-1}$ and Bob challenges with $c = 1$, then Mallory sends $r = m$ and Bob computes $r^2 \equiv m^2 \equiv m^2 t_A^{-1} t_A^1 \equiv w t_A^1 \pmod{n}$, and this passes step (4). However, if Mallory sends $w = m^2$ (guessing Bob will challenge with $c = 0$) and Bob challenges with $c = 1$, then Mallory cannot produce an r that will fool Bob. In this case, Mallory needs an r so that $r^2 \equiv w t_A \pmod{n}$ or $r^2 \equiv m^2 t_A \pmod{n}$. If he could find such an r, then $x \equiv r m^{-1} \pmod{n}$ would be a solution to $x^2 \equiv t_A \pmod{n}$, but we are assuming that he cannot solve this congruence. Also, if Mallory sends $w = m^2 t_A^{-1}$ and Bob challenges with $c = 0$, then Mallory needs an r so that $r^2 \equiv w \equiv m^2 t_A^{-1} \pmod{n}$. If he can find such an r, then $x \equiv mr^{-1} \pmod{n}$ would solve $x^2 \equiv t_A \pmod{n}$. Hence, if Mallory guesses which c Bob will send he can fool Bob for that round. But if he guesses wrong, he will not be able to provide an r that will pass step (4). Furthermore, Mallory cannot do any better than this. It is not hard to see that there is no w Mallory can send that will allow him to reply with an appropriate r, whether Bob sends $c = 0$ or $c = 1$. Suppose that there is such a w. Then if Bob sends $c = 0$, Mallory needs to have ready an $r = r_0$ so that $r_0^2 \equiv w \pmod{n}$, and in case Bob sends $c = 1$, he needs to have an $r = r_1$ with $r_1^2 \equiv w t_A \pmod{n}$. Yet, if r_0^{-1} exists, $(r_1 r_0^{-1})^2 \equiv t_A \pmod{n}$, so if Mallory really had this r_0 and r_1, he could compute $x \equiv r_1 r_0^{-1} \pmod{n}$, and he could solve $x^2 \equiv t_A \pmod{n}$, which we are assuming to be impossible. The bottom line is that when Mallory picks a witness, he can decide whether his witness will defeat $c = 0$ or whether it will defeat $c = 1$. He can always pick a witness that will work against whichever one of these he wants, but he cannot pick a witness that will work against both. Only Alice can do that. (See Exercise 2.2.)

The following protocol classification requires the introduction of another in our cast of cryptographic characters, *Judy, the adjudicator*.

◆ Adjudicated Protocols

This is actually a variation of an arbitrated protocol. If all entities follow the sequence of steps in the protocol and no entity believes that any other entity is cheating, then an adjudicator is not needed. However, if cheating is suspected by any entity, then Judy the adjudicator comes into play to analyze the dispute, render a ruling on who is right, and determine what punishment should be dispensed to the entity in the wrong. For instance, consider the "real life" scenario where Alice and Bob agree that Alice will sell her car to Bob. First, Alice gives the car keys to Bob, who gives her a cheque for the car. If the keys are fake or the cheque is fraudulent, then they go before Judy the adjudicator to present their cases. Judy rules on the evidence presented and the entity who cheated is fined or imprisoned. Thus, the penalty, if severe enough, may prevent cheating and the need for Judy. However, there is a means to work things out by another type of protocol that involves no third entity.

◆ Self-Enforcing Protocols

These are the most desirable protocols since they are designed to make cheating virtually impossible. We need neither Trent not Judy, since any cheating is immediately detected by other entities participating in the protocol. Cheaters gain no advantage by *not* following the protocol. Hence, the protocols are *self-enforcing*.

As an illustration, suppose that Alice and Bob were married but got a divorce, now live in different towns, and want to decide who gets the car by the flip of a coin over the telephone. Given that they no longer trust each other and have no trusted third party, such as Trent, they need to ensure that the protocol is fair and not open to either of them cheating. They do this as follows.

■ Coin Flipping by Telephone

There are numerous "coin flipping by telephone" protocols, the first being given by Blum in [30] (and we will re-visit the "coin flipping" notion several times throughout the text). However, we present one here that is different from that typically presented. The following uses what are called *Blum integers*, which are natural numbers of the form $n = pq$ where $p \equiv q \equiv 3 \,(\text{mod } 4)$ are primes. In the following, Bob is going to make a guess. If he is correct, he gets the car. If not, Alice gets the car. Also, in each congruence modulo n, we assume that the least positive residue is selected for the values discussed.

(1) Alice selects a Blum integer $n = pq$, and a random $x \in \mathbb{N}$ such that $\gcd(x, n) = 1$, then computes $x_0 \equiv x^2 \,(\text{mod } n)$ and $x_1 \equiv x_0^2 \,(\text{mod } n)$. Then Alice reveals n and x_1 to Bob.

(2) Bob guesses the parity of x and tells Alice his guess.

(3) Alice tells Bob what p, q, x, and x_0 happen to be.

(4) Bob verifies that n is a Blum integer, then computes both $x_0 \equiv x^2 \,(\text{mod } n)$ and $x_1 \equiv x_0^2 \,(\text{mod } n)$, and thereby is able to confirm that x is indeed what Alice selected.

Analysis: Let Q_n denote the set of all $x \in (\mathbb{Z}/n\mathbb{Z})^*$ (the group of units modulo n — see (C.12) in Appendix C on page 215), such that x is a quadratic residue modulo n. Thus, the conditions in step (1) mean that *both* $x_0 \in Q_n$ (the condition: $x_0 \equiv x^2 \,(\text{mod } n)$); *and* $x_1 \in Q_n$ (the condition: $x_1 \equiv x_0^2 \,(\text{mod } n)$). Exercise 2.1 shows that of the four square roots x_0 of x_1 in Q_n modulo n, *exactly one* of them is also in Q_n. This square root is often called the *principal square root of x_0 modulo n*. This guarantees the uniqueness of x. In fact, the function $f : Q_n \mapsto Q_n$, given by $f(y) \equiv y^2 \,(\text{mod } n)$ is a bijection with inverse,

$$f^{-1}(y) \equiv y^{((p-1)(q-1)+4)/8} \,(\text{mod } n).$$

If n is not a Blum integer, however, this uniqueness evaporates. In this case it is possible for Alice to find an integer x_2 such that $x_2^2 \equiv x^2 \equiv x_0 \,(\text{mod } n)$ with $x_2 \neq x$ but $x_2, x \in Q_n$. In this case, if the parity of x_2 and x differ, then Alice can cheat freely by sending x_2 and x_0 to Bob instead of x and x_0 in step (3). To see this, consider the following example.

Let $n = 65$ and $x = 4$. Then $x_0 \equiv 16 \equiv 4^2 \equiv x^2 \equiv x_2^2 \equiv 61^2 \,(\text{mod } 65)$. Also, both $x = 4$ and $x = 61$ are square roots of $16 = x_0$ modulo $n = 65$ and both are easily checked to be in Q_{65} as well. In step (1), Alice sends $(n, x_1) = (65, 61)$ to Bob. Suppose that Bob guesses (correctly) that x is even. Then Alice can send $x = 61$ and $x_0 = 16$ to Bob instead of $x = 4$ and $x_0 = 16$ in step (3). Bob checks that both $61^2 \equiv 16 \pmod{65}$ (verifying that $x_0 \in Q_n$), and

$$x_1 \equiv 61 \equiv 16^2 = x_0^2 \pmod{65}$$

(thereby verifying that that *false* x, namely, $x_2 = 61$, is indeed a square root of $x_0 = 16$ modulo n). Since they hold, Bob has been cheated without being any wiser. However, with Blum integers in place, this is a self-enforcing protocol that both Alice and Bob can be sure is a fair determination of who gets the car. This completes our discussion of protocols.

Exercises

2.1. Suppose that $a \in \mathbb{N}$ and $a \equiv z^2 \pmod{pq}$ where $p \equiv q \equiv 3 \pmod 4$ are primes. Prove that there are only four possible square roots of a modulo pq, and they are given as follows. For $x, y \in \mathbb{Z}$ given by the Euclidean Algorithm, such that $xp + yq = 1$ we have:

$$z = \pm(xpa^{(q+1)/4} + yqa^{(p+1)/4}), \text{ and } z = \pm(xpa^{(q+1)/4} - yqa^{(p+1)/4}).$$

2.2. Let $n = pq = 101 \cdot 757 = 76457$ in the Feige-Fiat-Shamir Protocol. Show how Alice can prove knowledge of $s_A = 3$ to Bob. Assume that $a = 2$, and that in the first round $m = 98$ and $c = 0$, whereas in the second round $m = 747$ and $c = 1$.

2.3. Given $n = pq = 2087 \cdot 487 = 1016369$, use the Feige-Fiat-Shamir Protocol to show how Alice can prove knowledge of a secret $s_A = 111$ to Bob. Assume that $a = 2$ and that on the first round $m = 21$ and $c = 0$, whereas on the second round $m = 12$ and $c = 1$.

2.4. Suppose that Alice and Bob each have computers that input bitstrings of length $m \in \mathbb{N}$ and output bitstrings of length $n \in \mathbb{N}$. Also, assume that they send each other copies of the circuits of their respective computers. The following coin tossing protocol is followed.

   (1) Bob chooses a random bitstring R of length m, which he then runs through the circuitry of both computers. The two outputs are added modulo 2 and the result is sent to Alice.

   (2) Alice guesses the parity of Bob's input (the number of 1's in it) and sends the guess to him.

   (3) Bob sends R to Alice.

   (4) Alice can verify that the modulo 2 addition of the outputs from the two circuits is indeed what Bob sent earlier, and determine the correctness of the guess.

   What two properties must hold for the above coin flipping algorithm to be fair? (*This problem is a precursor to the topic to be discussed in Section 3.1.*)

2.2 The Discrete Log Problem

The things I want to show are mechanical. Machines have less problems.
Andy Warhol (1927–1987) American Artist

The security of numerous algorithms, some of which we will study in Section 2.3, is predicated upon the difficulty of solving a problem that is the topic of this section. A general form of the problem may be stated as follows.

◆ **Generalized Discrete Log Problem**

Given a finite cyclic group G of order $n \in \mathbb{N}$, a generator α of G, and an element $\beta \in G$, find that unique nonnegative integer $e \leq n-1$ such that $\alpha^e = \beta$.

A specific instance of this occurs when $G = \mathbb{F}_p^*$, the multiplicative group of nonzero elements in $\mathbb{F}_p = \mathbb{Z}/p\mathbb{Z}$ for p an odd prime. Given a generator α (also called a *primitive root modulo p*, recall) of $\mathbb{F}_p^*$ and an element $\beta \in \mathbb{F}_p^*$, find the unique nonnegative integer $e \leq p - 2$ such that $\alpha^e \equiv \beta \pmod{p}$. This value e is called the *index of β to the base α modulo p* in elementary number theory. This methodology (see [163, Section 4.1, pp. 154–158], for instance, and see the index-calculus algorithm on page 43) is strikingly similar to that of logarithms. Thus, the problem of finding a unique nonnegative integer $e \leq p - 2$ such that $c \equiv m^e \pmod{p}$, given integers m, c and a prime p, is called the

Discrete Log Problem (DLP):

$$e \equiv \log_m(c) \pmod{p}, \qquad (2.2)$$

often called simply *discrete log*. If p is "properly chosen", this is a very difficult problem to solve. At the time of this writing, the consensus is that if p has more than 308 decimal digits, and $p - 1$ has at least one "large" prime factor, then p is properly chosen (see the discussion at the bottom of page 45). Why we insist upon $p - 1$ having at least one large prime factor is due to the following algorithm, which allows for efficient calculation of discrete logs when $p - 1$ has only small prime factors, an issue that will be discussed in more detail after presentation. This first appeared in [185].

◆ **Silver-Pohlig-Hellman Algorithm for Computing Discrete Logs**

Let α be a generator of $\mathbb{F}_p^*$ and let $\beta \in \mathbb{F}_p^*$, and assume that we have a factorization

$$p - 1 = \prod_{j=1}^{r} p_j^{a_j} \qquad a_j \in \mathbb{N},$$

where the p_j are distinct primes. The technique for computing $e = \log_\alpha \beta$ is to compute e modulo $p_j^{a_j}$ for $j = 1, 2, \ldots, r$, then apply the Chinese Remainder Theorem. To compute e modulo $p_j^{a_j}$ we need to determine e in its base p_j representation:

$$e = \sum_{i=0}^{a_j-1} b_i^{(j)} p_j^i \qquad \text{where } 0 \leq b_i^{(j)} \leq p_j - 1 \text{ for } 0 \leq i \leq a_j - 1.$$

To find these $b_i^{(j)}$, we proceed as follows. First, set $\beta = \beta_0$.

(1) Compute $b_0^{(j)}$.

   We have that $\beta_0^{(p-1)/p_j} \equiv \alpha^{(p-1)b_0^{(j)}/p_j} \pmod{p}$. Thus we compute $\alpha^{(p-1)k/p_j}$ modulo p for each $k = 0, 1, \ldots, p_j - 1$ until we get that

$$\alpha^{(p-1)k/p_j} \equiv \beta_0^{(p-1)/p_j} \pmod{p},$$

   in which case this k must be $b_0^{(j)}$.

(2) Compute $b_i^{(j)}$ for $i = 1, 2, \ldots, a_j - 1$ as follows. For each such i, recursively define

$$\beta_i = \beta\alpha^{-\sum_{k=0}^{i-1} b_k^{(j)} p_j^k}, \text{ and } x_i = \log_\alpha \beta_i.$$

   Then

$$x_i = \sum_{k=i}^{a_j-1} b_k^{(j)} p_j^k,$$

   since

$$\log_\alpha \beta_i = \log_\alpha (\beta\alpha^{-\sum_{k=0}^{i-1} b_k^{(j)} p_j^k}) = \log_\alpha \beta - \sum_{k=0}^{i-1} b_k^{(j)} p_j^k =$$

$$\sum_{k=0}^{a_j-1} b_k^{(j)} p_j^k - \sum_{k=0}^{i-1} b_k^{(j)} p_j^k = \sum_{k=i}^{a_j-1} b_k^{(j)} p_j^k.$$

By Exercise 2.5,

$$\beta_i^{(p-1)/p_j^{i+1}} \equiv \alpha^{(p-1)b_i^{(j)}/p_j} \pmod{p}, \tag{2.3}$$

so we compute $\alpha^{(p-1)k/p_j}$ modulo p for $k \geq 0$ until (2.3) occurs in which case k is $b_i^{(j)}$.

We now illustrate the Pohlig-Hellman algorithm with a very small example to get the flavour of the process without making the calculations onerous.

Example 2.4 Let $p = 73$. Then $\alpha = 5$ generates $\mathbb{F}_{73}^*$. Select $\beta_0 = \beta = 68$. We want to compute $e = \log_5(68)$ in $\mathbb{F}_{73}^*$. We have $p - 1 = 72 = 2^3 \cdot 3^2 = p_1^{a_1} p_2^{a_2}$. All congruences in the balance of this example are assumed to be modulo 73.

For $p_1 = 2$:

k	0	1
$\alpha^{(p-1)k/p_1}$	1	$5^{36} \equiv 72$

i	0	1	$2 = a_1 - 1$
β_i	68	$68 \cdot 5^{-1} \equiv 72$	$68 \cdot 5^{-1} \equiv 72$
$\beta_i^{(p-1)/p_1^{i+1}}$	$68^{36} \equiv 72$	$72^{18} \equiv 1$	$72^9 \equiv 72$
$b_i^{(1)}$	1	0	1

Thus, the base 2 representation of $\log_5(68)$ modulo 8 is:

$$\sum_{i=0}^{a_1-1} b_i^{(1)} p_1^i = 1 \cdot 2^0 + 0 \cdot 2^1 + 1 \cdot 2^2 \equiv 5 \,(\mathrm{mod}\,8). \tag{2.5}$$

For $p_2 = 3$:

k	0	1	2
$\alpha^{(p-1)k/p_2}$	1	$5^{24} \equiv 8$	$5^{24 \cdot 2} \equiv 64$

i	0	$1 = a_2 - 1$
β_i	68	$68 \cdot 5^{-1} \equiv 72$
$\beta_i^{(p-1)/p_2^{i+1}}$	$68^{24} \equiv 8$	$72^8 \equiv 1$
$b_i^{(2)}$	1	0

Thus, the base 3 representation of $\log_5(68)$ modulo 9 is:

$$\sum_{i=0}^{a_2-1} b_i^{(2)} p_2^i = 1 \cdot 3^0 + 0 \cdot 3^1 \equiv 1 \,(\mathrm{mod}\,9). \tag{2.6}$$

Solving (2.5)–(2.6) by the Chinese Remainder Theorem, we get that $e = \log_5 68 = 37$ in $\mathbb{F}_{73}^*$. (See Exercises 2.6–2.23.) There are efficient methods for implementing the Chinese Remainder Theorem (see [59], for instance).

The complexity of finding e in (2.2) when p has n digits is virtually the same as factoring an n-digit integer (see [178]). This tells us that computing discrete logs is roughly as hard as factoring (algorithms for which we will study in Chapter 5). To date, no tractable factorization algorithms are known, so factoring is assumed to be intrinsically difficult. Hence, cryptosystems based upon the computation of discrete logs are assumed to be intractable. However, it should be stressed that nobody has established nontrivial lower bounds for the complexity of integer factorization.

Remark 2.7 *If $n = p - 1$, then given a factorization of n, the running time of the Silver-Pohlig-Hellman Discrete Log Algorithm is*

$$O\left(\sum_{j=1}^{r} a_j \left(\ln n + \sqrt{p_j}\right)\right)$$

group multiplications. This implies that the Pohlig-Hellman Algorithm is only efficient if the prime divisors of $p-1$ are small. This is the reason why we talked about a proper choice of p in the beginning of this section for the intractability of the discrete log problem. It should also be noted that the above algorithm *makes use of what is known as the* baby-step giant-step *algorithm for computing discrete logs due to the late Dan Shanks, a pioneer in computational number theory. Lastly, recent techniques developed by Dan Bernstein* [24] *can speed up the above algorithm.*

Remark 2.7 motivates us to look at the baby-step giant-step method, which is a generic method in the sense that it can be used with any finite group. For simplicity of presentation, we state the algorithm for a cyclic group. The original idea is attributed to Dan Shanks[2.1] by Odlyzko [175] and Knuth [124].

◆ **Baby-Step Giant-Step Algorithm for Computing Discrete Logs**

Given a generator α of a cyclic group G of order n, and $\beta \in G$, the goal is to compute the discrete logarithm $x \equiv \log_\alpha \beta \pmod{n}$.

(1) Compute $s = \lfloor \sqrt{n} \rfloor$.

(2) **Baby-Step**: For $j = 0, 1, \ldots, s - 1$, compute $(j, \alpha^j \beta)$. Then sort the list by second component in ascending order.

(3) **Giant-Step**: For $i = 1, 2, \ldots, s$ compute (α^{is}, i) and sort by first component in ascending order.

(4) **Search and Compare**: Search the lists in Steps (2)–(3) to see if there is an $\alpha^j \beta$ from Step (2) and an α^{is} from step (3) such that $\alpha^j \beta = \alpha^{is}$. If so, then compute $x \equiv is - j \pmod{n}$, which is $\log_\alpha \beta \pmod{n}$.

Example 2.8 *Let $\alpha = 5$, $\beta = 71$, and $n = 167$. We want to determine*

$$x \equiv \log_5(71) \pmod{167}.$$

First, we calculate $s = \lfloor \sqrt{n} \rfloor = 12$. The baby-step is the computation of

$$(j, 5^j \cdot 71 \pmod{167}) \text{ for } j = 0, 1, \ldots, 11:$$

$(0, 71)$, $(1, 21)$, $(2, 105)$, $(3, 24)$, $(4, 120)$, $(5, 99)$, $(6, 161)$, $(7, 137)$, $(8, 17)$, $(9, 85)$, $(10, 91)$, $(11, 121)$. *Then we sort according to the second element:*

j	8	1	3	0	9	10
$5^j \cdot 71$	17	21	24	71	85	91

j	5	2	4	11	7	6
$5^j \cdot 71$	99	105	120	121	137	161

[2.1] Daniel Shanks (1917–1996) is responsible not only for this algorithm, but also for numerous others such as his SQUFOF algorithm for factoring (see [196]). His book [204] is a classic in number theory. He had not only a depth of intellect that contributed fundamental results, but also a rapier wit and warm humanity. He died on September 9, 1996 from a heart attack.

The giant-step *is the computation of*

$$(5^{12i} \pmod{167}, i) \ for \ i = 1, 2, \ldots, 12 :$$

$(152, 1)$, $(58, 2)$, $(132, 3)$, $(24, 4)$, $(141, 5)$, $(56, 6)$, $(162, 7)$, $(75, 8)$, $(44, 9)$, $(8, 10)$, $(47, 11)$, $(130, 12)$. *Then we order according to the first component:*

15^{12i}	8	24	44	47	56	58
i	10	4	9	11	6	2

15^{12i}	75	130	132	141	152	162
i	8	12	3	5	1	7

Then we search the two lists and find that

$$\alpha^3 \beta \equiv 24 \equiv \alpha^{4 \cdot 12} \pmod{167},$$

so $x = 4 \cdot 12 - 3 = 45$ *and indeed:*

$$\log_5(71) \equiv 45 \pmod{167} \ since \ 5^{45} \equiv 71 \pmod{167}.$$

The baby-step giant-step method presented above was first used by Shanks in August of 1968 to calculate the class number of an imaginary quadratic field (see [240]). The running time for the algorithm is $O(\sqrt{n})$ group operations and according to [156, Note 3.67(i), p. 109] is the same as the Silver-Pohlig-Hellman Algorithm if n is prime. Moreover, it uses $O(\sqrt{n})$ memory, so this deterministic algorithm has a runtime/memory trade-off. Shanks' method is a kind of square root method, of which Pollard provided other kinds such as his rho-method (see [163, pp. 127–130]). Such methods have also been used on elliptic curves (see [227]–[228]).

We now look at (arguably) the most potent and efficacious of the methods for computing discrete logs. In its general form, it bears a strong resemblance to some of the most powerful factoring algorithms (such as the *number field sieve* (see [165, Section 5.2, pp. 207–220]), which may be considered to be a variant of the following method). Although the following has a more general formulation for other cyclic groups, we restrict our attention to $\mathbb{F}_p^*$ for the sake of simplicity of presentation. The following is a subexponential time algorithm (see Appendix B).

◆ The Index-Calculus Algorithm for Computing Discrete Logs

We solve $\beta \equiv \alpha^x \pmod{p}$ where p is a large prime and α is a primitive root modulo p.

Precomputation stage:

(1) Select a *factor base*[2.2] (a set of "small primes" that will remain the primes under consideration for the duration of the algorithm): $\mathcal{B} = \{p_1, \ldots, p_B\}$

[2.2]The term *factor base* was coined by John Brillhart (see Footnote 4.5 on page 83).

consisting of the first B primes. (Here the choice for B should be made such that a "considerable number" of the elements of $\mathbb{F}_p^*$ can be expressed as products of powers of elements of $\mathcal{B}$.)

(2) Collect relations by choosing a random nonnegative integer $k \leq p - 2$ and compute the least positive residue of α^k modulo p, if possible, then its canonical prime factorization, $\prod_{j=1}^{B} p_j^{k_j}$ for $k_j \geq 0$. When such relations exist we may take logs and get

$$k \equiv \sum_{j=1}^{B} k_j \log_\alpha(p_j) \,(\mathrm{mod}\ p - 1). \tag{2.9}$$

Continue to choose (at least) B such k so that we are successful in securing B relations as in (2.9). Here we are trying to solve for $\log_\alpha(p_j)$ for $j = 1, 2, \ldots, B$.

Calculation of discrete logs stage:

(3) For each k in (2.9), determine the value of $\log_\alpha(p_j)$ for $1 \leq j \leq B$ by solving the B (modular) linear equations with unknowns $\log_\alpha(p_j)$.

(4) Select a random nonnegative integer $t \leq p - 2$ and compute $\beta\alpha^t$.

(5) If possible, *factor $\beta\alpha^t$ over* $\mathcal{B}$, namely, write

$$\beta\alpha^t = \prod_{j=1}^{B} p_j^{t_j} \qquad (t_j \geq 0). \tag{2.10}$$

If it is not possible to get (2.10), then go to step (4). If (2.10) is successfully obtained, then

$$\log_\alpha(\beta) + t \equiv \sum_{j=1}^{B} t_j \log_\alpha(p_j) \,(\mathrm{mod}\ p - 1),$$

from which we can calculate $\log_\alpha(\beta)$.

As usual, a small example will suffice to illustrate the algorithm.

Example 2.11 *Let $p = 3361$, $\alpha = 22$, and $\mathcal{B} = \{2, 3, 5, 7\}$. We wish to compute $\log_{22}(4)$ in $\mathbb{F}_{3361}^*$ using the index-calculus method. We choose randomly $k = 48, 100, 186, 2986$ and get*

$$22^{48} \equiv 2^5 \cdot 3^2 \,(\mathrm{mod}\ 3361), \qquad 22^{100} \equiv 2^6 \cdot 7 \,(\mathrm{mod}\ 3361),$$

$$22^{186} \equiv 2^9 \cdot 5 \,(\mathrm{mod}\ 3361), \qquad 22^{2986} \equiv 2^3 \cdot 3 \cdot 5^2 \,(\mathrm{mod}\ 3361).$$

Thus we get the system of four congruences in four unknowns:

$$48 \equiv 5\log_{22}(2) + 2\log_{22}(3) \,(\mathrm{mod}\ 3360),$$

$$100 \equiv 6\log_{22}(2) + \log_{22}(7) \,(\mathrm{mod}\ 3360),$$

$$186 \equiv 9\log_{22}(2) + \log_{22}(5) \,(\mathrm{mod}\ 3360)\ and,$$

$$2986 \equiv 3\log_{22}(2) + \log_{22}(3) + 2\log_{22}(5) \,(\mathrm{mod}\ 3360).$$

This completes the precomputation stage. Now we use this to compute:

$$\log_{22}(2) = 1100; \log_{22}(3) = 2314; \log_{22}(5) = 366;\ and\ \log_{22}(7) = 220.$$

Suppose that we now select $t = 754$ *at random and compute,*

$$\beta\alpha^t = 4 \cdot 22^{754} \equiv 2 \cdot 3^2 \cdot 5 \cdot 7 \,(\mathrm{mod}\ 3361).$$

Thus, we have,

$$\log_{22}(4) + 754 \equiv \log_{22}(2) + 2\log_{22}(3) + \log_{22}(5) + \log_{22}(7) \,(\mathrm{mod}\ 3360).$$

Hence, $\log_{22}(4) = 2200$, *and we check that indeed*

$$22^{2200} \equiv 4 \,(\mathrm{mod}\ 3361).$$

(See Exercise 2.24.)

The DLP will show up throughout the text. For instance, we will look at discrete logs and key exchange when we discuss public-key cryptography in Chapter 3. The DLP in subgroups of $\mathbb{F}_p^*$ is assumed intractable and is used, for instance, as the security of the American Government NIST Digital Signature Algorithm (see Chapter 7). In Chapter 7, we will see that the ElGamal signature scheme is based upon discrete log, and in Chapter 3 that the ElGamal encryption is also based upon the DLP. In fact, in Chapter 5, when we discuss elliptic curves, we will see that the DLP in elliptic curve groups appears to be several orders of magnitude more difficult than the DLP in the multiplicative group of a finite field of similar size. The DLP plays an active role in numerous cryptographic protocols. We begin in Section 2.3 to look at some of them.

It should be pointed out that in early 2001, Joux and Lercier [116] announced that they had set a new record for computing discrete logs modulo a prime of 120 decimal digits in ten weeks on a single 525 MHz quadri-processor Digital Alpha Server 8400 computer. Hence, today, for long-term security, one should choose a prime modulus of 1024 bits (308 decimal digits) for long-term security.

Exercises

2.5. Given $j \in \{1, 2, \ldots, r\}$, establish that for each $i = 0, 1, \ldots, a_j - 1$,

$$\beta_i^{(p-1)/p_j^{i+1}} \equiv \alpha^{(p-1)b_i^{(j)}/p_j} \pmod{p},$$

in the Silver-Pohlig-Hellman Algorithm for computing discrete logs, described on page 39.

In Exercises 2.6–2.23, assume α generates $\mathbb{F}_p^*$ for the given α and prime p in each case, and compute $\log_\alpha \beta$ for the given β using the Silver-Pohlig-Hellman Algorithm for computing discrete logs.)

2.6. Let $p = 2083$, $\alpha = 2$, and $\beta = 19$.

2.7. Let $p = 37$, $\alpha = 2$, and $\beta = 19$.

2.8. Let $p = 1483$, $\alpha = 2$, and $\beta = 21$.

2.9. Let $p = 1579$, $\alpha = 3$, and $\beta = 31$.

2.10. Let $p = 1637$, $\alpha = 2$, and $\beta = 5$.

2.11. Let $p = 1721$, $\alpha = 3$, and $\beta = 7$.

2.12. Let $p = 1759$, $\alpha = 6$, and $\beta = 13$.

2.13. Let $p = 1783$, $\alpha = 10$, and $\beta = 3$.

2.14. Let $p = 1801$, $\alpha = 11$, and $\beta = 2$.

2.15. Let $p = 1871$, $\alpha = 14$, and $\beta = 5$.

2.16. Let $p = 2039$, $\alpha = 7$, and $\beta = 15$.

2.17. Let $p = 2161$, $\alpha = 23$, and $\beta = 3$.

2.18. Let $p = 2287$, $\alpha = 19$, and $\beta = 10$.

2.19. Let $p = 2351$, $\alpha = 13$, and $\beta = 2$.

2.20. Let $p = 2521$, $\alpha = 17$, and $\beta = 2$.

2.21. Let $p = 2689$, $\alpha = 19$, and $\beta = 7$.

2.22. Let $p = 2999$, $\alpha = 17$, and $\beta = 7$.

2.23. Let $p = 3361$, $\alpha = 22$, and $\beta = 3$.

2.24. Go through Exercises 2.6–2.23 using the index-calculus method for computing discrete logs.

Genius is one per cent inspiration, ninety-nine per cent perspiration.

Thomas Alva Edison (1847–1931) American inventor

2.3 Exponentiation Ciphers and Diffie-Hellman

Nearly every man who develops an idea works it up to a point where it looks impossible, and then gets discouraged. That's not the place to become discouraged.

Thomas Alva Edison

As we shall see, modular exponentiation, $c \equiv m^e \,(\mathrm{mod}\ n)$ is at the heart of RSA and one of the most important operations in public-key cryptography. We begin by looking at some ciphers that use modular exponentiation. The first was introduced in 1978 in [185]. Encryption and decryption via the following are examples of the first kind of exponentiation algorithm, *fixed-exponent exponentiation*, where the exponent is fixed but the base may vary.

◆ **Pohlig-Hellman**[2.3] **Symmetric-Key Exponentiation Cipher**

(a) A secret prime p is chosen and a secret enciphering key $e \in \mathbb{N}$ with $e \le p-2$.

(b) A secret deciphering key d is computed via $ed \equiv 1 \,(\mathrm{mod}\ p-1)$.

(c) Encryption of plaintext message units m is accomplished via

$$c \equiv m^e \,(\mathrm{mod}\ p).$$

(d) Decryption is achieved via $m \equiv c^d \,(\mathrm{mod}\ p)$.

Since knowledge of e and p would allow a cryptanalyst to obtain d, then both p and e must be kept secret. The security of this cipher is based on the difficulty of solving the discrete log problem discussed in Section 2.2, namely, an adversary, without knowledge of e or d, would have to compute

$$e \equiv \log_m(c) \,(\mathrm{mod}\ p).$$

Example 2.12 *We will choose* $p = 104729$ *which is unrealistically small, but for pedagogical reasons we need to do this, as we will several times throughout the text. Let* $e = 1011$ *and* $d = 41539$, *observing that* $ed \equiv 1 \,(\mathrm{mod}\ p-1)$, *as required. Thus if* $m = 76$, *we encipher via*

$$c \equiv m^{1011} \equiv 26422 \,(\mathrm{mod}\ 104729),$$

and we see that deciphering is given by

$$m \equiv 26422^{41539} \equiv 76 \,(\mathrm{mod}\ 104729).$$

[2.3]Martin E. Hellman (1945–) received his Ph.D. in electrical engineering from Stanford University in 1969. After brief stints at IBM and MIT, he returned to Stanford in 1971 where he remained until 1996 when he became Professor Emeritus. He is credited (along with Diffie and Merkle — see Footnote 2.4) for discovery of the notion of public-key cryptography. It should be noted that (as we will discuss in Section 3.5) public-key cryptography was actually known (in the classified sector in England) several years before Diffie and Hellman published their result.

In Section 2.1, we saw how Alice and Bob solved a problem via coin flipping over a phone. We now show how they can solve a problem via a coin flipping protocol involving exponentiation, sometimes called *flipping coins into a well*.

◆ **Coin Flipping by Exponentiation**

Suppose that Alice and Bob want to make a decision based upon a random coin flip, but they are not in the same physical space. We now describe how they can do this over a channel remotely. The following is the cryptographic protocol for electronic coin flipping using exponentiation modulo a prime $p > 2$.

(1) Alice and Bob agree upon a prime p such that the factorization of $p - 1$ is known.

(2) Alice selects two generators $\alpha, \beta \in \mathbb{F}_p^*$ and sends them to Bob.

(3) Bob randomly chooses an integer x, relatively prime to $p - 1$, then he computes *one* of $y \equiv \alpha^x \pmod{p}$ or $y \equiv \beta^x \pmod{p}$, and sends y to Alice.

(4) Alice *guesses* whether y is a function of α or β and sends the guess to Bob.

(5) If Alice's guess is correct, then the result of the coin flip is deemed to be heads, and if incorrect, it is tails. Bob sends the result of the coin flip to Alice.

In order for Bob to cheat on the coin toss, two integers x_1 and x_2 must be known where $\alpha^{x_1} \equiv \beta^{x_2} \pmod{p}$. To compute x_2 given x_1, Bob must compute $\log_\alpha \beta^{x_2} \pmod{p}$, which is possible if Bob knows $\log_\alpha \beta$. However, Alice chooses α and β in step 2. Hence, Bob is in the position of having to compute a discrete log. (Thus, the coin flipping protocol relies on the difficulty of the DLP.) Also, Bob could cheat by choosing an x such that $\gcd(x, p-1) > 1$. That can be avoided by the following verification step added on the end, called the *verification protocol*.

(6) Bob reveals x to Alice. Then Alice computes $\alpha^x \pmod{p}$ and $\beta^x \pmod{p}$ to verify both the outcome of the coin toss, and that Bob has not cheated.

Step (6) ensures that Bob did not cheat at step (3) since Alice can then check $\gcd(x, p - 1)$.

Characteristics implicit in the above protocol are that neither Alice nor Bob learns about the coin flip at the same time. Also, at some point in the protocol, one of them knows the result of the coin flip, but cannot alter it. The reason this type of coin flipping protocol is called *flipping coins into a well* is that it is a metaphor for the following situation. Suppose that Bob is next to a well, and that Alice is physically removed from this well. Bob throws a coin into the well, and can see (but not reach) the coin at the bottom of it. Alice cannot see the result until Bob allows Alice to come to the well to have a look.

The following is an example of the second type of exponentiation, *fixed-base exponentiation*. Moreover, this was the first step in public-key cryptography, the main topic of this text. It was also the principal motivation for the intensive interest in discrete logs over the past quarter century.

◆ The Diffie-Hellman Key-Exchange Protocol

Suppose that Alice and Bob have not yet met nor exchanged keys, but they want to establish a shared secret key k by exchanging messages over an unsecured channel. First Alice and Bob agree on a large prime p and a generator α of $\mathbb{F}_p^*$ ($2 \leq \alpha \leq p - 2$). These need not be kept secret, so Alice and Bob can agree over an unsecured channel. Then the protocol proceeds as follows.

(1) Alice chooses a random (large) $x \in \mathbb{N}$ and computes the least positive residue X of α^x modulo p, then sends X to Bob (and keeps x secret).

(2) Bob chooses a random (large) $y \in \mathbb{N}$ and computes the least positive residue Y of α^y modulo p, then sends Y to Alice (and keeps y secret).

(3) Alice computes the least positive residue k of Y^x modulo p.

  (*Note that $Y^x \equiv \alpha^{yx} \pmod{p}$.*)

(4) Bob computes the least positive residue k of X^y modulo p.

  (Note that $X^y \equiv \alpha^{xy} \pmod{p}$.)

In the Diffie-Hellman protocol, k is the shared secret key independently generated by both Alice and Bob. The key exchange is complete, since Alice and Bob are in agreement on k. A cryptanalyst Eve listening to the channel would know p, α, X, and Y, but neither x nor y. Thus, Eve faces what is called the

Diffie-Hellman Problem (DHP):

find α^{xy} given α, α^x and α^y (but not x or y).

If Eve can solve the DLP, then she can clearly solve the DHP. Whether the converse is true or not is unknown. In other words, it is not known if it is possible for a cryptanalyst to solve the DHP without solving the DLP. Nevertheless, the consensus is that the two problems are equivalent. Thus, for practical purposes, one may assume that the Diffie-Hellman Key-Exchange Protocol is secure as long as discrete log is intractable. (The equivalence of the two problems has been proved for certain situations. See [39], [68], and [150]–[154].) Also, in [206], Shmuely showed that if $n = pq$ where p and q are odd primes, and the DHP in $(\mathbb{Z}/n\mathbb{Z})^*$ can be solved in polynomial time for a nontrivial magnitude of all generators (primitive roots) $\alpha \in (\mathbb{Z}/n\mathbb{Z})^*$, then n can be factored in expected polynomial time (see Appendix B).

Notice that the Diffie-Hellman Protocol differs from the Pohlig-Hellman Cipher in that the latter requires that both p and e be kept secret since d could be deduced from them, whereas in the former p and α may be made public due to the intractability of the DLP. The idea leading to the above protocol began in the 1974–1975 academic year with Whitfield Diffie[2.4] and Martin Hellman at Stanford University, along with Ralph Merkle at the University of California at Berkeley. The Diffie-Hellman idea, that the enciphering key could be made *public* since it is computationally infeasible to obtain the deciphering key from it, is at the heart of public-key cryptography, which we will begin to study in depth in Chapter 3 with Diffie-Hellman as the "door-opener" to the subject.

The Diffie-Hellman protocol [74] in 1976 was the first practical key-exchange technique to be published. It was a landmark in cryptographic development since this was the first protocol with public-key properties including the idea of a *trapdoor one-way function* (see Section 3.1), a partial solution to creating *public-key cryptosystems* (see Section 3.2), and *digital signatures* (see Chapter 7). (In Chapter 3 we will see how the authors of RSA took the ideas in [74] and created the first public-key cryptosystem.) We begin in the next chapter to travel the road paved by the appearance of this paper and explore all the ramifications that it brought to the cryptographic arena.

We end this section by looking at an algorithm that uses modular exponentiation to speed up calculations. This is an example of another (the last) kind of exponentiation algorithm, *basic exponentiation*, which can be used with any base m and exponent e.

Our task is to calculate the least positive residue of b^r modulo n. To do this in a single exponentiation would, for large values of r, overflow the memory of most computers. Even if we start with b and multiply by b, $r-1$ times reducing modulo n at each step, the process is too slow. Happily, there is an efficient process of squaring and reducing modulo n given in the following algorithm.

◆ The Repeated Squaring Method for Modular Exponentiation

We compute b^r modulo n for $b, r, n \in \mathbb{N}$ as follows. First write r in its binary representation:

$$r = \sum_{j=0}^{k} a_j 2^j.$$

We wish to calculate $c \equiv b^r \pmod{n}$ in a stepwise fashion as follows.

Initial Step: Set $b_0 = b$, and:

$$c = \begin{cases} 1 & \text{if } a_0 = 0, \\ b & \text{if } a_0 = 1. \end{cases}$$

[2.4]Whitfield Diffie (1944–) received his B.Sc. in mathematics from MIT in 1965. Later he worked on development of the mathematical symbolic manipulation package *Mathlab*, and at Stanford University on proofs of correctness of computer programs. For his involvement, along with Hellman and Merkle, in the discovery of the notion of public-key cryptography, he was awarded a Doctorate in Technical Sciences (Honoris Causa) by the Swiss Federal Institute of Technology in 1992. Prior to his current position as Distinguished Engineer at Sun Microsystems, in Palo Alto, California, assumed in 1991, he was Manager of Secure Systems Research for Northern Telecom. Since 1993, his primary focus has been on public policy concerning cryptography, taking a position against limitations on the use of cryptography by individuals and corporations, which included testifying before the House and the Senate. He has received numerous awards including the 1997 Louis E. Levy Medal from the Franklin Institute in Philadelphia.

Perform the following step for each $j = 1, 2, \ldots, k$:

j-th Step: Calculate the least nonnegative residue b_j of b_{j-1}^2 modulo n. If $a_j = 1$, then replace c by $c \cdot b_j$, and reduce modulo n. If $a_j = 0$ leave c unchanged. What is achieved at the jth step is the computation of

$$c_j \equiv b^{r_j} \pmod{n},$$

where c_j is the least nonnegative residue of b^{r_j} modulo n, and

$$r_j = \sum_{i=0}^{j} a_i 2^i.$$

Hence, at the kth step, we have calculated $c \equiv b^r \pmod{n}$.

Example 2.13 *Suppose that we wish to calculate the least positive residue of 3^{61} modulo 101. To do this in a single exponentiation would be an enormous task, but can be simplified by using the repeated squaring method. Since*

$$61 = 1 + 2^2 + 2^3 + 2^4 + 2^5,$$

then $k = 5$, $a_j = 1$ for $j = 0, 2, 3, 4, 5$ and $a_1 = 0$. Also, since $a_0 = 1$, we set $c = 3 = b = b_0$. The following are the jth steps for $j = 1, 2, 3, 4, 5$.

(1) $3^2 \equiv 9 \pmod{101}$, *so $b_1 = 9$, and since $a_1 = 0$, then $c = 3$ remains.*

(2) $b_2 \equiv 9^2 \pmod{101}$, *so $b_2 = 81$. Since $a_2 = 1$, then*

$$3 \cdot 81 \equiv 243 \equiv 41 \pmod{101}, \text{ *so c becomes* } c = 41.$$

(3) $b_3 \equiv 81^2 \equiv 97 \pmod{101}$, *so $b_3 = 97$. Since $a_3 = 1$, then*

$$c \cdot b_3 \equiv 41 \cdot 97 \equiv 38 \pmod{101}, \text{ *and c becomes* } c = 38.$$

(4) $b_4 \equiv 97^2 \equiv 16 \pmod{101}$, *so $b_4 = 16$. Since $a_4 = 1$, then*

$$c \cdot b_4 \equiv 38 \cdot 16 \equiv 2 \pmod{101}, \text{ *making* } c = 2.$$

(5) $b_5 \equiv 16^2 \equiv 54 \pmod{101}$, *so $b_5 = 54$. Also, since $a_5 = a_k = 1$, then*

$$c \cdot b_5 \equiv 2 \cdot 54 \equiv 7 \pmod{101}.$$

Hence, $3^{61} \equiv 7 \pmod{101}$.

Example 2.13 clearly shows the power of the repeated squaring method which we will have occasion to use throughout the text. In Chapter 3, we will begin with a mention of one such application. For the reader interested in an extremely fast (but more complicated) exponentiation algorithm that uses only multiplications, see Pippenger's developments in [180]–[182].

Exercises

In Exercises 2.25–2.30, find the least nonnegative residue modulo p for each given exponent using the repeated squaring method (see page 50).

2.25. 5^{51} modulo $p = 97$.

2.26. 7^{49} modulo $p = 103$.

2.27. 9^{71} modulo $p = 113$.

2.28. 10^{79} modulo $p = 151$.

2.29. 11^{22} modulo $p = 167$.

2.30. 13^{51} modulo $p = 179$.

In Exercises 2.31–2.34, decipher the cryptogram assuming that it was encrypted using the Pohlig-Hellman cipher with the given prime p and enciphering key e. Determine the deciphering key d to get the numerical equivalents of the plaintext, which will then be translated via Table 1.2 on page 3.

2.31. $p = 167$, $e = 69$ and cryptogram is $85, 50, 96, 96, 0, 27, 50$.

2.32. $p = 397$, $e = 95$ and cryptogram is $344, 107, 128, 107, 198, 339, 107, 49$.

2.33. $p = 1187$, $e = 107$ and cryptogram is $1084, 235, 904, 0, 18, 904$.

2.34. $p = 1481$, $e = 1201$ and cryptogram is $1175, 185, 1032, 293, 263, 185, 167$.

In Exercises 2.35–2.47, assume that the p, α, x, and y are the parameters in the Diffie-Hellman protocol. Determine the key k generated by Alice and Bob in each case.

2.35. $p = 3361$, $\alpha = 22$, $x = 2999$, $y = 3299$.

2.36. $p = 3529$, $\alpha = 17$, $x = 3400$, $y = 3501$.

2.37. $p = 3631$, $\alpha = 15$, $x = 2001$, $y = 3001$.

2.38. $p = 3881$, $\alpha = 13$, $x = 3801$, $y = 2799$.

2.39. $p = 3889$, $\alpha = 11$, $x = 2000$, $y = 3811$.

2.40. $p = 4051$, $\alpha = 10$, $x = 2051$, $y = 3000$.

2.41. $p = 4129$, $\alpha = 13$, $x = 1998$, $y = 3900$.

2.42. $p = 4201$, $\alpha = 11$, $x = 4200$, $y = 2111$.

2.43. $p = 4339$, $\alpha = 10$, $x = 2911$, $y = 3911$.

2.44. $p = 4391$, $\alpha = 14$, $x = 3311$, $y = 4131$.

2.45. $p = 4441$, $\alpha = 21$, $x = 4011$, $y = 3191$.

2.46. $p = 4657$, $\alpha = 15$, $x = 4111$, $y = 2229$.

2.47. $p = 5039$, $\alpha = 11$, $x = 3133$, $y = 4313$.

Chapter 3

Public-Key Cryptography

Eternal law has arranged nothing better than this, that it has given one way in to life, but many ways out.

Seneca ('the Younger') circa (4 B.C. – A.D. 65)
Roman philosopher and poet

3.1 One-Way Functions

In this chapter, we will be using numerous terms from number theory for which Appendix C will serve the reader as a quick reference.

In Section 2.3, we looked at the Pohlig-Hellman symmetric-key cipher and the Diffie-Hellman key-exchange protocol, both of which use modular exponentiation, which can be efficiently computed using, for instance, the repeated squaring method given on page 50. However, as discussed in Section 2.2, there is no known efficient algorithm for computing discrete logs. Hence, modular exponentiation is an example of the following type of function. The reader may want to refer to Appendix B as well as Remark 1.9 on page 7 for a discussion of terms used in the following, which is an informal rendering of the notion (see [99, pp. 32–43] for formal definitions and variations).

Definition 3.1 (One-Way Functions)

A one-to-one function f from a set $\mathcal{M}$ to a set $\mathcal{C}$ is called one-way *if $f(m)$ is "easy" to compute for all $m \in \mathcal{M}$, but for a randomly selected c in the image of f, finding an $m \in \mathcal{M}$ such that $c = f(m)$ is computationally infeasible. In other words, we can easily compute f, but it is computationally infeasible to compute f^{-1}.*

Diagram 3.2 One-Way Function

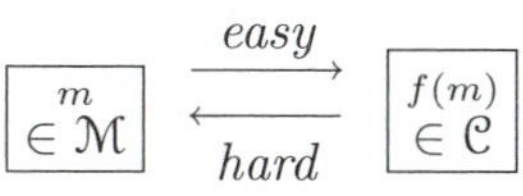

$$
\boxed{\begin{array}{c} m \\ \in \mathcal{M} \end{array}} \quad \overset{easy}{\underset{hard}{\overset{\longrightarrow}{\longleftarrow}}} \quad \boxed{\begin{array}{c} f(m) \\ \in \mathcal{C} \end{array}}
$$

It has not been rigorously proved that one-way functions actually exist since (as we have discussed in Remark 1.9 on page 7 and in Footnote B.9 on page 211) there is no rigorous definition of the terms "computationally easy" or "computationally infeasible". Nevertheless, from a pragmatic (working) viewpoint we have many cryptosystems basing their security upon the presumed computational infeasibility of integer factorization (see Chapter 5) or the DLP (see Section 2.2 on page 39), for instance. Thus, at the present state of mathematical development, one can only say that we have *conjectured* or *candidate* one-way functions, such as the latter two mentioned above. If we could prove the existence of one-way functions, a corollary of this would be $\mathbf{P} \neq \mathbf{NP}$ (see Appendix B). Nevertheless, we will take the pragmatic approach since, as we have seen, there are functions that are easy to evaluate, but for which there is no *known* efficient algorithm to invert them, so they can be used as one-way functions. A secure digital signature scheme which uses a one-way function will be studied in Chapter 7.

In Section 2.1, we described how to flip coins by telephone and in Section 2.3 how to flip coins via exponentiation. We now illustrate Definition 3.1 for yet another method of coin flipping.

◆ Coin Flipping Using One-Way Functions

(1) Alice and Bob both know a one-way function f but not the inverse f^{-1}.

(2) Bob selects an integer x at random and sends the value $f(x) = y$ to Alice.

(3) Alice makes a guess concerning a property of the number x that is valid fifty percent of the time, such as parity (whether x is odd or even) and sends the guess to Bob.

(4) Bob tells Alice whether or not the guess is correct.

(5) Bob sends Alice the value x.

(6) Alice confirms that $f(x) = y$ (verification step).

The security of this protocol relies on the choice of f, which must reliably produce, for instance, even and odd numbers with equal probability. If it does not, and were to produce say even numbers sixty percent of the time, then Alice could guess even every time in step (3) and have an advantage in the coin tosses. Also, of course, f must be one-to-one. If not then there could exist an x_1 even and an x_2 odd such that

$$f(x_1) = f(x_2) = y,$$

and Bob can cheat Alice every time.

The above protocol is another example of tossing coins into a well, which we first described on page 48. There is a related protocol for which we need the following concept.

Definition 3.3 (Hash Functions)

A hash function *is a computationally efficient function that maps bitstrings of arbitrary length to bitstrings of fixed length, called* hash values. *A* one-way hash function *is a hash function that satisfies Definition 3.1. The process of using a hash function on a message is called* hashing the message.

In order for a hash function to be cryptographically useful, it must be a one-way hash function to prevent easy unauthorized retrieval of the original bitstring. Thus, one-way hash functions are sometimes called *cryptographic hash functions*. We will mean a cryptographic hash function when we use the term *hash function* henceforth.[3.1] The hash values produced by such a hash function are used as a concentrated representative of the original bitstring, so they can be used as a unique identifier of it. Thus, this type of hash value is called a *message digest, imprint,* or *digital fingerprint*. In Chapter 7, we will look at applications of hash functions to message authentication and other aspects. For now, we look at some illustrations of one-way functions that build upon previous illustrations. We set the stage as follows. The problem is that Bob wants to commit to a prediction (of either a 0 or a 1), and does not want to reveal this prediction to Alice until sometime later (where *commit* means that Bob cannot change the choice later on). Alice, on the other hand, wants to ensure that Bob cannot change the bit prediction once made.

Now we exhibit a bit commitment scheme based upon our main topic.

◆ **Bit Commitment Protocol Using One-Way Functions**

Suppose that Alice and Bob agree upon a hash function h. Then the following steps are executed.

(1) Alice generates a random random bitstring R_1 and sends it to Bob.

(2) Bob generates a random bitstring R_2, and creates a message consisting of R_1, R_2 and the bit b to which he wants to commit, producing (R_1, R_2, b).

(3) Bob sends the pair $(h(R_1, R_2, b), R_1)$ to Alice. This concludes the commitment portion of the protocol. Later, when Bob is ready to reveal the committed bit, we continue as follows.

(4) Bob sends Alice the original message (R_1, R_2, b).

(5) Alice hashes the message and compares it, and R_1, with $h(R_1, R_2, b)$ and R_1. If they match, the bit is valid.

Bob cannot cheat since he is using a one-way hash function so it is not possible to send a message (R_1, R_2, b_1), for $b_1 \neq b$ in step (4), with $h(R_1, R_2, b) = h(R_1, R_2, b_1)$. Also, since Alice chooses R_1 in step (1), he cannot anticipate the

---

[3.1]There is a special terminology for one-way hash functions. A *hash collision* for a hash function h occurs if there exist $x_1 \neq x_2$ such that $h(x_1) = h(x_2)$. If it is computationally infeasible to find values x_1, x_2 with $x_1 \neq x_2$ and $h(x_1) = h(x_2)$, we call h *collision-free* (sometimes called *strongly* collision-free).

protocol and begin a brute force attack in advance to find a hash collision (which is possible if *he* generates R_1) and which he might find under optimal conditions if given the chance, and the time (see the *birthday attack* in Section 9.2).

The bitstrings that Bob sends to Alice to commit to a bit are typically called *blobs* or *envelopes*. Blobs have these properties: (1) Bob, by committing to a blob, commits to a bit. (2) Bob can open any committed blob, and once opened can convince Alice of the value of the committed bit (called *binding*), but Alice cannot open the blob (called *concealing*). (3) Blobs do not carry any information other than Bob 's committed bit, and the blobs themselves. The means by which Bob commits to and opens them are not related to any other data that Bob might want to keep secret from Alice. This method for enciphering a blob is called a *bit commitment scheme*, namely, a two-entity protocol involving the above defined requirements of *binding* and *concealing*.

Historically, the term "bit commitment" developed as follows. In 1982, Manuel Blum introduced the problem of fair electronic coin flipping in [30]. He solved his problem using a bit commitment protocol in the fashion described below.

(1) Bob commits to a random bit using the above bit commitment protocol.

(2) Alice makes the guess.

(3) Bob reveals the committed bit to Alice, who wins if the predicted bit is correct.

To be "fair," this protocol must ensure that a guess is correct fifty percent of the time, that Bob cannot change the bit once it is committed, and Alice cannot know the predicted bit in advance of the guess. The protocol for coin flipping via exponentiation, given on page 48 is one way of ensuring that the fairness properties are satisfied (see Exercise 2.4 on page 38).

Now we describe two other bit commitment schemes using one-way functions. The first ensures that Bob commits to a bit that is unconditionally concealing to Alice but binding to Bob only under the assumption of the infeasibility of the DLP in $(\mathbb{Z}/p\mathbb{Z})^*$. For this, we need our trusted third party, Trent.

We assume that we are given a prime p, a generator α of $(\mathbb{Z}/p\mathbb{Z})^*$, and a random $x \in (\mathbb{Z}/p\mathbb{Z})^*$.

(1) Let $\beta \in (\mathbb{Z}/p\mathbb{Z})^*$ be arbitrarily chosen by Trent and let b be the bit to which Bob will commit.

(2) Bob enciphers b by computing $f(b, x) \equiv \beta^b \alpha^x \pmod{p}$.

(3) Bob commits to the blob $y = f(b, x)$.

This is unconditionally concealing to Alice. However, it is binding to Bob if and only if it is infeasible for Bob to compute $\log_\alpha(\beta)$.

The following is another method of using one-way functions, this time based upon an RSA modulus (see page 35) and modular square roots (see Exercise 2.1 on page 38). Again, we need Trent.

Let $n = pq$ be an RSA modulus, m a quadratic residue modulo n, and $x \in (\mathbb{Z}/n\mathbb{Z})^*$ arbitrarily chosen.

(1) We entrust p, q, and m to Trent, and let b be the bit to which Bob will commit.

(2) Bob enciphers by computing $f(b, x) \equiv m^b x^2 \pmod{n}$.

(3) Bob commits to the blob $y = f(b, x)$.

In this case, the concealing property is unconditional since given a quadratic residue $y \equiv x^2 \pmod{n}$ and $m \equiv z^2 \pmod{n}$, we have,

$$y \equiv f(0, x) \equiv f(1, xz^{-1}) \pmod{n},$$

so any quadratic residue is an encryption of both 0 and 1. Yet for binding, Bob can open a blob (as both a 0 and a 1) if and only if Bob can compute a square root of m modulo n. Hence, binding is based upon the infeasibility of computing a modular square root of m.

From the discussion above, the commitment schemes discussed are the digital analogues of opaque sealed envelopes, since by sealing a message in this envelope an entity commits to the substance of the message while keeping it secret.

We began the discussion in this section by talking about exponentiation as a motivator for the topic of one-way functions via such protocols as Diffie-Hellman key-exchange. However, what we have not yet addressed is the very important requirement that the intended receiver of the message, enciphered using a one-way function, must have some additional information for inverting the function. For instance, the sender could write the message on a piece of paper and then burn it, which is certainly a one-way function, but not a very useful one since the receiver has little hope of retrieving the message. Hence, we need a special kind of one-way function.

Definition 3.4 (Trapdoor One-Way Functions)

A trapdoor one-way function or *public-key enciphering function* *is a one-way function*

$$f : \mathcal{M} \mapsto \mathcal{C}$$

satisfying the additional property that there exists information, called trapdoor *information, or simply* trapdoor, *that makes it feasible to find* $m \in \mathcal{M}$ *for a given* $c \in \text{img}(f)$ *such that* $f(m) = c$, *but without the trapdoor this task becomes infeasible.*

The very heart of the Diffie-Hellman idea is the use of trapdoor one-way functions. The Diffie-Hellman Protocol, discussed in Section 2.3, allows for entities who have never met or exchanged information to establish a shared secret key by exchanging messages over an unsecured channel.

Diagram 3.5 Trapdoor One-Way Function

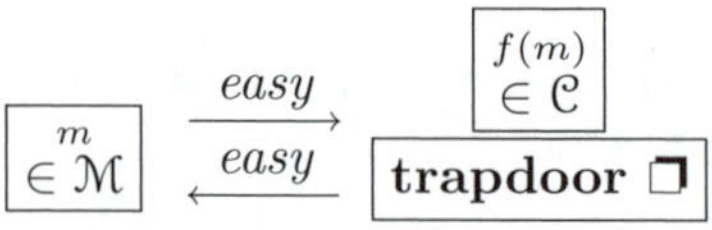

$$\boxed{\substack{m \\ \in \mathcal{M}}} \xrightarrow[\xleftarrow{\quad easy \quad}]{\quad easy \quad} \begin{array}{c} \boxed{\substack{f(m) \\ \in \mathcal{C}}} \\ \boxed{\textbf{trapdoor } \square} \end{array}$$

The most relevant trapdoor function for our purposes is illustrated as follows.

Example 3.6 *Let* $f(x) \equiv x^e \,(\mathrm{mod}\ n)$ *where* $n = pq$ *with* $p \neq q$ *primes and*

$$ed \equiv 1 \,(\mathrm{mod}\ (p-1)(q-1)).$$

We know that modular exponentiation is easy to compute since it can be done in polynomial time. However, $f^{-1}(x^e) \equiv f^{-1}(f(x)) \equiv x \,(\mathrm{mod}\ n)$ *is difficult to compute,* unless *the trapdoor* d *is known, since* $x^{ed} \equiv x \,(\mathrm{mod}\ n)$.

We now engage in a discussion of the reasons for the difficulty in computing the inverse. When we say that two algorithms are "computational equivalent", or "polynomially equivalent", we will mean that one can be obtained from the other in polynomial time. For the formal definition of "computational equivalence", also called "computational indistinguishability", see [98] or [156].

We now show that computing $(p-1)(q-1)$ is computationally equivalent to factoring n. If we have $(p-1)(q-1)$, then we can find p and q by successively computing,

$$p + q = n - (p-1)(q-1) + 1 \text{ and } p - q = \sqrt{(p+q)^2 - 4n}, \qquad (3.6)$$

so we get $p = \frac{1}{2}[(p+q) + (p-q)]$ and $q = \frac{1}{2}[(p+q) - (p-q)]$. Conversely, if we can factor n as pq, we can immediately compute $(p-1)(q-1)$, and we can use the Euclidean algorithm to find d from e (in computationally feasible time). Thus, the two are computationally equivalent. In Section 3.2, we will prove that computing the exponent d is computationally equivalent to factoring $n = pq$ (see Exercise 3.14). The above argument shows that knowing how to factor n allows us to compute d. So our task in Section 3.2 will be to prove that being able to compute d can be converted into an algorithm for factoring n.

We introduced the Diffie-Hellman key-exchange protocol on page 49, wherein we may now observe that the trapdoor in that protocol is either of the pair (x, y) without which a cryptanalyst is reduced to solving the DLP, which we know is intractable.

In the next section, we will introduce the notion of public-key cryptosystems for which the above paves the way. We will see that the existence of one-way functions is the basic underlying assumption given that knowledge of the encryption key e does not allow for computation of the decryption key d.

Exercises

In Exercises 3.1–3.10, refer to Example 3.6 for the setup, and find x and d in each case.

3.1. $p = 101$, $q = 167$, $e = 7$, and $x^e = 10303$.

3.2. $p = 157$, $q = 173$, $e = 5$, and $x^e = 111777$.

3.3. $p = 179$, $q = 191$, $e = 7$, and $x^e = 2828$.

3.4. $p = 211$, $q = 1481$, $e = 11$, and $x^e = 68641$.

3.5. $p = 881$, $q = 1231$, $e = 7$, and $x^e = 23921$.

3.6. $p = 1483$, $q = 1759$, $e = 7$, and $x^e = 5111$.

3.7. $p = 1753$, $q = 2081$, $e = 7$, and $x^e = 3269$.

3.8. $p = 2087$, $q = 2383$, $e = 5$, and $x^e = 67851$.

3.9. $p = 1753$, $q = 2699$, $e = 5$, and $x^e = 155515$.

3.10. $p = 2707$, $q = 3361$, $e = 13$, and $x^e = 31313$.

3.11. Prove that the choice of d in Example 3.6 actually does decrypt to give x.

3.12. If, in Example 3.6, p and q are chosen close together in the sense that $(p+q)/2$ is only slightly bigger than $\sqrt{pq}$, then it becomes easier to factor n, since we need only look at values $x^2 - n$ for $x > \sqrt{n}$ until a value

$$y^2 = x^2 - n$$

is achieved. Show that, whether p and q are close or not, for such a value of y, we get that $\gcd(x \pm y, n)$ are factors of n. Moreover, show that if

$$x \not\equiv \pm y \,(\mathrm{mod}\ n),$$

then $x+y$ and $x-y$ are nontrivial factors of n. Illustrate the problem with the closeness of p and q by performing such a calculation for the values in Exercise 3.2.

3.13. Suppose that, in Example 3.6, Alice sends $x^e \equiv c \,(\mathrm{mod}\ n)$ to Bob. However, Mallory intercepts c and selects a random $y \in (\mathbb{Z}/n\mathbb{Z})^*$, then sends $cy^e \equiv c' \,(\mathrm{mod}\ n)$ to Bob. Not knowing this, Bob computes

$$x' = c'^d \,(\mathrm{mod}\ n)$$

and sends it to Alice. How can Mallory retrieve x from x' if he intercepts it?

3.14. Prove that knowledge of the equations (3.6) allows us to factor n. In other words, knowledge of pq and $p + q$ allows us to find both p and q.

3.2 Public-Key Cryptosystems and RSA

Yet some there be that by due steps aspire to lay their just hands on that golden key that opes the place of eternity

John Milton (1608–1674) English poet

We have built the foundation for the introduction of the notion of public-key cryptography. As we have seen in Section 2.3, with the advent of the Diffie-Hellman key-exchange protocol arose the notion of a *public* enciphering key, since it is computationally infeasible to obtain the deciphering key from it. In Chapter 1, we studied symmetric-key cryptography, wherein both the enciphering key and the deciphering key had to be kept secret, since it is computationally easy to obtain one from the other. Now we are ready for the modern cryptographic offspring.

Definition 3.7 (Public-Key Cryptosystems)

A cryptosystem consisting of a set of enciphering transformations $\{E_e\}$ *and a set of deciphering transformations* $\{D_d\}$ *is called a* Public-key Cryptosystem *or an* Asymmetric Cryptosystem *if, for each key pair* (e, d), *the enciphering key* e, *called the* public key, *is made publicly available, while the deciphering key* d, *called the* private key, *is kept secret. The cryptosystem must satisfy the property that it is computationally infeasible to compute* d *from* e.

We use the convention that the term *private key* is reserved for use in association with public-key cryptography, whereas the term *secret key* is reserved for use in association with symmetric-key cryptosystems. This convention is used in the cryptographic community because it takes two or more entities to share a secret, but a key is truly private when only one entity knows about it.

A standard analogy for public-key cryptography is given as follows. Suppose that Bob has a wall safe with a secret combination lock known only to him, and the safe is left open and made available to passers-by. Then anyone, including Alice, can put messages in the safe and lock it. However, only Bob can retrieve the message, since even Alice, who left a message in the box, has no way of retrieving the message.

Diagram 3.8 (Asymmetric (Public-Key) Cryptosystems)

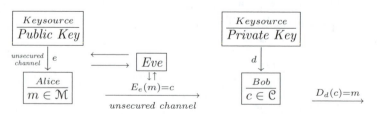

The key-exchange protocol devised by Diffie and Hellman [74] did not provide a complete solution to the notion given above of a public-key cryptosystem. The first to (publicly) do this were Rivest,[3.2] Shamir,[3.3] and Adleman,[3.4] for which their names are attached to the cryptosystem which we now describe.

◆ **The RSA Public-Key Cryptosystem**

We break the algorithm into two parts with the underlying assumption that Alice wants to send a message to Bob.

(I) RSA Key Generation

(1) Bob generates two large, random primes $p \neq q$ of roughly the same size.

(2) He computes both $n = pq$ and

$$\phi(n) = (p-1)(q-1).$$

The integer n is called his *(RSA) modulus*.

(3) He selects a random $e \in \mathbb{N}$ such that $1 < e < \phi(n)$ and $\gcd(e, \phi(n)) = 1$. The integer e is called his *(RSA) enciphering exponent*.

(4) Using the extended Euclidean algorithm (see Appendix C), he computes the unique $d \in \mathbb{N}$ with $1 < d < \phi(n)$ such that

$$ed \equiv 1 \,(\text{mod } \phi(n)).$$

(5) Bob publishes (n, e) in some public database and keeps d, p, q, and $\phi(n)$ private.[3.5] Thus, Bob's *(RSA) public-key* is (n, e) and his *(RSA) private key* is d. The integer d is called his *(RSA) deciphering exponent*.

[3.2] Ronald L. Rivest obtained his Ph.D. in computer science from Stanford University in 1974. He is an inventor of the RSA public-key cryptosystem and a founder of RSA Data Security (now called RSA Security after being bought by Security Dynamics). He is currently the Viterbi Professor of Computer Science in the Department of Electrical Engineering and Computer Science at MIT. Among other duties, he is also leader of Cryptography and Information Security Group of MIT's Laboratory for Computer Science. Along with Shamir and Adleman, he was awarded the 2000 IEEE Koji Kobayashi Computers and Communications Award, as well as the Secure Computing Lifetime Achievement Award. He is a widely recognized expert in cryptographic design and cryptanalysis, but has also done significant work in the areas of computer algorithms, machine learning, and VLSI design.

[3.3] Adi Shamir received his Ph.D. in computer science from Stanford University in 1977. He is an inventor of the RSA cryptosystem, the Feige-Fiat-Shamir Identification Protocol (see page 34), as well as various other cryptanalytic schemes. He is currently a Professor in the Department of Applied Mathematics and Computer Science at the Weizmann Institute of Science in Israel. His research interests lie in cryptography, cryptanalysis, complexity theory, and algorithms.

[3.4] Leonard Adleman was born on December 31, 1945 in San Francisco, California. He received his Ph.D. in computer science from UC Berkeley in 1976. He is currently Distinguished Professor in the Department of Computer Science at the University of Southern California, a title awarded in 2000. He is an inventor of the RSA cryptosystem. His research interests include algorithms, computational complexity, cryptography, DNA computing, immunology, molecular biology, number theory, and quantum computing.

[3.5] Note that the four items $d, p, q, \phi(n)$ form the trapdoor and knowledge of any one of them reveals the remaining three items. In other words, they are not independent items.

(II) RSA Public-Key Cipher[3.6]

enciphering stage:

In order to simplify this stage, we assume that the plaintext message $m \in \mathcal{M}$ is in numerical form with $m < n$. Also, $\mathcal{M} = \mathcal{C} = \mathbb{Z}/n\mathbb{Z}$, and we assume that $\gcd(m, n) = 1$.

(1) Alice obtains Bob's public-key (n, e) from the database.

(2) She enciphers m by computing $c \equiv m^e \pmod{n}$ using the repeated squaring method given on page 50.

(3) She sends $c \in \mathcal{C}$ to Bob.

deciphering stage:

Once Bob receives c, he uses d to compute $m \equiv c^d \pmod{n}$.

Moreover, decryption is unique in that we always recover the intended plaintext (see Exercise 3.16).

Example 3.9 *Suppose that Bob chooses* $(p, q) = (1759, 7487)$. *Then* $n = 13169633$ *and* $\phi(n) = 13160388$. *If Bob selects* $e = 5$, *then by solving* $1 = 5d + \phi(n)x$ *he gets* $d = 7896233$ *(for* $x = -3$). *Thus,* $(13169633, 5)$ *is his public key and* $d = 7896233$ *is his private key. Alice obtains Bob's public key and wishes to send the message* $m = 7115697$. *She enciphers using Bob's public key to get*

$$c \equiv m^5 \equiv 10542186 \pmod{n},$$

which she sends to Bob. He uses his private key d *to decipher via*

$$c^d \equiv 10542186^{7896233} \equiv 7115697 \equiv m \pmod{n}.$$

We exchange roles if Bob wants to send a message to Alice. In this case, the above key generation is performed by Alice to generate her own RSA public and private keys, and Bob performs the enciphering stage sending his message to her for deciphering using her private key.

We have not addressed the issue of what occurs if the plaintext message unit is a numerical value $m \geq n$. In this case, we must subdivide the plaintext numerical equivalents into blocks of equal size, a process called *message blocking*. Suppose that we are dealing with numerical equivalents of the plaintext in base N integers for some fixed $N > 1$. Message blocking may be achieved by choosing that unique integer ℓ such that such that $N^\ell < n < N^{\ell+1}$ (see Exercise 3.15), then writing the message as blocks of ℓ-digit, base N integers (with zeros packed to the right in the last block if necessary), and encipher each separately. In this way, since each block of plaintext corresponds to an element of $\mathbb{Z}/n\mathbb{Z}$ given that $N^\ell < n$; and since $n < N^{\ell+1}$, then each ciphertext message unit can be uniquely written as an $(\ell + 1)$-digit, base N integer in $\mathcal{C} = \mathbb{Z}/n\mathbb{Z} = \mathcal{M}$.

[3.6]To ensure a secure cryptosystem, there must be "preprocessing" of plaintext message units before the enciphering stage. We will not discuss the means here but leave this for later in Section 6.1 where we discuss several security issues involving the implementation of RSA.

Example 3.10 *Suppose that Bob chooses* $n = 1943 = 29 \cdot 67$. *Then* $\phi(n) = 1848$, *and if he chooses* $e = 701$, *then* $d = 29$ *is a solution of* $1 = 5d + 1848x$ *with* $x = -11$. *Therefore,* $(n, e) = (1943, 701)$ *is his public key and* $d = 29$ *is his private key. Now suppose that Alice wants to send the message* **power** *to Bob. She must first convert this to numerical equivalents. She chooses Table 1.2 on page 3, so we are using base* 26 *integers. Since* $26^2 < n < 26^3$, *then* $\ell = 2$ *and we write the message as 2-digit, base* 26 *integers. First, via Table 1.2, we get the base* 26 *equivalents for the plaintext as* $15, 14, 22, 4, 17$, *so we break this up as follows:* $m_1 = \mathbf{po} = 15 \cdot 26 + 14 = 404$; $m_2 = \mathbf{we} = 22 \cdot 26 + 4 = 576$; *and* $\mathbf{ra} = 17 \cdot 26 + 0 = 442$, *with the* $\mathbf{a} = 0$ *packed to the right in the last block. Thus, Alice enciphers:*

$$404^{701} \equiv 1419 \,(\mathrm{mod}\ 1943), \quad 576^{701} \equiv 344 \,(\mathrm{mod}\ 1943),$$

$$and \quad 442^{701} \equiv 210 \,(\mathrm{mod}\ 1943).$$

Bob recovers the plaintext via his private key:

$$1419^{29} \equiv 404 \,(\mathrm{mod}\ 1943); \quad 344^{29} \equiv 576 \,(\mathrm{mod}\ 1943);$$

$$and \quad 210^{29} \equiv 442 \,(\mathrm{mod}\ 1943).$$

He then rewrites each deciphered block as 2-digit base 26 *integers and recovers the English plaintext via Table 1.2. Note that in Example 3.9,* **power** *was enciphered using only one block since we had the use of a much longer modulus, hence a longer block, given that* $\ell = 5$ *in that case.*

Now let us look at what would happen if Bob did not do any message blocking. Then since **power** *may be represented as the 5-digit, base* 26 *integer* m *as follows:* $m = 15 \cdot 26^4 + 14 \cdot 26^3 + 22 \cdot 26^2 + 4 \cdot 26 + 17 = 7115697$, *a single enciphering would yield* $m^{701} = 7115697^{701} \equiv 1243 \,(\mathrm{mod}\ 1943)$, *which is* $1 \cdot 26^2 + 21 \cdot 26 + 21$ *as a base* 26 *integer and via Table 1.2, this yields* **BVV**, *having nothing to do with the original plaintext. Too much information is lost. Hence, the message blocking above is necessary. Moreover, the choice of* ℓ *is maximal (and therefore optimal for encryption), as well as necessary for a unique decryption (see Exercise 3.17).*

We now look at a slightly more complicated example (see [9]).[3.7]

[3.7]On April 2, 1994, the authors of this paper factored the RSA-129 challenge number, for which RSA had offered $100.00(US) as a prize in 1977. (RSA challenge numbers, for factoring algorithms, are denoted by RSA-n for $n \in \mathbb{N}$, which are n-digit integers that are products of two primes of approximately the same size. These RSA challenge numbers are published on the web and one may send a request for a copy of the list to: challenge-rsa-list@rsa.com.) The plaintext they deciphered is the one given in the example, with of course, a much different modulus. They factored the number using a variation of the Multiple Polynomial Quadratic Sieve (MPQS) (see Chapter 5), which took eight months and more than 600 researchers from more than twenty countries around the globe. This method is called *factoring by electronic mail*, a term used by Lenstra and Manasse in [139] to mean the distribution of the quadratic sieve operations to hundreds of physically separated computers all over the world. The unit of time measurement for factoring is called a *mips year*, which is defined to being equivalent

Example 3.11 *Suppose that Alice wants to send the message*

The magic words are squeamish ossifrage,

and that Bob has chosen $n = 131369633 = 57593 \cdot 2281$, *and* $e = 7$, *from which we get* $d = 112551223$ *via* $7d + 131309760x = 7d + \phi(n)x = 1$ *with* $x = -6$. *Since* $26^5 < n < 26^6$, *then we choose* $\ell = 5$, *so the message will be blocked using 5-digit, base* 26 *integers via Table 1.2 as follows.*

$$\mathbf{thema} = 19 \cdot 26^4 + 7 \cdot 26^3 + 4 \cdot 26^2 + 12 \cdot 26 + 0 = 8808592,$$

$$\mathbf{gicwo} = 6 \cdot 26^4 + 8 \cdot 26^3 + 2 \cdot 26^2 + 22 \cdot 26 + 14 = 2884402,$$

$$\mathbf{rdsar} = 17 \cdot 26^4 + 3 \cdot 26^3 + 18 \cdot 26^2 + 0 \cdot 26 + 17 = 7833505,$$

$$\mathbf{esque} = 4 \cdot 26^4 + 18 \cdot 26^3 + 16 \cdot 26^2 + 20 \cdot 26 + 4 = 2155612,$$

$$\mathbf{amish} = 0 \cdot 26^4 + 12 \cdot 26^3 + 8 \cdot 26^2 + 18 \cdot 26 + 7 = 216795,$$

$$\mathbf{ossif} = 14 \cdot 26^4 + 18 \cdot 26^3 + 18 \cdot 26^2 + 8 \cdot 26 + 5 = 6726413,$$

$$\mathbf{ragea} = 17 \cdot 26^4 + 0 \cdot 26^3 + 6 \cdot 26^2 + 4 \cdot 26 + 0 = 7772752.$$

Now enciphering is accomplished by the following where each congruence is understood to mean modulo n:

$$8808592^7 \equiv 56806804; \quad 2884402^7 \equiv 65895615; \quad 7833505^7 \equiv 45842787;$$

$$2155612^7 \equiv 43647783; \quad 216795^7 \equiv 123817334; \quad 6726413^7 \equiv 110825702;$$

$$\text{and} \quad 7772752^7 \equiv 48882513.$$

Then Alice converts to 6-digit, base 26 *integers and produces ciphertext via Table 1.2 as follows.*

$$56806804 = 4 \cdot 26^5 + 20 \cdot 26^4 + 8 \cdot 26^3 + 1 \cdot 26^2 + 19 \cdot 26 + 2 = \mathbf{EUIBTC},$$

$$65895615 = 5 \cdot 26^5 + 14 \cdot 26^4 + 5 \cdot 26^3 + 4 \cdot 26^2 + 18 \cdot 26 + 19 = \mathbf{FOFEST},$$

$$45842787 = 3 \cdot 26^5 + 22 \cdot 26^4 + 8 \cdot 26^3 + 6 \cdot 26^2 + 20 \cdot 26 + 3 = \mathbf{DWIGUD},$$

$$43647783 = 3 \cdot 26^5 + 17 \cdot 26^4 + 13 \cdot 26^3 + 9 \cdot 26^2 + 18 \cdot 26 + 23 = \mathbf{DRNJSX},$$

$$123817334 = 10 \cdot 26^5 + 10 \cdot 26^4 + 24 \cdot 26^3 + 17 \cdot 26^2 + 19 \cdot 26 + 4 = \mathbf{KKYRTE},$$

$$110825702 = 9 \cdot 26^5 + 8 \cdot 26^4 + 13 \cdot 26^3 + 13 \cdot 26^2 + 9 \cdot 26 + 0 = \mathbf{JINNJA},$$

$$48882513 = 4 \cdot 26^5 + 2 \cdot 26^4 + 25 \cdot 26^3 + 5 \cdot 26^2 + 10 \cdot 26 + 17 = \mathbf{ECZFKR},$$

which Alice sends to Bob who may decipher using his private key d. *The reader may verify this by doing the computations.* (*See Exercises 3.26–3.31.*)

---

to the computational power of a computer rated at one million instructions per second (mips) and used for one year, which is tantamount to approximately $3 \cdot 10^{13}$ instructions. The RSA-129 challenge number took 5000 mips years. In Chapter 5, we will return to a discussion of these numbers in more detail.

We began a discussion of the security of RSA at the end of Section 3.1. There we showed that knowing how to factor n allows us to compute d (see Exercises 3.18–3.25). Suppose that we know d (as well as e). We now show that this allows us (with arbitrarily high probability) to factor n. Clearly having both e and d allows us to compute

$$ed - 1 = 2^k s$$

where $k \in \mathbb{N}$ and s is odd. Since $ed \equiv 1 \,(\mathrm{mod}\ \phi(n))$, then there exists an $a \in (\mathbb{Z}/n\mathbb{Z})^*$ such that $a^{2^k s} \equiv 1 \,(\mathrm{mod}\ n)$. If $j \in \mathbb{N}$ is the least value such that $a^{2^j s} \equiv 1 \,(\mathrm{mod}\ n)$, then $j \leq k$. If

$$\text{both } a^{2^{j-1}s} \not\equiv 1 \,(\mathrm{mod}\ n) \quad \text{and} \quad a^{2^{j-1}s} \not\equiv -1 \,(\mathrm{mod}\ n), \tag{3.11}$$

then $b = a^{2^{j-1}s}$ is a *nontrivial square root of* 1 *modulo* n. In other words,

$$n \mid (b+1)(b-1) \text{ with } n \nmid (b+1), \text{ and } n \nmid (b-1).$$

Therefore, we can factor n since

$$\gcd(b+1, n) = p, \text{ or } \gcd(b+1, n) = q.$$

Hence, (3.11) is what we need to ensure that we can factor n. It can be shown (by a method we will learn in Chapter 5, called the *universal exponent method*) that the probability that (3.11) occurs can be made to approach 1 (for a sufficiently large number of trials testing the values of $a \in (\mathbb{Z}/n\mathbb{Z})^*$). Hence, knowledge of d can be converted into an algorithm for factoring n (with arbitrarily small probability of failure to do so). Of course all of this is predicated upon the RSA cryptosystem being *properly* set up, and we will discuss more aspects of this in Chapter 6 when we return to a discussion of the security of RSA.

Essentially what we have been skirting around is the following.

◆ The RSA Conjecture[3.8]

Cryptanalyzing RSA must be as difficult as factoring.

However, there is no known proof of this conjecture, although the general consensus is that it is valid. The reason for the consensus is that the only known method for finding d given e is to apply the extended Euclidean algorithm to e and $\phi(n)$. Yet to compute $\phi(n)$, we need to know p and q, namely, to cryptanalyze the RSA cryptosystem, we must be able to factor n.

[3.8]There are numerous cryptosystems which are called *equivalent to the difficulty of factoring*. For instance, see [238], where RSA-like public-key cryptosystems are studied whose difficulty to break is as hard as factoring the modulus. It can be shown that any cryptosystem for which there is a constructive proof of equivalence to the difficulty of factoring, is vulnerable to a chosen-ciphertext attack (see page 26 and [236, p. 729]). For a nice discussion of the security of such cryptosystems, see [169]. We have already seen that factoring an RSA modulus allows the breaking of the cryptosystem, but the converse is not known. In other words, it is not known if there are other methods of breaking RSA. For a discussion of this and related ideas, see [197].

Exercises

3.15. Show that for given $n, N \in \mathbb{N}$, with $N > 1$ and $n > N$, there exists a unique integer $\ell \geq 0$ such that $N^{\ell} \leq n < N^{\ell+1}$.

3.16. Prove that the decryption described in the RSA public-key cipher uniquely recovers the intended plaintext m.

3.17. Show that the choice of ℓ in the discussion of message blocking on page 62 is maximal for unique decryption when the plaintext numerical equivalents m are bigger than n. In other words, if $k > \ell$ is chosen as the blocklength when $m > n$, then decryption will not be unique in the RSA cipher. Illustrate by using Example 3.10 with $k = \ell + 1 = 3$.

In Exercises 3.18–3.25, find the primes p and q assuming $n = pq$. Hint: See Exercise 3.14.

3.18. $n = 10765283$ and $\phi(n) = 10758720$.

3.19. $n = 11626579$ and $\phi(n) = 11619760$.

3.20. $n = 14785217$ and $\phi(n) = 14777520$.

3.21. $n = 23268953$ and $\phi(n) = 23259300$.

3.22. $n = 26264851$ and $\phi(n) = 26254600$.

3.23. $n = 36187829$ and $\phi(n) = 36175776$.

3.24. $n = 65827471$ and $\phi(n) = 65811232$.

3.25. $n = 87732199$ and $\phi(n) = 87713464$.

In Exercises 3.26–3.31, assume the given cryptogram was enciphered using the RSA cryptosystem with $n = 10765283$, $e = 11$, $\phi(n) = 10758720$, $\ell = 4$, with numerical values from Table 1.2 on page 3.

3.26. **HNCLGMQIDO**

3.27. **EGSIOXEWXGDPXMA**

3.28. **VISYCMNUEVALGLE**

3.29. **FENFLPLNMZXLMPS**

3.30. **SZAZUHAPSB**

3.31. **BOTDTICBYJ**

3.3   ElGamal Cryptosystems

A cryptographic scheme has to operate in a maliciously selected environment which typically transcends the designer's view.

Oded Goldreich
On the foundation of modern cryptography
invited lecture, CRYPTO 1997

In the last section, we introduced public-key cryptography and looked at the best-known example, the RSA cipher. In this chapter, we will concentrate on another public-key cipher given in the title.[3.9] The reader should be familiar with number-theoretic concepts in Appendix C before proceeding.

Recall that the RSA public-key cipher has security essentially based on the difficulty of integer factoring. The following public-key cryptosystem bases its security upon the intractability of the discrete log problem, DLP, see page 39, and the Diffie-Hellman problem, DHP, see page 49.

The following cipher first appeared in 1985 [80].

◆ **The ElGamal[3.10] Cryptosystem**

The following is performed assuming that Alice wants to send a message m to Bob, and $m \in \{0, 1, \ldots, p-1\}$ (equivalent to the actual plaintext).

(I) ElGamal Key Generation

(1) Bob chooses a large random prime p (with a prescribed number of bits) and a primitive root[3.11] α modulo p.

(2) Bob then chooses a random integer a with $2 \leq a < p-1$ and computes α^a (mod p).

[3.9]There is another well-known public-key cryptosystem, called the knapsack cipher. However, all known versions of this cryptosystem have been broken. Until recently, the only one based on the subset sum problem (see Exercises 1.59–1.60 on page 24), still unbroken was the Chor-Rivest cryptosystem introduced in 1984, [56], and refined in 1988, [57]. However, in 2001, this last standing knapsack cipher was cryptanalyzed (see [232]), wherein there are also directions given for the breaking of the related *powerline system* devised by H. Lenstra in 1991, [137]. Thus, we do not cover knapsack ciphers in this text, since they have been essentially abandoned by the cryptographic community for all but passing interest. Even before the breaking of Chor-Rivest, knapsack ciphers were generally suspected to be weaker than RSA and ElGamal, which we now study. (More detail on knapsack ciphers may be found in [165, Section 3.4, pp. 153–160].)

[3.10]Taher ElGamal was born in Cairo, Egypt on August 18, 1955. In 1984, he obtained his Ph.D. at Stanford University under the supervision of Martin Hellman (see Footnote 2.3). While at Stanford, he helped to pioneer digital signatures. Later, he became Director of Engineering at *RSA Data Security Inc.*, where he developed RSA cryptographic toolkits. He also was the Chief Scientist of *Netscape Communications* where he pioneered internet security technology such as *Secure Sockets Layer* (SSL) — see page 183, the standard for web security. Other accomplishments include development of internet credit card payment schemes. He founded *Security Inc.* which later became the *Kroll-O'Gara Information Security Group* of which he became president. ElGamal is a recognized leader in the cryptographic community and the information security industry.

[3.11]See Appendix C for definitions. Note as well that, although preferable, one need not choose a primitive root, as long as one chooses an element $\alpha \in (\mathbb{Z}/p\mathbb{Z})^*$ whose order is close to the size of p; i.e. the smallest $r \in \mathbb{N}$ such that $\alpha^r \equiv 1 \pmod{p}$ must be nearly as large as p. Such α are called *near-primitive roots*. In the case of a primitive root, $r = p-1$.

(3) Bob's public key is (p, α, α^a) and his private (session) key is a.

(II) ElGamal Public-Key Cipher
enciphering stage:

(1) Alice obtains Bob's public-key (p, α, α^a).

(2) She chooses a random natural number $b < p - 1$.

(3) She computes $\alpha^b \pmod{p}$ and $m\alpha^{ab} \pmod{p}$.

(4) Alice then sends the ciphertext $c = (\alpha^b, m\alpha^{ab})$ to Bob.

deciphering stage:

(1) Bob uses his private key to compute $(\alpha^b)^{-a} \equiv (\alpha^b)^{p-1-a} \pmod{p}$.

(2) Then he deciphers m by computing $(\alpha^b)^{-a}m\alpha^{ab} \pmod{p}$.

The reason Bob's deciphering stage works is due to the following.

$$(\alpha^b)^{-a}m\alpha^{ab} \equiv m\alpha^{ab-ab} \equiv m \pmod{p}.$$

The ElGamal cipher is tantamount to the Diffie-Hellman key-exchange protocol. To see this, suppose that Eve wants to get m and she can solve the DHP. Eve seeks to get m from $c = (\alpha^b, m\alpha^{ab})$. Since she can solve the DHP, she can determine $\beta \equiv \alpha^{ab} \pmod{p}$ from α^a and α^b, and reconstruct the message $m = \beta^{-1}m\alpha^{ab} \pmod{p}$; i.e., if Eve can break the Diffie-Hellman key-exchange protocol, she can break the ElGamal cipher.

Now let's assume that Eve can cryptanalyze the ElGamal cipher above. Then she can obtain any message m from knowledge of $p, \alpha, \alpha^a, \alpha^b$, and $m\alpha^{ab}$. If Eve wants to get α^{ab} from $p, \alpha, \alpha^a, \alpha^b$, she computes $(m\alpha^{ab})m^{-1} \equiv \alpha^{ab} \pmod{p}$. In other words, we have shown that cryptanalyzing ElGamal is equivalent to cryptanalyzing Diffie-Hellman. In fact, the ElGamal cipher may be viewed as a Diffie-Hellman key exchange on $k = \alpha^{ab}$, which is used to encipher m in step (3). Thus, we have shown that although Diffie-Hellman is not itself a public-key cryptosystem, it is the basis for (and has difficulty equivalent to) the ElGamal public-key cryptosystem. Moreover, as noted on page 49, if Eve can solve the DLP, she can solve the DHP. The converse is not known, but is generally assumed to be true. Hence, we assume that the security of the ElGamal cipher is *based* upon the DLP.

It is generally acknowledged that, as with RSA, a modulus of 1024 bits is recommended for long-term security. However, as usual, we will illustrate with small unrealistic parameters for pedagogical reasons.

Example 3.12 *Suppose that Alice wants to send the message $m = 696$ to Bob using the ElGamal cipher. Bob chooses $p = 3361$, $\alpha = 22$, and $a = 5$, his private key. He computes computes $\alpha^a \equiv 1219 \,(\mathrm{mod}\ p)$. Bob's public key is therefore $(p, \alpha, \alpha^a) = (3361, 22, 1219)$, which Alice obtains. She chooses $b = 56$ and computes both $\alpha^b \equiv 2904 \,(\mathrm{mod}\ p)$ and $m\alpha^{ab} \equiv 696 \cdot 1219^{56} \equiv 609 \,(\mathrm{mod}\ p)$. Thus, she sends $c = (2904, 609)$ to Bob, who uses his private key to compute $(\alpha^b)^{p-1-a} \equiv 2904^{3355} \equiv 2882 \,(\mathrm{mod}\ p)$ and*

$$(\alpha^b)^{-a} m\alpha^{ab} \equiv 2882 \cdot 609 \equiv 696 \,(\mathrm{mod}\ p),$$

thereby recovering m. See Exercises 3.32–3.35.

Notice that in Example 3.12, the ciphertext is roughly twice as long as the plaintext. This is a phenomenon called *message expansion* and is one disadvantage of the ElGamal cipher. Later, when we look at a generalization of ElGamal, called the ElGamal elliptic curve cryptosystem, we will see that this message expansion goes up by a factor of four. When we study elliptic curve versions of ElGamal, we will also see that there is a procedure for changing a cryptosystem based upon the DLP to one using elliptic curves.

The simplicity of Example 3.12 does not address the problem of message blocking. If $m > p$, then we break the message into smaller blocks as we did in Section 3.2 for the RSA cipher. For instance, if we maintain the setup in Example 3.12, but change the message, we may illustrate as follows.

Example 3.13 *Suppose that Alice wants to send the message:* **big block** *with the parameters given in Example 3.12. Since the base 26 equivalents for the letters are $1, 8, 6, 1, 11, 14, 2, 10$, and since $26^2 < 3361 < 26^3$, we choose message blocks of length 2. Thus, the message blocks are* **bi** $= m_1 = 1 \cdot 26 + 8 = 34$, **gb** $= m_2 = 6 \cdot 26 + 1 = 157$, **lo** $= m_3 = 11 \cdot 26 + 14 = 300$, *and* **ck** $= m_4 = 2 \cdot 26 + 10 = 62$. *Thus, Alice enciphers each block individually. For instance,*

$$m_1 \alpha^{ab} \equiv 870 \,(\mathrm{mod}\ p).$$

Thus, the first ciphertext Alice would send is $c_1 = (2904, 870)$, and Bob would decipher via his private key $a = 5$, $(\alpha^b)^{p-1-a} \equiv 2882 \,(\mathrm{mod}\ p)$, as in Example 3.12, and

$$(\alpha^b)^{-a} m_1 \alpha^{ab} \equiv 2882 \cdot 870 \equiv 34 \,(\mathrm{mod}\ p),$$

recovering m_1. Now Bob would have to go through the process again of choosing a new private session key (and so new public key) for each of the message blocks, which the reader may try as an exercise.

The above cryptosystem can be generalized to any (appropriately chosen) finite cyclic group G. The choice of G must be made to ensure the intractability of the DLP therein (for security reasons), and the group operation must be relatively easy (for reasons of efficiency). In what follows, we assume that such an *appropriate choice* for G has been made.

☞ **Generalized ElGamal Public-Key Cryptosystem**

The following is performed assuming that Alice wants to send a message m to Bob, where m is an element of a cyclic group of order n having generator α.

(I) Generalized ElGamal Key Generation

(1) Bob selects a random integer a with $2 \leq a < n - 1$ and computes the group element α^a.

(2) Bob's public key is (α, α^a) and his private key is a.

(II) Generalized ElGamal Public-Key Cipher

<u>enciphering stage</u>:

(1) Alice obtains Bob's public key (α, α^a).

(2) She chooses a random natural number $b < n - 1$.

(3) She computes α^b and $m\alpha^{ab}$.

(4) Alice then sends the ciphertext $c = (\alpha^b, m\alpha^{ab})$ to Bob.

<u>deciphering stage</u>:

(1) Bob uses his private key to compute $(\alpha^b)^{-a}$.

(2) Then he deciphers m by computing $(\alpha^b)^{-a} m\alpha^{ab}$.

Example 3.14 *Let G be the multiplicative group of the finite field $\mathbb{F}_{5^3}$, whose elements we will represent as polynomials, with degree at most 2, over $\mathbb{F}_5$. We observe that $|G| = 124$, and if $r(x) = x^3 + x + 1$, then $\mathbb{F}_{5^3} \cong (\mathbb{Z}/5\mathbb{Z})[x]/(r(x))$, so multiplication in $\mathbb{F}_{5^3}$ may be performed modulo $r(x)$, a polynomial which we assume both Bob and Alice have in common. Furthermore, for the sake of notational convenience, we represent the polynomials by their coefficients; for instance, a generator of G is $2x^2 + 2$, represented by $\alpha = (2, 0, 2)$. If Bob chooses $a = 9$ as his private key, then he computes $\alpha^a = \alpha^9 = (1, 2, 3)$. Thus, Bob's public key is $(\alpha, \alpha^a) = ((2, 0, 2), (1, 2, 3))$.*

Suppose that Alice wants to send the message $(4, 3, 2)$ to Bob. She selects at random $b = 96$ and computes $\alpha^b = \alpha^{96} = (1, 4, 2)$ and $m\alpha^{ab} = (4, 3, 2)(1, 2, 3)^{96} = (4, 4, 2)$. She sends $c = (\alpha^b, m\alpha^{ab}) = ((1, 4, 2), (4, 4, 2))$ to Bob. Upon receipt, he computes $(\alpha^b)^{-a} = (1, 4, 2)^{-9} = (2, 4, 3)$ and uses this to compute $\alpha^{-ab} m\alpha^{ab} = (2, 4, 3)(4, 4, 2) = (4, 3, 2) = m$, recovering the plaintext. (See Exercises 3.36–3.40.)

We conclude this section with a cryptosystem that allows us to use the setting in Example 3.14 for an illustration of it. However, the following is unique in that it involves *neither* public keys *nor* shared private keys, so is not a public-key cipher, but is of interest in its own right. The basic idea behind what follows was credited to Shamir by Konheim [121, p. 345], though never published. Initially, it was set up to establish a session key for use by two entities, so it is called either *Shamir's three-pass protocol* or *Shamir's no-keys protocol*. However, there was later an independent refinement. According to Massey [146, p. 35], Omura later proposed an exponentiation-based cipher that was based on Shamir's original idea.

☞ **Massey-Omura Cryptosystem**

The following is performed assuming that Alice wants to send a message m to Bob, where p is prime, $n \in \mathbb{N}$, and m is an element of the multiplicative group of $\mathbb{F}_{p^n}$.

(1) Alice and Bob independently select random integers e_A and e_B, respectively, with $2 \leq e_A, e_B < p^n - 1$ and $\gcd(e_A, p^n - 1) = 1 = \gcd(e_B, p^n - 1)$.

(2) Alice and Bob compute $d_A \equiv e_A^{-1} \pmod{p^n - 1}$ and $d_B \equiv e_B^{-1} \pmod{p^n - 1}$, respectively, using the extended Euclidean algorithm.

(3) Alice sends m^{e_A} to Bob.

(4) Bob sends $m^{e_A e_B}$ back to Alice.

(5) Alice sends $m^{e_A e_B d_A} = m^{e_B}$ to Bob.

(6) Bob computes $m^{e_B d_B} = m$.

Notice that in step (3), Bob is unable to decipher m^{e_A} not having access to d_A, and in step (4), Alice cannot decipher $m^{e_A e_B}$ not having access to d_B. This is because they are faced with the DLP, which we are assuming is intractable in the group under consideration. However, in step (6), Bob can indeed decipher m^{e_B} since he has his key d_B. Hence, no public keys are necessary, no shared secret key is necessary, and new keys are used each time. We now use the setting in the previous example to illustrate as promised earlier.

Example 3.15 *Suppose that Alice wishes to send the message $m = (1, 2, 3)$ as an element of $\mathbb{F}_{5^3}$ (see Example 3.14), using the Massey-Omura cryptosystem. Suppose further that Alice has generated her keys as $e_A = 25$ and $d_A = 5$, while Bob has independently generated his keys as $e_B = 3$ and $d_B = 83$. Alice sends $m^{e_A} = (1, 2, 3)^{25} = (3, 3, 1)$. Bob replies with $m^{e_A e_B} = (3, 3, 1)^3 = (4, 0, 3)$, and Alice sends $m^{e_A e_B d_A} = m^{e_B} = (4, 0, 3)^5 = (2, 2, 0)$. Bob now easily decrypts with the computation $m^{e_B d_B} = (2, 2, 0)^{83} = (1, 2, 3)$.*

One drawback is that three separate transmissions are required, which could have disastrous consequences under some circumstances (see Exercise 3.42). Furthermore, as the cipher stands, without any further protection, it is vulnerable to the man-in-the-middle attack (page 27). See Exercise 3.41.

Exercises

In Exercises 3.32–3.35, assume that the given ciphertext $c = (\alpha^b, m\alpha^{ab})$ was encrypted using the ElGamal cryptosystem with the parameters given. Find the numerical plaintext.

3.32. $c = (8, 90)$; $p = 173$; $\alpha = 2$; $a = 7$; $b = 3$.

3.33. $c = (32, 12)$; $p = 409$; $\alpha = 21$; $a = 6$; $b = 2$.

3.34. $c = (614, 115)$; $p = 659$; $\alpha = 21$; $a = 6$; $b = 95$.

3.35. $c = (512, 303)$; $p = 941$; $\alpha = 2$; $a = 14$; $b = 9$.

In Exercises 3.36–3.40, assume that the given ciphertext $c = (\alpha^b, m\alpha^{ab})$ was encrypted using the Generalized ElGamal cryptosystem with the given value of a and the assumption that all the other parameters are the same as those given in Example 3.14. Find the numerical plaintext.

☆ 3.36. $c = ((1, 4, 2), (3, 1, 4))$; $a = 15$.

☆ 3.37. $c = ((3, 3, 2), (0, 2, 1))$; $a = 44$.

☆ 3.38. $c = ((3, 4, 0), (1, 4, 2))$; $a = 100$.

☆ 3.39. $c = ((0, 3, 1), (4, 0, 4))$; $a = 24$.

☆ 3.40. $c = ((0, 2, 0), (2, 0, 2))$; $a = 121$.

3.41. Assume that Bob and Alice are conversing over a channel using the Massey-Omura cryptosystem and that Mallory is listening to the channel. Describe how Mallory can impersonate Bob and recover m using a man-in-the-middle attack.

3.42. Assume that in the Massey-Omura cryptosystem, Alice and Bob decide to substitute exponentiation with the Vernam cipher (see page 9). Explain how Eve, listening to the three transmissions can decipher and recover m with no additional knowledge.

☆ 3.43. Suppose that Alice wants to send the message **today** to Bob using the ElGamal cryptosystem. Describe how she does this using the prime $p = 15485863$, $\alpha = 6$ a primitive root modulo p, and her choice of $b = 69$. Assume that Bob has private key $a = 6$. How does Bob recover the message?

3.4 Symmetric vs. Asymmetric Cryptosystems

Symmetry is a vast subject, significant in art and nature. Mathematics lies at its root, and it would be hard to find a better one on which to demonstrate the working of the mathematical intellect.

Hermann Weyl (1885–1955)
mathematician and pioneer of modern quantum theory

Now that we have been introduced to both symmetric-key and public-key cryptosystems, it is time to compare and contrast.

◆ **Advantages of Public-Key Cryptosystems**

(1) **Security**: Only the private key needs to be kept a secret.

(2) **Longevity**: Key pairs may be used without change in most cases over long periods of time – years in some situations.

(3) **Key Management**: If a multi-user large network is being used (without a key server — see page 164) then fewer private keys will be required with a public-key cryptosystem than with a symmetric-key cryptosystem. For instance, if $n \in \mathbb{N}$ entities are communicating, using DES, say (see page 22), then the number of keys required to allow any two entities to communicate is $n(n-1)/2$ (see Exercise 3.45). Also, every user on the system has to store $n-1$ keys. This is called *key predistribution*. With a public-key cryptosystem, only n keys are required for any two entities to communicate since only one (public) key for each entity has to be stored.

(4) **Key-Exchange**: In a public-key cryptosystem, no key-exchange between communicating entities is necessary. (Note that this tells us that the Diffie-Hellman key exchange protocol, discussed on page 49, is not a public-key cryptosystem, although it contained the basic original ideas for it. See the remarks on page 50.)

Some further comments on advantage (3) are in order. It must be noted that if the large network, employing a symmetric-key cryptosystem, has a trusted arbitrator, such as Trent, then each user only needs a single key that is shared with Trent. However, for Alice and Bob to communicate with assurance not only of privacy of communication (security), but also of each other's identity (authentication), Trent must not only be unconditionally trusted, but also available 24/7, namely, at all times. The reason is that Alice, wanting to communicate with Bob (at any time), sends a message to Trent with such a request, enciphered using a key k_A, shared with Trent. Trent then generates a key, k_T, enciphered using k_A, which he sends to Alice for her conversation with Bob. Trent also sends k_T to Bob encrypted using the key, k_B, that he shares with Bob. Then Alice and Bob can use this k_T to communicate. (This description of how Alice and Bob can agree on a key using DES keys is the basis for *Kerberos* which is a key distribution protocol developed at MIT, and named after the three-headed dog of Greek mythology that stood guard at the gates of Hades — see page 166.) Public-key cryptosystems virtually eliminate the need for Trent.

There is another advantage not listed above, namely, digital signatures and general authentication, about which we will learn in Chapter 7. This may be, arguably, the greatest advantage of public-key cryptosystems since they offer virtually the only means for providing digital signatures. However, we have enough now to proceed and much more to digest before we get there.

◆ **Disadvantages of Public-Key Cryptosystems**

(1) **Efficiency**: Public-key cryptosystems are slower than their symmetric-key counterparts. For instance, the RSA cryptosystem is roughly a thousand times slower than the DES symmetric-key cryptosystem (see the bottom of page 22).

(2) **Key sizes**: The key sizes for a public-key cryptosystem are significantly larger than that required for a symmetric-key cryptosystem. For instance, the private key in the RSA cryptosystem should be 1024 bits, whereas with a symmetric-key cipher, generally 128 bits will suffice. Usually, private keys are ten times larger than secret keys.

◆ **Analysis and Summary**

The primary advantage of public-key cryptosystems is increased security and convenience. One might add the disadvantage to the list that no public-key cryptosystem has been proven secure. However, other than the Vernam Cipher (see page 10), neither has any symmetric-key cryptosystem been so proven. Public-key cryptography, given its disadvantages, is not meant for enciphering the bulk of a given communication. In other words, public-key cryptography is not meant to replace symmetric-key cryptography, but rather to supplement it for the goal of achieving more security.

The idea behind modern cryptographic usage is to employ public-key cryptography to obtain keys, which are then used in a symmetric-key cryptosystem. Such cryptosystems are called *hybrid cryptosystems* or *digital envelopes*, which have the advantages of both types of cryptosystems. Here is how they work in practice.

Alice and Bob have access to a symmetric-key cryptosystem, which we will call S. Also, Bob has a public-private key pair (e, d). Alice wishes to send a message m to Bob. Alice first generates a symmetric key, called a *session key* or *data encryption key*, k to be used only once. She enciphers m using k and S obtaining $c = k(m)$, the ciphertext. Then Alice obtains Bob's public key e and uses it to encrypt k, yielding $e(k) = k'$. Both of these encryptions are fast since S is efficient in the first enciphering, and the session key is small in the second enciphering. Then Alice sends c and $e(k)$ to Bob. Bob deciphers k with his private key d, via $d(e(k)) = k$ from which he easily deduces the symmetric deciphering key k^{-1}. Then he deciphers: $k^{-1}(c) = k^{-1}(k(m)) = m$.

Hence, the public-key cryptosystem is used only for the sending of the session key, which provides a digital envelope that is both secure and efficient — an elegant solution to the above problems.

Diagram 3.16 (Digital Envelope — Hybrid Cryptosystem)

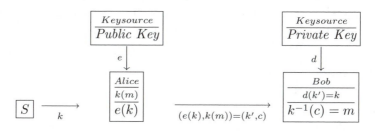

Example 3.17 *Suppose that the symmetric-key cryptosystem, S, that Alice and Bob agree to use is a permutation cipher (see Definition 1.28 on page 21), with $r = 6$, $\mathcal{M} = \mathcal{C} = \mathbb{Z}/26\mathbb{Z}$, and key $k = (5, 2, 3, 1, 6, 4)$ from which one easily deduces the deciphering key $k^{-1} = (4, 2, 3, 6, 1, 5)$. Further suppose that Alice wants to send $m =$ **quiver** to Bob, who has set up his RSA keys as follows. He chooses $n = pq = 1759 \cdot 5023 = 8835457$, with public key $e = 11$, and private key $d = 802607$ determined from $11d + \phi(n)x = 11d + 8828676x = 1$ with $x = -1$. Since $10^6 < n < 10^7$, then $\ell = 6$. So Alice proceeds as follows.*

She converts m to numerical equivalents via Table 1.2 on page 3 to get $m = (16, 20, 8, 21, 4, 17)$ to which she applies k to get

$$c = k(m) = (4, 20, 8, 16, 17, 21).$$

She then proceeds to encipher k using Bob's public key as follows.

Since $\ell = 6$, then we may encipher the key k as a 6-digit, base 10 integer:

$$k = 5 \cdot 10^5 + 2 \cdot 10^4 + 3 \cdot 10^3 + 1 \cdot 10^2 + 6 \cdot 10 + 4 = 523164. \qquad (3.17)$$

She then enciphers k as

$$k' = k^e = 523164^{11} \equiv 6013077 \,(\text{mod } 8835457),$$

and sends the pair $(k', c) = (k^e, k(m))$ to Bob. Bob receives the pair and makes the following calculations.

He computes $(k')^d = (k^e)^d = k^{ed} \equiv k \equiv 523164 \,(\text{mod } n)$. He then converts this back to its original format via (3.17), and is able to easily deduce k^{-1} which he applies to c to get

$$k^{-1}(k(m)) = m = (16, 20, 8, 21, 4, 17) = \textbf{quiver}.$$

Exercises

3.44. Suppose that Alice and Bob use the RSA cryptosystem but they choose the same RSA modulus n, and enciphering (public) keys e_A and e_B, respectively, with $\gcd(e_A, e_B) = 1$. Suppose that Eve intercepts two cryptograms m^{e_A} and m^{e_B}, enciphering the same message m, sent by a third entity to Alice and Bob, respectively. Show how Eve can determine m without knowledge of the factorization of n or of either of Alice or Bob's (private) decryption keys. (This is called *common modulus protocol failure.*)

3.45. Explain why, in the use of a symmetric-key cryptosystem with n users, there is the requirement that $n(n-1)/2$ keys be stored.

In Exercises 3.46–3.53, assume that we are in the situation developed in Example 3.17 with the same r, $\mathcal{M}, \mathcal{C}, n, e, d$, but with k' and c as given in each of the exercises. Use the methodology of that example to find k as a permutation, and m in English plaintext via Table 1.2.

3.46. $(k', c) = (950866, (11, 8, 21, 17, 4, 18))$.

3.47. $(k', c) = (4019872, (8, 11, 21, 4, 17, 18))$.

3.48. $(k', c) = (5790603, (19, 17, 4, 12, 8, 18))$.

3.49. $(k', c) = (1525853, (9, 0, 1, 1, 14, 3))$.

3.50. $(k', c) = (1420735, (17, 7, 4, 0, 10, 2))$.

3.51. $(k', c) = (7155548, (3, 17, 4, 8, 5, 13))$.

3.52. $(k', c) = (688724, (0, 2, 17, 19, 8, 6))$.

3.53. $(k', c) = (371155, (14, 12, 8, 13, 18, 4))$.

3.54. $(k', c) = (6604589, (8, 7, 18, 22, 2, 19))$.

3.55. $(k', c) = (8182887, (1, 2, 17, 0, 0, 8))$.

3.56. $(k', c) = (1662406, (19, 17, 19, 4, 7, 8))$.

3.57. $(k', c) = (4125753, (8, 12, 13, 19, 0, 7))$.

3.58. $(k', c) = (5607923, (19, 11, 6, 7, 18, 8))$.

3.59. $(k', c) = (1968543, (17, 3, 17, 0, 10, 4))$.

3.60. $(k', c) = (187177, (17, 6, 4, 4, 21, 8))$.

3.61. $(k', c) = (7066510, (17, 25, 0, 4, 18, 1))$.

3.62. $(k', c) = (995088, (4, 11, 25, 8, 25, 5))$.

3.5 Secret History of Public-Key Cryptography

If men could learn from history, what lessons it might teach us! But passion and party blind our eyes, and the light of experience gives us a lantern on the stern, which shines only on the waves behind us!

Samuel Taylor Coleridge (1772–1834)
English poet, critic, and philosopher

Earlier in this chapter, we learned about the pioneering efforts of Diffie, Hellman, and Merkle in establishing public-key cryptography. However, it is now *public* knowledge that the notion had already been discovered years earlier by British cryptographers, but not *officially* released until relatively recently. Here are the known facts.

Public-key methodologies were first discovered by the Communications-Electronics Security Group (CESG) in the early 1970s. The function of CESG, as a branch of the British Government Communications Headquarters (GCHQ), is to ensure information security for the British government, which regards CESG as their technical authority on official cryptographic applications.

In December of 1997, five papers, [58], [82]–[83], [242]–[243], were released by CESG, which the reader may download from:

http://www.cesg.gov.uk/publications/index.htm#nsecret.

In January of 1970, J. Ellis[3.12] established the fundamental ideas behind public-key cryptography in [82]. He called his method *non-secret encryption* (NSE). Hence, the discovery of the idea of public-key cryptography predated Diffie, Hellman, and Merkle by more than a half dozen years. In [58], dated November 20, 1973, C. Cocks[3.13] essentially describes what we now call the RSA cryptosystem, with any differences being entirely superficial. In [242], dated January 24, 1974, Williamson[3.14] describes what we now call the Diffie-Hellman key-

---

[3.12]James H. Ellis (1924–1997) was born in Australia. However, his parents returned to London, England when he was still a baby. After graduating from the University of London, he was employed at the Post Office Research Station at Dollis Hill. In 1965, the cryptography section at Dollis Hill moved to join the (newly formed) CESG in GCHQ. There Ellis became a leading figure in British cryptography. Fellow workers turned to him for advice and ideas. The British government asked Ellis in the 1960s to investigate the key distribution problem since management of large amounts of key material needed for secure communication was a headache for the military. Shortly after his death in 1997, GCHQ/CESG released the five publications cited above. According to a spokesman for the British government, the release of the papers was a "pan governmental drive for openness" by the Labour party.

[3.13]Clifford Cocks joined CESG in September of 1973, where he became acquainted with Ellis's ideas for NSE. He naturally moved to the idea of a one-way function since he had studied number theory at the University of Cambridge as a student. Cocks claimed that it took him only a half hour to invent the notion in [58].

[3.14]Malcolm Williamson joined CESG in September of 1974. He learned from Cocks about the NSE idea, but found it difficult to believe. By trying to disprove the existence of NSE, he discovered a notion equivalent to the Diffie-Hellman key-exchange protocol. This means that the discovery of (a notion equivalent to) RSA preceded that of (a notion equivalent to) Diffie-Hellman, which is the opposite of what occurred in the public domain.

exchange protocol. In [243], dated August 10, 1976, Williamson improved upon the ideas [242] he put forth in 1974.

In [83], published (internally) in CESG in 1987, Ellis describes the history of NSE. In this paper, he says: "The task of writing this paper has devolved to me because NSE was my idea and I can therefore describe their developments from personal experience." Also, in this paper Ellis cites the 1944 publication [230] (by an unknown author for Bell Laboratories) which he describes as an ingenious idea for secure conversation over a telephone. This was his inspiration for NSE. Ellis states, in the aforementioned paper, that this is how the idea was born, that secure communication was possible if the recipient took part in the encryption process. At the end of his paper Ellis concludes that the Diffie-Hellman idea: "was the start of public awareness on this type of cryptography and subsequent rediscovery of the NSE techniques I have described."

In an interview in the the *New York Times* in December of 1997, Williamson said that he felt badly knowing that others were taking credit for solutions found at CESG. However, he concluded that this was just one of the restrictions to which you agree and accept when you work for a government agency on secrecy projects. On the other hand, Hellman has said that these things are like stubbing your toe on a gold nugget left in the forest: "If I'm walking in the forest and stub my toe on it, who's to say I deserve credit for discovering it?" Hellman's philosophical bent here is that of a *Platonist* in the sense that all discoveries are assumed to be just that — *discoveries*, rather than creations. Hellman also stated that he, Diffie, and Merkle were all "working in a vacuum". He claimed that if they had had access to the classified documents over the previous three decades, it would have been a great advantage. Diffie commented that the history of ideas is hard to write because people find solutions to different problems and later find out that they have discovered the same thing as someone else. It is up to historians to sort out the details and the claims, but it is certain that the ideas for public-key cryptography were known (in the classified domain) well in advance of the (publicly acknowledged) efforts of Diffie, Hellman, and Merkle. CGHQ/CESG have stated that more documents are scheduled for release.

Chapter 4

Probabilistic Primality Tests

Probable impossibilities are to be preferred to improbable possibilities.

Aristotle (384–322 B.C.) Greek philosopher

4.1   Introduction

For long-term security of the RSA cipher, the primes in the modulus should be 154 digits (512 bits) each ensuring a modulus of 308 digits (1024 bits). Hence, it is essential to generate large "random primes" (see Remark 1.13). To do this, we generate large random numbers and test them for primality using primality testing algorithms. The fastest known such algorithms are probabilistic algorithms (those that use random numbers — see the discussion of randomized algorithms on page 205). This is in contrast to *deterministic algorithms*, also discussed on page 205. (See Exercises 4.1–4.5, where some deterministic primality testing algorithms, also called *primality proofs* may be found). Moreover, there are much faster (compared to deterministic) probabilistic algorithms for primality testing. There exist several general types of probabilistic algorithms, which we now describe. (If necessary, the reader may review complexity theory in Appendix B before proceeding.)

◆ Types of Probabilistic Algorithms

The following probabilistic algorithms are based upon decision problems — those problems whose solution is either "yes" or "no" (see the discussion of problem classification on page 207). In each of the following algorithms, there is an assigned *error probability* or *failure probability* (computed over all possible random choices made by the algorithm when it is run with a given input), $\alpha \in \mathbb{R}$ such that $0 \leq \alpha < 1$. This means that the algorithm will give an *incorrect* answer with probability at most α. However, by running the algorithm a sufficient number of times, we may reduce α to a value as small as desired.

We will discuss this in more detail once we have the specific primality tests at our disposal.

❶ Monte Carlo Algorithms

Monte Carlo algorithms are those probabilistic polynomial-time algorithms that answer correctly at least fifty percent of the time. More precisely, A *yes-biased* Monte Carlo algorithm is a probabilistic algorithm for which a "yes" answer is always correct, but a "no" answer may be incorrect. A *no-biased* Monte Carlo algorithm is one for which a "no" answer is always correct, but a "yes" answer may be incorrect. Hence, either form of Monte Carlo yields that half its answers are certain to be correct. The term "Monte Carlo" algorithm has been used since the turn of the twentieth century (see [119]).

❷ Atlantic City Algorithms

An Atlantic City algorithm is a a probabilistic polynomial-time algorithm that answers correctly at least seventy-five percent of the time. In other versions of the algorithm, the value 3/4 may be replaced by any value greater than 1/2. The term "Atlantic City" was first introduced in 1982 by J. Finn in an unpublished manuscript entitled *Comparison of probabilistic tests for primality.*

❸ Las Vegas Algorithms

A Las Vegas algorithm is a probabilistic expected polynomial-time algorithm that either produces no answer whatsoever, or it produces a correct answer. A Las Vegas algorithm is often viewed as a combination of both the yes-biased and no-biased Monte Carlo algorithms. The term "Las Vegas" algorithm was introduced by L. Babai in 1982 (see [114]).

As noted above, the error probability in the above algorithms may be reduced to an acceptably small value by running the algorithm a "sufficient number" of times. So, how many integers will have to be tested to determine one that is prime? The Prime Number Theorem (see Theorem C.29 on page 218) tells us that if $n \in \mathbb{N}$ is fixed then the number of primes less than n is around $n/\ln n$. One may therefore conclude that if m is a randomly chosen (odd) integer, its probability of being prime is approximately $2/\ln m$. For instance, we stated above that the RSA modulus should have primes p of 512 bits each, so

$$2/\ln p \cong 2/355, \text{ since } \ln(2^{512}) \cong 355.$$

Hence, for every 355 odd integers searched two are likely be prime. This is not unreasonable. In Section 4.4, we will begin to look in detail at some of the probabilistic primality tests that are most utilized.

4.2 Pseudoprimes and Carmichael Numbers

Time's glory is to calm contending kings,
To unmask falsehood, and bring truth to light.

William Shakespeare (1564–1616)

In Section 4.1, we looked at various types of randomized primality testing algorithms and alleged that they are faster than deterministic types. We now discuss this in further detail with an eye to setting up some notions essential for the two types of probabilistic primality tests that we will study in Sections 4.3–4.4.

There do not exist any practical primality tests that encompass the three desired properties of: (1) generality; (2) speed; and (3) correctness. In fact, in each test that has been discovered, exactly one of these properties is sacrificed. For instance, there are deterministic polynomial time algorithms, but they lack generality. A case in point is Pepin's test presented in Exercise 4.5, which works well on Fermat numbers

$$\mathfrak{F}_n = 2^{2^n} + 1.$$

On the other hand, there are deterministic tests that allow for general input, and are guaranteed to produce a proof of primality when a prime is input. However, in general, these tests require factorization, and there are no known polynomial time algorithms to do this, as we have already discussed in Chapter 3 and will discuss in more detail in Chapter 5. For instance, there is Pocklington's Theorem given in Exercise 4.1, that relies on a partial factorization of $n - 1$ given input n. This is a deterministic test since it produces a proof of primality. There is also the improvement by Kraitchik and Lehmer given in Exercise 4.2, and that by Brillhart and Selfridge in Exercise 4.3, all requiring some knowledge of factorization. In fact, the results in Exercises 4.2–4.3 rely on Fermat's Little Theorem, C.10 on page 214, the converse of which does not hold without some additional information. The tests in Exercises 4.2–4.3 actually use the converse of Fermat's Little Theorem by giving a primality test with the additional data input as stated in each Exercise. This motivates the following notion.

Definition 4.1 (Pseudoprimes)[4.1]

If $n \in \mathbb{N}$ is composite and there exists an $a \in \mathbb{N}$ such that

$$a^{n-1} \equiv 1 \,(\mathrm{mod}\ n), \tag{4.1}$$

then n is called a pseudoprime to base a. *The set of all pseudoprimes to base a is denoted by* $\mathrm{psp}(a)$.[4.2]

[4.1]In [84], Erdös states that the term "pseudoprime" is due to D.H. Lehmer (see Footnote 4.3). The notion has been generalized several times over. For instance, see [135] and [159]. In this text, we do not allow (as is sometimes done in the literature) for a pseudoprime to be prime.

[4.2]Although some sources do not use this symbol for the *set* of such pseudoprimes, we adopt the convention as it is used in [15] since we find it to be a convenient usage.

We see why we would call such entities "pseudoprimes" since (4.1) is the property held by primes in Fermat's Little Theorem. In fact, the reader can go to Exercise 4.6 for more connections in this regard.

Example 4.2 *Consider:*
$$2^{560} \equiv 1 \,(\text{mod } 561).$$

Also
$$3^{8910} \equiv 1 \,(\text{mod } 8911).$$

However,
$$561 = 3 \cdot 11 \cdot 17 \; and \; 8911 = 7 \cdot 19 \cdot 67.$$

Thus, 561 *and* 8911 *are pseudoprimes to base* 2 *and* 3, *respectively. However,* 561 *and* 8911 *are much more and in fact are examples of a very important kind of composite number.*

Definition 4.3 (Carmichael Number)
  A composite $n \in \mathbb{N}$ *satisfying*

$$a^{n-1} \equiv 1 \,(\text{mod } n) \; for \; each \; a \in \mathbb{N} \; with \; \gcd(a, n) = 1$$

is called a Carmichael *number.*

In 1922, the American mathematician, Robert Carmichael conjectured that there are infinitely many Carmichael numbers. This was proved some seventy years later by Alford, Granville, and Pomerance (see [3]). The reader may verify that both 561 and 8911 are indeed Carmichael numbers using Exercise 4.7. In fact, 561 is the smallest Carmichael number. Definition 4.3 tells us that all Carmichael numbers are pseudoprimes to base a for all $a \in \mathbb{N}$ such that $\gcd(a, n) = 1$, so Carmichael numbers are sometimes called *absolute pseudoprimes*. We see that the converse of Fermat's Little Theorem fails in the extreme for Carmichael numbers. However, what Exercises 4.1–4.3 demonstrate is that the converse of Fermat's Little Theorem can be used to prove primality of n if we can find an element of order $n - 1$ in $(\mathbb{Z}/n\mathbb{Z})^*$, namely, a primitive root modulo n (see Proposition C.22 on page 216).

In Sections 4.3–4.4, we introduce variants of the notion in Definition 4.1 in the presentation of two probabilistic primality tests. These tests sacrifice correctness, but are extremely fast and allow for general input. Moreover, if the input is prime, we are guaranteed that the test will declare it to be so. However, the uncertainty comes attached to the fact that a composite number might also be deemed to be prime. Nevertheless, we can reduce this probability to acceptably low levels.

Exercises

4.1. Let $n = ab + 1$ ($a, b \in \mathbb{N}$, $b > 1$). Suppose that for any prime divisor q of $b > 1$, there exists an integer m with $\gcd(m^{(n-1)/q} - 1, n) = 1$, and $m^{n-1} \equiv 1 \pmod{n}$. Prove that $p \equiv 1 \pmod{b}$ for every prime p dividing n. Also, if $b > \sqrt{n} - 1$, then n is prime. (*This is known as* Pocklington's Theorem, *which is a* deterministic primality test (*see page 124*).)

4.2. Given $n \in \mathbb{N}$ with $n \geq 3$, prove that n is prime if and only if there is an $m \in \mathbb{N}$ such that $m^{n-1} \equiv 1 \pmod{n}$, but $m^{(n-1)/q} \not\equiv 1 \pmod{n}$ for any prime $q \mid (n-1)$. (*This is another deterministic primality test due to Lehmer*[4.3] *and Kraitchik*[4.4], *according to* [15, p. 267] — *see* [125, p. 135].)

4.3. Prove that $n \geq 3$ is prime if and only if for each prime $q \mid (n-1)$, there exists an $m_q \in \mathbb{Z}$ with both $m_q^{n-1} \equiv 1 \pmod{n}$ and $m_q^{(n-1)/q} \not\equiv 1 \pmod{n}$. (*This deterministic test is due to Brillhart*[4.5] *and Selfridge* (*see Footnote 4.10*) *in* [47], *improving upon the result in Exercise 4.2.*)

4.4. Let $\mathfrak{F}_n = 2^{2^n} + 1$ for $n \in \mathbb{N}$ be the n-th Fermat number. Prove that if $\mathfrak{F}_n$ is prime, then 3 is a primitive root modulo $\mathfrak{F}_n$.

4.5. Let $\mathfrak{F}_n = 2^{2^n} + 1$ be the n-th Fermat number. Prove that $\mathfrak{F}_n$ is prime if and only if $3^{(\mathfrak{F}_n - 1)/2} \equiv -1 \pmod{\mathfrak{F}_n}$. (*This is called* Pepin's primality test, *another deterministic test.*)

4.6. If n is an odd composite number satisfying (4.1), sometimes called *Fermat's Primality test*, for some integer a, then a is called a *Fermat Liar* (to the primality of n). Thus, the elements of psp(a) (see Definition 4.1) are sometimes called *ordinary pseudoprimes* or *Fermat pseudoprimes*. Find all Fermat liars for $n = 65$.

4.7. Let $n = \prod_{j=1}^{r} p_j$ where $r \geq 2$ and the p_j are distinct odd primes. Prove that n is a Carmichael number if and only if $(p_j - 1) \mid (n - 1)$ for $j = 1, 2, \ldots, r$. (See Definition 4.3.)

[4.3]Derrick Henry Lehmer (1905–1991) was born in Berkeley, California on February 23, 1905. He obtained his Ph.D. from Brown University in 1930. After some brief stints elsewhere, he took a position at UC Berkeley in 1940, remaining there until he retired in 1972. He was a truly great pioneer in computational number theory. See a collection of his selected works [133] for evidence of his genius.

[4.4]Maurice Borisovich Kraitchik (1882–1957) obtained his Ph.D. from the University of Brussels in 1923. He worked as an engineer in Brussels and later as a Director at the Mathematical Sciences section of the Mathematical Institute for Advanced Studies there. From 1941–1946, he was Associate Professor at the New School for Social Research in New York. In 1946, he returned to Belgium where he died on August 19, 1957. His work over thirty-five years on factoring methods stands tall today because he devised and used a variety of practical techniques that are found today in computer methods such as the Quadratic Sieve (see Chapter 5). He is also the author of the popular book *Mathematical Recreations* [126].

[4.5]John Brillhart is a Professor Emeritus at the University of Arizona at Tucson. He obtained his Ph.D. under the supervision of Dick Lehmer (see Footnote 4.3) at UC Berkeley. He is a pioneer in computational number theory, and a co-inventor of the continued fraction factoring algorithm (see Chapter 5). His seminal work in primality testing and factoring is part of the structure of the computational edifice we enjoy today.

4.3   Solovay-Strassen Test

Idleness is only the refuge of weak minds.

Lord Chesterfield
(Philip Dormer Stanhope, Earl of Chesterfield) (1694–1773)
English writer and politician

To prepare for this section, the reader must be familiar with the Jacobi symbol and Euler's criterion, for which Appendix C will provide a quick reference (see the discussion starting on page 217). The probabilistic primality test in the title of this section was the first such test that became prominent due to the arrival of public-key cryptography upon the cryptographic scene. We present it here for both historical and motivational reasons in comparison to the stronger test to be presented in Section 4.4 since it will provide us with a means of introducing new concepts for later use.

The following is our first example of one of the types of algorithms presented in Section 4.1. It is a Monte Carlo algorithm for compositeness, namely, a yes-biased Monte Carlo algorithm for the decision problem: "Is the input composite?" (See the discussion of problem classification in Appendix B starting on page 207.) Thus, if the algorithm answers that input $n \in \mathbb{N}$ is composite, then indeed it is, whereas if it answers that n is prime, we only have some (good) evidence of primality. This is due to the fact that the test is based upon the Euler criterion (see Corollary C.27 on page 218), which is a necessary, but not sufficient, test for primality in terms of the Jacobi symbol. The following is named after the discoverers who published the result [224] in 1977. Also, due to the use of Euler's criterion, the test is also called the *Euler pseudoprimality test*. This test was modified in 1982 by Atkin and Larson [8].

◆ The Solovay-Strassen Primality Test

Let $n \in \mathbb{N}$, $n > 1$ and select at random r positive random integers a_j with $a_j < n$ and $\gcd(a_j, n) = 1$ for $j = 1, 2, \ldots, r \in \mathbb{N}$. For each such j compute both:

$$a_j^{(n-1)/2} \text{ and } \left(\frac{a_j}{n}\right) \text{ modulo } n$$

until one of the following occurs.

(1) For some $j \leq r$,

$$a_j^{(n-1)/2} \not\equiv \left(\frac{a_j}{n}\right) \pmod{n},$$

in which case, terminate the algorithm with "n is definitely not prime".

(2) For all $j = 1, 2, \ldots, r$,

$$a_j^{(n-1)/2} \equiv \left(\frac{a_j}{n}\right) \pmod{n} \tag{4.4}$$

in which case, terminate the algorithm with "n is probably prime".

Any composite integer n that is declared to be "probably prime" to base a by the Solovay-Strassen test is said to be a *base-a Euler pseudoprime*,[4.6] and a is called an *Euler liar*[4.7] (to the primality of n). The set of Euler psuedoprimes to base a is denoted by epsp(a). The set of Euler liars, $E(n)$, is defined in Exercise 4.8, which shows that $E(n)$ actually is a subgroup of $(\mathbb{Z}/n\mathbb{Z})^*$. When

$$a \in (\mathbb{Z}/n\mathbb{Z})^*, \text{ but } a \notin E(n),$$

then a is called an *Euler witness* (to the compositeness of n). Exercise 4.13 tells us that at least half of the integers a with $1 < a < n$ are Euler witnesses. Moreover, by Exercise 4.12:

$$n \text{ is an odd prime if and only if } E(n) = (\mathbb{Z}/n\mathbb{Z})^*,$$

a foundational fact upon which the Solovay-Strassen test is built.

If n is declared to be "definitely not prime" in step (1) of the test, then we may conclude *with certainty* that n is composite since Euler's criterion tells us that all primes *necessarily* satisfy (4.4), so failure to do so in step (1) is a guarantee that the number cannot be prime. Moreover, if n is indeed prime then the test will declare that it is since it will pass Euler's criterion, necessarily. However, if n is composite, then there is no certainty that the test will declare it to be so. Nevertheless, by Exercises 4.12–4.13, the probability that an odd composite integer will be declared to be prime by the Solovay-Strassen test is less than $(1/2)^r$ for randomly chosen r. Hence, the larger the parameter r, the greater the probability that a composite number will not be declared to be prime by the Solovay-Strassen test.

Example 4.5 *Let $n = 8911$ and choose $a = 2$. Then*

$$2^{(n-1)/2} \equiv 2^{4455} \equiv 6364 \not\equiv \left(\frac{2}{n}\right) \equiv 1 \,(\mathrm{mod}\ n).$$

hence, n is declared to be composite by the Solovay-Strassen test. Indeed, $8911 = 7 \cdot 19 \cdot 67$.

In terms of complexity, it can be shown that for a given base a, in the Solovay-Strassen test, each of the computations takes $O((1 + \lfloor \log_2 |n| \rfloor)^3)$ bit operations (see [15, Theorem 9.4.2, p. 280]). In Section 4.4, we will introduce a test which also uses this number of bit operations, but which is both more efficient and more often correct in its outcome.

[4.6]The term "Euler pseudoprime" was introduced by Shanks [204] in 1978.

[4.7]In the literature, the term "false witness" is sometimes used instead of "liar".

Exercises

4.8. Let $n \in \mathbb{N}$ be odd and composite. Define $E(n)$ by

$$\left\{ a \in (\mathbb{Z}/n\mathbb{Z})^* : \left(\frac{a}{n} \right) \equiv a^{(n-1)/2} \,(\mathrm{mod}\ n) \right\}.$$

Prove that this is a subgroup of $(\mathbb{Z}/n\mathbb{Z})^*$, called the *group of Euler liars* for n. (*Hint: It is sufficient to prove that $ab^{-1} \in E(n)$ whenever $a, b \in E(n)$.*)

4.9. Prove that if $p > 2$ is prime, $a \in \mathbb{N}$, then

$$x^2 \equiv 1 \,(\mathrm{mod}\ p^a) \text{ if and only if } x \equiv \pm 1 \,(\mathrm{mod}\ p^a).$$

4.10. Prove that if n is a Carmichael number, then n is squarefree.

4.11. Prove that 8911 and 10585 are Carmichael numbers, and that 10585 is also an Euler pseudoprime to base 2.

Exercises 4.12–4.13 are in reference to Exercise 4.8.

4.12. Prove that any odd natural number $n \geq 3$ is prime if and only if $E(n) = (\mathbb{Z}/n\mathbb{Z})^*$. (*Hint: Use Exercise 4.10.*)

4.13. Prove that for composite n, $|E(n)| \leq \phi(n)/2$.

4.14. Prove that 15841 is an Euler pseudoprime to base 2.

4.15. Prove that 29341 is an Euler pseudoprime to base 2.

4.16. Prove that 41041 is an Euler pseudoprime to base 2.

4.17. Prove that 62745 is an Euler pseudoprime to base 2.

4.18. Prove that the values in Exercises 4.14–4.17 are all Carmichael numbers. (*Hint: Use Exercise 4.7.*)

4.19. Find an Euler pseudoprime to base 3 that is also a Carmichael number.

☆ 4.20. Let n be an odd natural number. With reference to Exercise 4.8, prove that if

$$n = \prod_{j=1}^{r} p_j^{m_j}$$

where the p_j are distinct primes, then

$$|E(n)| = d_n \prod_{j=1}^{r} \gcd\left(\frac{n-1}{2}, p_j - 1 \right),$$

where $d_n \in \{1/2, 1, 2\}$.

(*Hint: Use Theorem C.28 on page 218.*)

4.4 Miller-Selfridge-Rabin Test

Beauty is the first test: there is no permanent place in the world for ugly mathematics.   **G. H. Hardy**[4.8]

We may now provide our second example of a Monte Carlo algorithm for compositeness (see Section 4.1).

There are methods for proving that a number is composite without finding any factors of that number. For instance, consider the following simple illustration.

Example 4.6 *It is easy to prove that* $n = 77$ *is composite without finding a factor, since by Fermat's Little Theorem C.10 on page 214, 77 cannot be prime, given that*

$$2^{n-1} \not\equiv 1 \,(\text{mod } n).$$

(The reader may show this to be true in incremental steps. Consider:

$$2^8 \equiv 25 \,(\text{mod } 77); \qquad 2^{12} \equiv 2^4 \cdot 2^8 \equiv 16 \cdot 25 \equiv 15 \,(\text{mod } 77);$$

$$2^{16} \equiv 25^2 \equiv 9 \,(\text{mod } 77); \quad 2^{32} \equiv 9^2 \equiv 4 \,(\text{mod } 77); \quad 2^{64} \equiv 16 \,(\text{mod } 77);$$

so $2^{76} \equiv 2^{64} \cdot 2^{12} \equiv 16 \cdot 15 \equiv 9 \,(\text{mod } 77).)$

This idea is used in a more general fashion as follows.

◆ **The Miller[4.9]-Selfridge[4.10]-Rabin[4.11] Primality Test**

Let $n - 1 = 2^t m$ where $m \in \mathbb{N}$ is odd and $t \in \mathbb{N}$. The value n is the input to be tested by executing the following steps, where all modular exponentiations are done using the repeated squaring method described on page 50.

(1) Choose a random integer a with $2 \leq a \leq n - 2$.

[4.8]Godfrey Harold Hardy was born in Cranleigh, Surrey on February 7, 1877. He taught at Trinity College, Cambridge from 1906 to 1919. In 1919, he was appointed to the Savilian Chair of Geometry at Oxford University. In 1931, he was appointed Sadlerian Professor of Pure Mathematics at Cambridge, where he died on December 1, 1947. Hardy is famous, not only for his numerous contributions to number theory, but also for his book, first published in 1940, *A Mathematician's Apology*, from which this quote is taken. This book took the stand that number theory is "useless" in the sense of having no real-world applications. Hardy perhaps foresaw that some applications might come to the fore when he said in his book: "Time may change all this." Indeed, he would be greatly surprised to see the impact of number theory on cryptography and its varied applications.

[4.9]Gary Miller obtained his Ph.D. in Computer Science from UC Berkeley in 1974. He is currently a Professor in Computer Science at Carnegie-Mellon University. His expertise lies in computer algorithms.

[4.10]This test is most often called the *Miller-Rabin Test* in the literature. However, John Selfridge was using the test in 1974 before Miller first published the result, so we credit Selfridge here with this recognition. John Selfridge was born in Ketchican, Alaska, on February 17, 1927. He received his doctorate from U.C.L.A. in August of 1958, and became a Professor at Pennsylvania State University six years later. He is a pioneer in computational number theory.

[4.11]See Footnote B.11 on page 211.

(2) Compute:
$$x \equiv a^m \, (\text{mod } n).$$

If
$$x \equiv \pm 1 \, (\text{mod } n),$$

then terminate the algorithm with:

"*n* is probably prime".

If $t = 1$, terminate the algorithm with

"*n* is definitely not prime."

Otherwise, set $j = 1$ and go to step (3).

(3) Compute:
$$x \equiv a^{2^j m} \, (\text{mod } n).$$

If $x \equiv 1 \, (\text{mod } n)$, then terminate the algorithm with

"*n* is definitely not prime."

If $x \equiv -1 \, (\text{mod } n)$, terminate the algorithm with

"*n* is probably prime."

Otherwise set $j = j + 1$ and go to step (4).

(4) If $j = t - 1$, then go to step (5). Otherwise, go to step (3).

(5) Compute:
$$x \equiv a^{2^{t-1} m} \, (\text{mod } n).$$

If $x \not\equiv -1 \, (\text{mod } n)$, then terminate the algorithm with

"*n* is definitely not prime."

If $x \equiv -1 \, (\text{mod } n)$, then terminate the algorithm with

"*n* is probably prime."

If n is declared to be "probably prime" with base a by the Miller-Selfridge-Rabin test, then

n is said to be a *strong pseudoprime to base a.*

Thus, the above test is often called the *strong pseudoprime test*[4.12] in the literature. The set of all pseudoprimes to base a is denoted by spsp(a).

[4.12]The term "strong pseudoprime" was introduced by Selfridge in the mid-1970s, but he did not publish this reference. However, it did appear in a paper by Williams [235] in 1978.

Let us look a little closer at the above test to see why it it is possible to declare that "n is definitely not prime" in step (3). If $x \equiv 1 \pmod{n}$ in step (3), then for some j with $1 \leq j < t - 1$:

$$a^{2^j m} \equiv 1 \pmod{n}, \text{ but } a^{2^{j-1} m} \not\equiv \pm 1 \pmod{n}.$$

Thus, by Exercise 4.23 (a special case of which appears in Exercise 3.12) $\gcd(a^{2^{j-1} m} - 1, n)$ is a nontrivial factor of n. Hence, if the Miller-Selfridge-Rabin test declares in step (3) that n is "definitely not prime", then indeed it is composite. Another way of saying this is that if n is prime, then Miller-Selfridge-Rabin will declare it to be so. However, if n is composite, then it can be shown that the test fails to recognize n as composite with probability at most $(1/4)$. This is why the most we can say is that "n is probably prime". However, if we perform the test r times for r large enough, this probability $(1/4)^r$ can be brought arbitrarily close to zero. Moreover, at least in practice, using the test with a single choice of a base a is usually sufficient.

Also, in step (5), notice that we have not mentioned the possibility that

$$a^{2^{t-1} m} \equiv 1 \pmod{n}$$

specifically. However, if this did occur, then that means that in step (3), we would have determined that

$$a^{2^{t-2} m} \not\equiv \pm 1 \pmod{n},$$

so by Exercise 4.9, n cannot be prime. Furthermore, by the above method, we can factor n since $\gcd(a^{2^{t-2} m} - 1, n)$ is a nontrivial factor. This final step (4) is required since, if we get to $j = t - 1$, with $x \not\equiv \pm 1 \pmod{n}$ for any $j < t - 1$, then simply invoking step (3) again would dismiss those values of $x \not\equiv \pm 1 \pmod{n}$, and this would not allow us to claim that n is composite in those cases. Hence, it allows for more values of n to be deemed composite, with certainty, than if we merely performed step (3) as with previous values of j.

The above discussion contains a fundamental principle that is worth discussion. A basic point made in Exercise 4.23 is that if we have a natural number $n > 1$ such that

$$x^2 \equiv y^2 \pmod{n} \text{ with } x \not\equiv \pm y \pmod{n} \text{ for integers } x, y, \tag{4.7}$$

then n is necessarily composite since $\gcd(x - y, n)$ is a nontrivial factor of n. Although Gauss knew this, the basic idea goes back to Fermat who, in 1643, developed a method of factoring that was based upon a simple observation. If $n = rs$ is an odd natural number with $1 < r < \sqrt{n}$, then

$$n = a^2 - b^2 \text{ where } a = (s + r)/2 \text{ and } b = (s - r)/2.$$

Therefore, in order to find a factor of n, we need only look at values $x = y^2 - n$ for

$$y = \lfloor \sqrt{n} \rfloor + 1, \lfloor \sqrt{n} \rfloor + 2, \ldots, (n - 1)/2$$

until a perfect square is found. We call this *Fermat's difference of squares method*, which has been rediscovered numerous times and is the underpinning of many algorithms for factoring today. (The basic idea is attributed also to Legendre.) We will discuss this in more detail in Chapter 5. Exercise 3.12 illustrates how this idea of Fermat can be used to factor an RSA modulus, $n = pq$, when p and q are "close together". This is one of the reasons for ensuring that the primes are "properly" chosen in an RSA modulus, a topic that we will revisit in Chapter 6. Now we return to the Miller-Selfridge-Rabin test with an illustration.

Example 4.8 *Let $n = 1105$. Since $n - 1 = 2^t m = 2^4 \cdot 69$, we choose $a = 2$ and compute*

$$x \equiv 2^m \equiv 967 \,(\mathrm{mod}\ n)$$

in step (2). Then we go to steps (3)–(4), compute

$$x \equiv 2^{2^j m} \,(\mathrm{mod}\ n) \text{ for } j = 1, 2,$$

for which $x \not\equiv \pm 1 \,(\mathrm{mod}\ n)$. Since $j = 3 = t - 1$ in step (4), we go to step (5) and compute:

$$x \equiv 2^{2^3 m} \equiv 1 \,(\mathrm{mod}\ n).$$

Thus, we conclude with "n is definitely not prime".

Example 4.8 shows that 1105 is *not* a strong pseudoprime to base 2. However,

$$2^{1104} \equiv 1 \,(\mathrm{mod}\ 1105),$$

so 1105 *is* a pseudoprime to base 2. In general, there are far fewer strong pseudoprimes than pseudoprimes (see Exercise 4.21). Moreover, 1105 is a Carmichael number (see Definition 4.3). By Exercise 4.27, a Carmichael number n cannot be a strong pseudoprime to *every* base a with $\gcd(a, n) = 1$. Also, Carmichael numbers provide infinitely many counterexamples to the converse of Fermat's Little Theorem. However, there exists an additional condition that makes the converse of Fermat's theorem work as a primality test — see Exercise 4.28.

Exercise 4.22 provides another characterization of strong pseudoprimes. Therein, we define a *strong witness* to the compositeness of n. It can be shown that there are at most $(n - 1)/4$ natural numbers (prime to, and less than n) which are *not* strong witnesses for the compositeness of n, namely, if n is composite, then in the notation of Exercise 4.22,

$$|S(n)| \leq (n - 1)/4.$$

In fact, it is known that for *any* natural number $n \neq 9$,

$$|S(n)| \leq \phi(n)$$

(see [15] and [156]). This gives further evidence for the earlier allegation that the probability of a composite number n failing detection by the strong pseudoprimality test is less than $1/4$. It can also be shown, as well, that the Miller-Selfridge-Rabin test uses $O((1 + \lfloor \log_2 |n| \rfloor)^3)$ bit operations (see [15, Theorem 9.4.5, p. 282]). Also, the following may be of interest to the reader with knowledge of the generalized Riemann hypothesis (GRH). Under the assumption of GRH, it has been shown that for every composite integer n, there exists a base a with

$$1 < a \leq 2(\log_2^2 n),$$

such that n fails the Miller-Selfridge-Rabin test to base a. This was done by Bach [13]–[14], based upon the earlier work of Miller [158], which in turn appealed to the ideas of Ankeny [4] (see [239, p. 351]). Lastly, it has been shown, under the assumption of the GRH, that there exists a deterministic polynomial time algorithm for primality testing (see [156] and [245]).

Now we have a look at comparisons among the tests we have discussed in this chapter. We know that both of the Solovay-Strassen and Miller-Selfridge-Rabin tests use $O((1 + \lfloor \log_2 |n| \rfloor)^3)$ bit operations for input n. However, Miller-Selfridge-Rabin is computationally less expensive to implement than the Solovay-Strassen test given that the latter requires computation of Jacobi symbols. Furthermore, for a given number, r, of iterations of either algorithm, the error probability of Solovay-Strassen is at most $(1/2)^r$, while that for Miller-Selfridge-Rabin is at most $(1/4)^r$. Also, by Exercise 4.35, strong liars must be Euler liars, so Miller-Selfridge-Rabin can never err more than Solovay-Strassen. In other words, Miller-Selfridge-Rabin is always *at least as correct* as Solovay-Strassen. By Exercises 4.6 and 4.35–4.36, we have the following diagram.

Figure 4.8: **Hierarchy of Pseudoprimality**

This diagram encapsulates the practical fact that one should never use either of Fermat or Solovay-Strassen over Miller-Selfrige-Rabin. The former two were presented for motivational, historical, and pedagogical reasons.

In Chapter 5, we move up to the next cryptographic step by looking at factoring methods, which will lead us into Chapter 6 where we discuss security issues surrounding RSA and public-key cryptography in general.

Exercises

4.21. For given $a \in \mathbb{N}$, prove that $\mathrm{spsp}(a) \subseteq \mathrm{psp}(a)$. (See Definition 4.1.)

4.22. Assume that $n = 1 + 2^t m$ is composite, where $m \in \mathbb{N}$ is odd, and $t \in \mathbb{N}$. Let $S(n)$ denote the following set:

$$\{a \in (\mathbb{Z}/n\mathbb{Z})^* : a^m \equiv 1 \,(\mathrm{mod}\ n) \text{ or } a^{2^j m} \equiv -1 \,(\mathrm{mod}\ n) \text{ for } 0 \le j < t\},$$

called the *set of strong liars* (to primality) for n. If $a \in (\mathbb{Z}/n\mathbb{Z})^*$ and $a \notin S(n)$ then a is called a *strong witness* (to compositeness) for n. Prove that $a \in S(n)$ if and only if n is a strong pseudoprime to base a.

4.23. Let $n \in \mathbb{N}$ and $x, y \in \mathbb{Z}$ with $x^2 \equiv y^2 \,(\mathrm{mod}\ n)$ and $x \not\equiv \pm y \,(\mathrm{mod}\ n)$. Prove that n is composite and $\gcd(x - y, n)$ is a nontrivial factor of n.

4.24. Prove that if $S(n) = (\mathbb{Z}/n\mathbb{Z})^*$, defined in Exercise 4.22, then n is a Carmichael number.

4.25. Prove that each of 15841, 29341, 52633, and 252601 is both a Carmichael number *and* a strong pseudoprime to base 2.

4.26. Prove that the only possible even Carmichael number is $n = 2$. (This leads to the trivial statement that $a \equiv 1 \,(\mathrm{mod}\ 2)$ for all odd integers a. Thus, we always assume that a Carmichael number is odd.)

4.27. Prove that a Carmichael number n cannot be a strong pseudoprime to *every* base a with $\gcd(a, n) = 1$.

4.28. Let g be a primitive root modulo an odd prime p. Prove that if a prime $q \mid (p - 1)$, then $g^{(p-1)/q} \not\equiv 1 \,(\mathrm{mod}\ p)$.

4.29. Use the Miller-Selfridge-Rabin test to prove that $n = 561$ is a composite number (in fact, as noted in Section 4.2, the smallest Carmichael number).

In Exercises 4.30–4.34, use the Miller-Selfridge-Rabin test to determine that the given value of n is a strong pseudoprime to the given base(s) a.

4.30. $n = 2047$ and $a = 2$. (*This is the smallest strong pseudoprime to base 2.*)

4.31. $n = 121$ and $a = 3$. (*This is the smallest strong pseudoprime to base 3.*)

4.32. $n = 781$ and $a = 5$. (*This is the smallest strong pseudoprime to base 5.*)

4.33. $n = 25$ and $a = 7$. (*This is the smallest strong pseudoprime to base 7.*)

4.34. $n = 3215031751$ and $a = 2, 3, 5, 7$. (*This is the smallest strong pseudoprime to all bases $2, 3, 5, 7$. This discovery was published in [190].*)

4.35. Prove that for $a \in \mathbb{N}$, $\mathrm{epsp}(a) \subseteq \mathrm{psp}(a)$.

☆ 4.36. Prove that $\mathrm{spsp}(a) \subseteq \mathrm{epsp}(a)$. (*Hint: Use the general theory of orders of integers and properties of the Jacobi symbol given in Appendix C.*)

4.37. Prove that for any prime power $p^b \in \mathrm{spsp}(a)$ if and only if $p^b \in \mathrm{psp}(a)$.

Chapter 5

Factoring

The problem of distinguishing prime numbers from composite numbers and of resolving the latter into their prime factors is known to be one of the most important and useful in arithmetic.

                                                                    C. F. Gauss[5.1]

5.1 Universal Exponent Method

The problem of primality testing covered in Chapter 4 is less difficult, in general, than the topic of this chapter. We have already learned that public-key cryptosystems such as RSA, base their security upon the presumed difficulty of integer factorization, a term that we now make precise.

◆ **The Integer Factorization Problem (IFP)**

Given $n \in \mathbb{N}$ find primes p_j for $j = 1, 2, \ldots, r \in \mathbb{N}$ with $p_1 < p_2 < \cdots < p_r$ such that $n = \prod_{j=1}^{r} p_j^{e_j}$.

Essentially what the conclusion of the IFP says is the content of the Fundamental Theorem of Arithmetic (see Theorem C.6 on page 213). A subset of the IFP (which is actually the same as the IFP in the case of an RSA modulus) is the notion of *splitting* $n \in \mathbb{N}$, by which we mean that we find $r, s \in \mathbb{N}$ such that $1 < r \leq s < n$ and $n = rs$.

[5.1]Carl Friederich Gauss (1777–1855) is perhaps the best-known and possibly the greatest mathematician of all time. Evidence of his genius emerged at age three when he found an error in his father's bookkeeping. By age eight he had astonished his teacher, Büttner, by rapidly calculating $\sum_{j=1}^{100} j$ with the simple observation that the fifty pairs $(j + 1, 100 - j)$ for $j = 0, 1, \ldots, 49$ each sum to 101 for a total of 5050. Ultimately, he entered the Collegium Carolinum with funding from the Duke of Brunswick, to whom he dedicated his magnum opus *Disquisitiones Arithmeticae* [95], published in 1801, from which the above quote was taken. By 1795, he entered Göttingen University and received his doctorate at the age of twenty. His accomplishments are too numerous to list here (see [164] for more of them), but it is safe to say that his work continues to have influence to this day. Gauss remained a professor at Göttingen until he died in his sleep on February 23, 1855. His awareness over two centuries ago of the importance of factoring methods was prescient given today's applications to public-key cryptography.

Of course, the oldest method of splitting n is *trial division* by which we mean to divide n by all primes up to $\sqrt{n}$. For $n < 10^8$ or so, this is not unreasonable. However, since the running time of trial division can be shown to be $O(\max(p_{r-1}, \sqrt{p_r}))$ (see [123, p. 381] and [245, Theorem 2.3.1, p. 197]), then we need more sophisticated methods since we already know that for long-term security an RSA modulus should have 1024 bits. In fact, at the end of Section 3.2 when we were discussing RSA moduli, we promised the following method.

First, we need a new concept. For a given modulus $n \in \mathbb{N}$, an exponent $e \in \mathbb{N}$ is called a *universal exponent* if $x^e \equiv 1 \pmod{n}$ for all $x \in \mathbb{N}$ with $\gcd(x, n) = 1$.

◆ Universal Exponent Factorization Method

Let e be a universal exponent for $n \in \mathbb{N}$ and set $e = 2^b m$ where $b \geq 0$ and m is odd. Execute the following steps.

(1) Choose a random base a such that $1 < a < n - 1$. If $\gcd(a, n) > 1$, then we have a factor of n, and we may terminate the algorithm. Otherwise go to step (2).

(2) Let $x_0 \equiv a^m \pmod{n}$. If $x_0 \equiv 1 \pmod{n}$, then go to step (1). Otherwise, compute $x_j \equiv x_{j-1}^2 \pmod{n}$ for all $j = 1, \ldots, b$. If

$$x_j \equiv -1 \pmod{n},$$

then go to step (1). If

$$x_j \equiv 1 \pmod{n}, \text{ but } x_{j-1} \not\equiv \pm 1 \pmod{n},$$

then $\gcd(x_{j-1} - 1, n)$ is a nontrivial factor of n so we may terminate the algorithm.

If one compares the above with the Miller-Selfridge-Rabin probabilistic primality test in Section 4.4, striking similarities will be seen. However, the primality test is not guaranteed to have a value such that $x_j \equiv 1 \pmod{n}$ as we have in the universal exponent method (due to the existence of the exponent e).

When $n = pq$ is an RSA modulus, a universal exponent is sometimes taken to be $\mathrm{lcm}(p - 1, q - 1)$ instead of $\phi(n) = (p - 1)(q - 1)$. Yet these two values will be roughly the same since $\gcd(p - 1, q - 1)$ has an expectation of being small when p and q are chosen arbitrarily. Furthermore, recalling the discussion at the end of Section 3.2 on page 65, since $de - 1$ is a multiple of $\phi(n)$, then $de - 1$ is a universal exponent, and the above method can be used to factor n as alleged in Section 3.2. This is the major value of our universal exponent test since actually finding e is difficult in practice.

Example 5.1 *Let $n = 3763$ and suppose that we know $e = 3640$ is a universal exponent for n. Since $e = 2^3 \cdot 455$, then we first choose $a = 2$ as a base and compute $2^{455} \equiv 924 \equiv x_0 \pmod{n}$. Then $x_1 \equiv 924^2 \equiv 3338 \pmod{n}$, and $x_2 \equiv 3338^2 \equiv 1 \pmod{n}$. Since $x_1 \not\equiv \pm 1 \pmod{n}$, then $\gcd(x_1 - 1, n) = \gcd(3337, 3763) = 71$ is a factor of n. Indeed $n = 53 \cdot 71$.*

Exercises

5.1. If $n = 2^{a+2}$ for $a \in \mathbb{N}$, prove that 2^a is a universal exponent modulo n.

5.2. Let $n = 2^a \prod_{j=1}^{k} p_j^{a_j}$, where $a \geq 0$, $a_j \in \mathbb{N}$, and the p_js are distinct odd primes. The following is called the *Carmichael function* (see page 82):

$$\lambda(n) = \begin{cases} \phi(n) & \text{if } n = 2^a, \text{ and, } 1 \leq a \leq 2, \\ 2^{a-2} = \phi(n)/2 & \text{if } n = 2^a, a > 2, \\ \text{lcm}(\lambda(2^a), \phi(p_1^{a_1}), \ldots, \phi(p_k^{a_k})) & \text{if } k \geq 1. \end{cases}$$

Prove that $\lambda(n)$ is a universal exponent for n.

5.3. Prove that for a given $n \in \mathbb{N}$, $\lambda(n)$ is the *minimal universal exponent* for n (see Exercise 5.2).

5.4. Prove that for any $n \in \mathbb{N}$, $\lambda(n) \mid \phi(n)$.

In Exercises 5.5–5.13, use the universal exponent method to find a factor of each given n assuming that the given e is a universal exponent modulo n.

5.5. $n = 15841$ and $e = 12960$.

5.6. $n = 46657$ and $e = 41472$.

5.7. $n = 107381$ and $e = 106572$.

5.8. $n = 2919533$ and $e = 2915328$.

5.9. $n = 14660479$ and $e = 14652760$.

5.10. $n = 32774813$ and $e = 1170120$.

5.11. $n = 47440937$ and $e = 447426$.

5.12. $n = 61429883$ and $e = 30707100$.

5.13. $n = 76859539$ and $e = 12807000$.

5.14. Factor $n = 4294967297$ assuming that $e = 33502080$ is a universal exponent. (*Hint: The task will be made easier and more transparent in this case if Exercise 4.4 on page 83 is employed.*)

5.15. Use the technique of Exercise 5.14 to factor $n = 18446744073709551617$ with $e = 72057331223781120$.

Suddenly Christopher Robin began to tell Pooh about some of the things People called Kings and Queens and something called Factors....

from **The House on Pooh Corner**

by **A. A. Milne (1882 – 1956)**

5.2   Pollard's $p-1$ Method

*I could be bounded in a nut-shell, and count myself a king of infinite space,
were it not that I have bad dreams.*

from **Hamlet** — by **Shakespeare**

The algorithm presented herein can quickly find certain primes p dividing n
when we have that $p-1$ has only "small" prime factors. In order to describe
the algorithm, we need the following notion.

Definition 5.2 (Smoothness)
*If $B, n \in \mathbb{N}$ and n is divisible only by primes $p \leq B$, then n is said to be
B-smooth and B is called a* smoothness bound.

Given that deciding whether a given integer is composite or prime is easier
in general than factoring, we will assume that n has been checked and found
to be composite before applying any factoring algorithm. Moreover, we may
assume that we have checked that n is not a perfect power, for if $n = m^e$
where $m, e \in \mathbb{N}$ and $e > 1$, then we can actually determine m and e as follows.
For each prime $p \leq \log_2 n$, do a binary search for $r \in \mathbb{N}$ satisfying $n = r^p$,
restricting attention to the range $2 \leq r \leq 2^{\lfloor (\log_2 n)/p \rfloor + 1}$. This calculation may
be accomplished in $O((\log_2 n)^3 (\log_2(\log_2(\log_2 n))))$ bit operations. Thus, this is
a reasonable pre-test for ensuring that we do not have such a power.

The following was known to the Lehmers in the 1930s, but they did not publish it since it was not useful with hand calculations, and they had no computer
on which to implement it. However, when it was discovered by Pollard [187] in
1974, the computational value of the technique came to light.

◆ Pollard's $p-1$ Factoring Method

We assume that n is known to be composite, but not a perfect power, in the
application of this algorithm, which attempts to split n.

(1) Select a random $a \in \mathbb{Z}/n\mathbb{Z}$, $a \geq 2$, and if $\gcd(a, n) > 1$, then terminate
the algorithm, since we have split n. Otherwise go to step (2).

(2) Select a smoothness bound B, let $a_1 = a$, and set $j = 2$.

(3) Compute $a_j \equiv a_{j-1}^j \pmod{n}$ using the repeated squaring method.

(4) Compute $g = \gcd(a_j - 1, n)$.

(5) If $g > 1$, terminate the algorithm, since we have split n. If $g = 1$, or n, set
$j = j + 1$. If $j \leq B$, go to step (3). Otherwise, terminate with "failure".

Remark 5.3 *The smoothness bound needs to be small enough to ensure that
the algorithm runs quickly, but large enough to ensure some reasonable chance
of success. Clearly we do not want to choose B anywhere near $\sqrt{n}$ or we will*

have no advantage over trial division. Also, we do not want it to be so small that there is no hope of finding a nontrivial factor. More precisely, since the running time of Pollard's method can be shown to be $O((B \ln B \ln^2 n + \ln^3 n)$ *modular multiplications, then for* $B = O(\ln^m n)$, *for some fixed* $m \in \mathbb{N}$, *the algorithm is polynomial time. However, in this case,* B *is too small to ensure any reasonable hope of success. Pollard's method will find at least one out of every* N *primes in time*[5.2] $N^{1+o(1)}$ *if* n *has* N *bits, where* $2 \ln^2 N = \ln B \ln \ln B$. *In practice a bound for* B *is taken in the range* $10^5 < B < 10^6$. *If the algorithm ends in failure, one variant is to look at some primes* $j > B$ *and try steps* (3)–(4) *for those values. There is also a second phase of Pollard's* $p - 1$ *method that is not included in a typical description of the method. The interested reader may consult* [196, p. 176] *for details.*

The Pollard method works when some prime factor p of n satisfies the property that $p - 1$ has only small factors since it is then likely that $B!$ will be divisible by $p - 1$. If so, then $B! = (p - 1)m$ for some $m \in \mathbb{N}$ and since the algorithm computes, ultimately, $a_B \equiv a^{B!} \pmod{n}$, we know that

$$a^{B!} \equiv (a^{p-1})^m \equiv 1 \pmod{p},$$

from Fermat's Little Theorem. Thus, $p \mid \gcd(a_B - 1, n)$.

To thwart this method of attack on RSA, for example, one has to construct primes for which the RSA modulus does not suffer from this vulnerability. For instance, if r is a large prime such that $p = 2r + 1$ is also prime, then the RSA modulus will be resistant to the $p - 1$ attack. We will revisit this and surrounding issues in Chapter 6 (see the discussion of *strong primes* on page 120).

Example 5.4 *Let* $n = 8051$ *and select* $a = 29$, $B = 10$. *Then the values* a_j *for* $1 \leq j \leq 7$ *all satisfy* $\gcd(a_j - 1, n) = 1$. *However,* $a_8 \equiv 61^{108} \equiv 7179 \pmod{n}$, *and* $\gcd(7178, 8051) = 97$. *Indeed,* $n = 8051 = 83 \cdot 97$. *Also,* $96 = 2^5 \cdot 3$ *is B-smooth.*

Example 5.5 *Let* $n = 8549$ *and select* $a = 50$, $B = 17$. *Then the reader may verify that for* $j = 1, 2, \ldots, 16$, $\gcd(a_j - 1, n) = 1$. *However,* $\gcd(6592, n) = 103$ *where* $a_{17} \equiv a_{16}^{17} \equiv 4426^{17} \equiv 6593 \pmod{n}$. *In fact,* $n = 103 \cdot 83$. *Notice that* $102 = 2 \cdot 3 \cdot 17$ *is B-smooth, but* $82 = 2 \cdot 41$ *is not so Pollard's method achieved the larger prime factor since it excels at getting those where* $p - 1$ *is B-smooth.*

The $p - 1$ idea by Pollard has an analogue called the $p + 1$ factoring method designed by Hugh Williams in [237], where the assumption is that there is a prime factor p of n with $p + 1$ being B-smooth. Moreover, Pollard's $p - 1$ idea generalizes to the much more powerful elliptic curve algorithm about which we will learn in Section 5.3.

[5.2]The reader is reminded that $f(n) = o(g(n))$ means that $\lim_{n \to \infty} f(n)/g(n) = 0$. Thus, $o(1)$ is used to symbolize a function whose limit as n approaches infinity is 0.

Exercises

In Exercises 5.16–5.19, use Pollard's $p-1$ method to find a nontrivial factor of the given n with the selected smoothness bound B and the value of a chosen in each case.

5.16. $n = 53203$, $B = 10$, and $a = 35$.

5.17. $n = 12091$, $B = 10$, and $a = 37$.

5.18. $n = 11663$, $B = 5$, and $a = 41$.

5.19. $n = 37151$, $B = 9$, and $a = 51$.

☆ 5.20. The following is a version of *Dixon's algorithm*, Dixon [76], which is based upon Fermat's difference of squares method discussed on pages 89–90. The goal is to find a nontrivial factor of a given composite $n \in \mathbb{N}$.

We are given a set $\mathcal{S}_r = \{p_1, p_2, \ldots, p_r\}$ consisting of the first r primes, which is called the *factor base* (see the discussion on page 43). Execute the following steps with initial value $i = 1$.

(1) Randomly choose $a_i \in \mathbb{Z}/n\mathbb{Z}$ and compute $b_i \equiv a_i^2 \pmod{n}$.

(2) Trial divide b_i by the elements of $\mathcal{S}_r$ to determine if b_i is p_r-smooth. If not, then discard it, and return to step (1) with the same value of i. If it is, then save it as

$$b_i = \prod_{j=1}^{r} p_j^{e_{i,j}},$$

set $i = i+1$, and return to step (1) if $i < r+1$. Otherwise go to step (3).

(3) Select a subset $\mathcal{T} \subseteq \{1, 2, \ldots, r\}$ such that

$$\prod_{i \in \mathcal{T}} b_i \equiv y^2 \pmod{n}.$$

If we set

$$x = \prod_{i \in \mathcal{T}} a_i,$$

then $x^2 \equiv y^2 \pmod{n}$. If $x \not\equiv \pm 1 \pmod{n}$, then by Exercise 4.23 on page 92, $\gcd(x - y, n)$ is a nontrivial factor of n.

Explain why such a subset $\mathcal{T}$ in step (3) must exist. (*Hint: Determine a means for the r-tuples $(e_{i,1}, e_{i,2}, \ldots, e_{i,r})$ to be vectors over $\mathbb{Z}/2\mathbb{Z}$.*)

5.21. Use Dixon's method in Exercise 5.20 to find a nontrivial factor of $n = 8549$ using the factor base $\mathcal{S}_4 = \{2, 3, 5, 7\}$.

(*Exercises 5.20–5.21 are a precursor to the topic to be studied in Section 5.4.*)

5.3 ☞ Lenstra's Elliptic Curve Method

*Mathematics, rightly viewed, possesses not only truth, but supreme beauty —
a beauty cold and austere, like that of sculpture.*

Bertrand Russell (1872–1970)
British philosopher and mathematician

Knowledge of this optional section is is not necessary to move on to Section 5.4. This section is intended for the reader with prior knowledge of basic elliptic curve theory such as that given in [165, Section 6.1, pp. 221–235]. While we assume some such familiarity, we begin by summarizing the key facts about elliptic curves that we will use.

● **Elliptic Curve Facts**

We assume that $E(\mathbb{Q})$ is an elliptic curve over $\mathbb{Q}$ given by $y^2 = x^3 + ax + b$ where $a, b \in \mathbb{Z}$, and $\mathfrak{o}$ denotes the point at infinity.

(1) (**Addition of points**): For any two points $P = (x_1, y_1)$ and $Q = (x_2, y_2)$ on E, with $P, Q \neq \mathfrak{o}$ and $P \neq -Q$, define

$$P + Q = (x_3, y_3) = (m^2 - x_1 - x_2, m(x_1 - x_3) - y_1), \qquad (5.6)$$

where

$$m = \begin{cases} m_1/m_2 = (y_2 - y_1)/(x_2 - x_1) & \text{if } P \neq Q, \\ m_1/m_2 = (3x_1^2 + a)/(2y_1) & \text{if } P = Q, \end{cases} \qquad (5.7)$$

and

if $P = \mathfrak{o}$, for instance, then $P + Q = Q$ for all points Q on E,

and

if $P = -Q$, then $P + Q = \mathfrak{o}$.

(2) (**Reduction modulo n**): Let $n > 1$ be given and fixed with $\gcd(n, 6) = 1$, and $\gcd(4a^3 + 27b^2, n) = 1$. Then we refer to E reduced modulo n when the coefficients a, b are reduced modulo n, and each point P on E is reduced modulo n in the following fashion. If $P = (r_1/r_2, s_1/s_2)$ where

$$\gcd(r_1, r_2) = \gcd(s_1, s_2) = \gcd(r_2 s_2, n) = 1,$$

then

$$P = (t_1, t_2), \text{ where } t_1 \equiv r_1 r_2^{-1} \pmod{n} \text{ and } t_2 \equiv s_1 s_2^{-1} \pmod{n},$$

with r_2^{-1} and s_2^{-1} being the multiplicative inverses of r_2 and s_2 modulo n, respectively (see Exercise 5.22). We denote the reduced curve by $E(\mathbb{Z}/n\mathbb{Z})$, and if n is a prime, then this is a group.

(3) (**Modular group law**): Suppose that P_1, P_2 are points on $E(\mathbb{Q})$ where $P_1 + P_2 \neq \mathfrak{o}$ and the denominators of P_1, P_2 are prime to n. Then $P_1 + P_2$ has coordinates having denominators prime to n if and only if there does not exist a prime $p \mid n$ such that $P_1 + P_2 = \mathfrak{o} \pmod{p}$ on the elliptic curve $E(\mathbb{Z}/p\mathbb{Z})$.

Fact (3) above is the underlying key to the following algorithm.

In 1985, H. W. Lenstra[5.3] discovered the *elliptic curve method* or ECM (see [136]). The ECM is actually a generalization of Pollard's $p-1$ method, studied in Section 5.2, in the following sense. Pollard's method relies upon the existence of a prime divisor p of n, which has the property that $p - 1$ (the order of $(\mathbb{Z}/p\mathbb{Z})^*$) is B-smooth for a suitable bound B. If no such prime exists, then the method fails. Suppose that instead of $\mathbb{Z}/p\mathbb{Z}$, we choose an elliptic curve group over $\mathbb{Z}/p\mathbb{Z}$. Since it is known that the order, g, of such a group is of the form $p + 1 - t$ for some integer t with $|t| < 2\sqrt{p}$ (see [165, Theorem 6.27, p. 237]), then the selection of g being smooth, with respect to some bound B, will find a nontrivial factor of n with high probability. If g is not B-smooth, then repeated selection of a new elliptic curve group may succeed, an option not available with the Pollard method.

◆ Lenstra's Elliptic Curve Method (ECM)

In this algorithm, $n \in \mathbb{N}$ is assumed to be composite, prime to 6, and not a perfect power (see the discussion on page 96), and $r \in \mathbb{N}$ is a parameter. The goal is to split n.

(1) (**Select and elliptic curve**): Choose a random pair (E, P) where $E = E(\mathbb{Z}/n\mathbb{Z})$ is an elliptic curve:

$$y^2 = x^3 + ax + b \text{ and } P \text{ is a point on } E.$$

Check that $\gcd(n, 4a^3 + 27b^2) = 1$.[5.4] If not, then we have split n if $1 < g < n$, and we may terminate the algorithm. Otherwise, we select another (E, P) pair.

(2) (**Choosing bounds**): Select $M \in \mathbb{N}$ and bounds $A, B \in \mathbb{N}$ such that the canonical prime factorization for M is $M = \prod_{j=1}^{\ell} p_j^{a_{p_j}}$ for small primes $p_1 < p_2 < \cdots < p_\ell \leq B$ where $a_{p_j} = \lfloor \ln(A)/\ln(p_j) \rfloor$ is the largest exponent such that $p_j^{a_j} \leq A$. Set $j = k = 1$.

[5.3] Hendrik Willem Lenstra, Jr. (1949–) was born in Zaandam, the Netherlands. At the age of twenty-eight, he was appointed full professor at the University of Amsterdam, where he had obtained his doctorate under the direction of Frans Oort. By 1987, he was appointed as full professor at UC Berkeley, where he currently resides. His considerable achievements are reflected in his recognition including: the 1985 Fulkerton Prize; the 1995 honourary doctorate from the Université de Franche-Comté, Besançon; and the 1995 honour as Kloosterman-lecturer at the University of Leiden.

[5.4] Recall that the discriminant of an elliptic curve cannot be 0 over the field of definition, so the *gcd* condition must hold over $\mathbb{Z}/n\mathbb{Z}$. The nonzero discriminant guarantees that the elliptic curve has no multiple roots.

(3) (**Calculating multiple points**):[5.5] Using (5.6)–(5.7), compute $p_j P$.

(4) (**Computing the gcd**):

  (a) If $p_j P \not\equiv \mathfrak{o} \pmod{n}$, then set $P = p_j P$, and reset k to $k + 1$.
     (i) If $k \leq a_{p_j}$, then go to step (3).
     (ii) If $k > a_{p_j}$, then reset j to $j + 1$, and reset k to $k = 1$. If $j \leq \ell$, then go to step (3). Otherwise go to step (5).

  (b) If $p_j P \equiv \mathfrak{o} \pmod{n}$, then compute $\gcd(m_2, n)$ for m_2 in (5.7). If $n > g$, terminate the algorithm, since we have split n. If $g = n$, go to step (5).

(5) (**Selecting a new pair**): Set $r = r - 1$. If $r > 0$, go to step (1). Otherwise, terminate with "failure".

One of the beauties of the ECM is that its running time is highly reliant on the factor, $p \mid n$, found. Hence, one of the most useful means of employing the ECM is for finding "small" prime factors in a number n which is too large to find *all* its factors. The reasons behind this are as follows. Assuming that p is the smallest prime dividing n, the expected running time of the ECM is known (under certain plausible assumptions) to be

$$O(\exp(\sqrt{(2 + o(1)) \ln p(\ln \ln p)}) \cdot \ln^2 n).$$

This may be used in practice to select a smoothness bound B in step (2) of the algorithm as:

$$B = \exp(\sqrt{\ln p(\ln \ln p)/2}). \tag{5.8}$$

Since we do not know p in advance, we may nevertheless select (for p) the value $\lfloor \sqrt{n} \rfloor$. In this case, it is estimated that one out of every B iterations will be successful in splitting n.

The worst-case scenario for the ECM is when n is an RSA modulus, in which case we have that the expected running time is:

$$O\left(\exp(\sqrt{(2 + o(1)) \ln n(\ln \ln n)})\right) = O\left(n^{\sqrt{(2 + o(1))(\ln \ln n)/\ln n}}\right).$$

With this being said, it is not surprising that ECM is most successful at splitting *non*-RSA moduli, *usually* finding prime factors of less than 40 decimal digits in large composite numbers. The current record is the finding of a 47-digit prime in a large composite number. Perhaps the most notable is the finding, in 1995 by Richard Brent, of a 40-digit prime dividing $\mathfrak{F}_{10}$, thereby leading to the complete factoring of the tenth Fermat number (see page 81).

It is time for a very small example (of course, not realistic) to illustrate the basic ideas behind the ECM.

[5.5]Recently, Richard Schroeppel announced that *halving* as an elliptic curve operation on points dramatically speeds up cryptographic operations such as key exchange, typically by a factor of 2.4. Point halving was independently discovered by Knudsen [122].

Example 5.9 *Let* $n = 923$ *and select* $(E, P) = (y^2 = x^3 + 2x + 9, (0, 3))$. *Then* $\gcd(4 \cdot 2^3 + 27 \cdot 9^2, 923) = 1$, *so we choose* $B = 4$, *based upon* (5.8), *and let* $A = 3, M = 6 = 2 \cdot 3 = p_1 \cdot p_2$. *Now, using* (5.6)–(5.7), *with* $p_1 = 2$, *we calculate*

$$p_1 P = 2(0, 3) \equiv (9^{-1}, -82 \cdot 27^{-1}) \equiv (718, 373) \not\equiv \mathfrak{o} \,(\mathrm{mod}\ n).$$

Thus we set $P = (718, 373)$ *and compute*

$$p_2 P = 3P \equiv 2P + P \equiv (505, 124) + (718, 373) \equiv \mathfrak{o} \,(\mathrm{mod}\ n).$$

Thus, we have that a denominator in (5.7) *is not prime to* n. *In fact, the calculation of* m *for* $4P + 2P$ *yields* $m = (124 - 373)/(505 - 718) = 83/71$, *and* $\gcd(923, 71) = 71$. *Indeed,* $n = 13 \cdot 71$, *and we have split* n.

What Example 5.9 illustrates is that the failure of the existence of a modular inverse for some m in the calculations may lead to a factor of n. Another way of saying this is that the group law for multiplication actually fails in $\mathbb{Z}/n\mathbb{Z}$ since n is not prime and this allows us to get the factor. Indeed, it is somewhat inaccurate in the ECM algorithm to say that $p_j P \equiv \mathfrak{o} \,(\mathrm{mod}\ n)$, when in fact it is $p_j P \equiv \mathfrak{o} \,(\mathrm{mod}\ p)$ where p is he factor for which we were searching. However, this is legitimate since we were, in a sense, assuming n to be prime and doing the calculations as if it were so, in the *hope* that the calculations would "break down" with an undefined denominator for some value of m in (5.7).

Some concluding comments on complexity and the relationship between ECM and the topic of Section 5.4 are in order. In the discussion, we will be using, for the sake of convenience, the notation introduced in Appendix B (see equation (B.5) on page 206). In this notation, the expected running time of ECM may be stated as $L_p(1/2, \sqrt{2})$ to find a factor p of n. Again, this shows, as mentioned earlier, that the ECM depends very much on the size of the prime factors of n so it will find the small ones first. In the case of an RSA modulus, the expected running time of ECM may be stated as $L_p(1/2, 1)$, and this is the same as the (ordinary) *quadratic sieve*, a generalization of which we will study in the next section. However, even the ordinary quadratic sieve is much more efficient at factoring RSA moduli. The reason, as we will see, is that even the ordinary quadratic sieve involves single precision operations, while elliptic curve operations are more computationally intensive.

A summary of what we have learned thus far is this. Some algorithms operate more efficiently on integers of a special form, such as Pollard's $p - 1$ method. Thus an overall strategy for splitting n would be to use trial division first, up to some reasonable bound, then use Fermat's difference of squares method, after which one could employ algorithms for small prime factors such as Pollard's $p - 1$ methods. Then the ECM can be used if such methods as $p - 1$ fail. When all this fails to get a nontrivial factor, then we can bring out the bigger guns such as the quadratic sieve or its stronger cousin, the multipolynomial quadratic sieve, which is our next topic of study.

Exercises

5.22. Show that $P = (9/4, 29/8)$ is a point on the elliptic curve $E(\mathbb{Q})$ given by $y^2 = x^3 - x + 4$. Find the reduction of P modulo 15.

5.23. If $E(\mathbb{Z}/n\mathbb{Z})$ is the elliptic curve over $\mathbb{Z}/n\mathbb{Z}$ for $n = 3749$ given by $y^2 = x^3 - 6x + 6$ and $P = (1, 1)$ on E, find $6P$ on $E(\mathbb{Z}/n\mathbb{Z})$.

5.24. Using the curve and the point in Exercise 5.23, find $6P$ if $n = 4727$.

5.25. Suppose that $n > 1$ and P is a point on the elliptic curve $E(\mathbb{Z}/n\mathbb{Z})$. Prove that there exist $j, k \in \mathbb{N}$ such that $jP = kP$.

5.26. Prove that if $n > 1$ and there are $j, k \in \mathbb{N}$ with $j \leq k$ such that $jP = kP$ for a point P on the elliptic curve $E(\mathbb{Z}/n\mathbb{Z})$, then $(k - j)P = \mathfrak{o}$.[5.6]

5.27. With reference to Exercise 5.26, prove that if k is the order of P on $E(\mathbb{Z}/n\mathbb{Z})$ and $\ell P = \mathfrak{o}$, then $k \mid \ell$.

5.28. Suppose that for $k \in \mathbb{N}$, $kP = \mathfrak{o}$ for some point P on $E(\mathbb{Z}/n\mathbb{Z})$ where $n > 1$. Furthermore, assume that

$$kP = \mathfrak{o} \text{ but } (k/p)P \neq \mathfrak{o} \text{ for any prime divisor } p \text{ of } k. \qquad (5.10)$$

Prove that k is the order of P (see Exercise 5.26).[5.7]

In Exercises 5.29–5.34, use the elliptic curve and point in Example 5.9 to factor the given n.

5.29. $n = 3977$.

5.30. $n = 25973$.

5.31. $n = 17821$.

5.32. $n = 18323$.

5.33. $n = 38411$.

5.34. $n = 1884257$.

[5.6]The smallest value $r \in \mathbb{N}$ such that $rP = \mathfrak{o}$ is called the *order of P*. In general elliptic curves, such points are called *torsion points*. (The value $\mathfrak{o}$ is called the *trivial torsion point*.) Thus, elliptic curve groups over finite fields consist only of torsion points. Such groups are called *torsion groups*. A famous theorem by B. Mazur [155] says that if E is an elliptic curve over $\mathbb{Q}$, then the torsion subgroup of $E(\mathbb{Q})$ is either $\mathbb{Z}/n\mathbb{Z}$ for $n \in \{1, 2, 3, 4, 5, 6, 7, 8, 9, 10, 12\}$ or is of the form $\mathbb{Z}/2\mathbb{Z} \oplus \mathbb{Z}/2n\mathbb{Z}$ where $n \in \{1, 2, 3, 4\}$. For elliptic curves over arbitrary number fields, this remains an open problem.

[5.7]This is related to primality testing using elliptic curves. A result of Goldwasser and Killian says that if there is prime divisor p of k with $p > (n^{1/4} + 1)^2$ and there is a point satisfying (5.10), then n is prime (see [101]). There are also elliptic curve cryptosystems, which we will not study here. These cryptosystems are based upon the discrete log problem for elliptic curves which is to find $x \in \mathbb{Z}$ such that $P = xQ$ for a given $P, Q \in E(F)$, where $E(F)$ is an elliptic curve over a field F. (See [165, Chapter 6, pp. 243–251].)

5.4 Multipolynomial Quadratic Sieve

The elliptic curve method is one of the fastest integer factorization methods that is currently used in practice. The quadratic sieve algorithm seems to perform better on integers that are built up from two prime numbers of the same order of magnitude; such integers are of interest in cryptography.

Hendrik Lenstra

We describe a generalization of the quadratic sieve method named in the title. The ordinary quadratic sieve and its generalization go back to Seelhoff [201] in 1886, and was later used by Kraitchik [125] (see Footnote 4.4 on page 83) who developed it further. Also, it was first analyzed and modernized by Pomerance [188]. (For a further in-depth discussion of the history, see [239].) The version that we study herein, using multiple polynomials, was independently proposed in [66] and [167]. The idea is based at its foundation upon Fermat's differences of squares method (see the discussion on pages 89–90).

That basic idea was developed by Dixon as seen in Exercise 5.20 on page 98. With the ordinary quadratic sieve, in order to split $n \in \mathbb{N}$, one looks at a polynomial

$$g(x) = (x + \lfloor \sqrt{n} \rfloor)^2 - n$$

for $x \in (-n^\epsilon, n^\epsilon)$. Then one builds a set of integers $i \in \mathcal{T}$ so that $g(x_i)$ factors over the factor base and

$$\prod_{x_i \in \mathcal{T}} b_i \equiv \prod_{x_i \in \mathcal{T}} g(x_i) \equiv y^2 \, (\text{mod } n)$$

as in the exercise. The problem, however, is that for a single choice of $g(x)$, there is an unreasonable amount of time required to generate a sufficiently large enough set $\mathcal{T}$ over which $g(x)$ will factor. The reason is that for large n, the interval $(-n^\epsilon, n^\epsilon)$ is also large since $g(x) = O(n^{1/2+\epsilon})$ is large as well, and so we will probably not be able to factor most of the $g(x)$ over a small set of primes. The multipolynomial version solves this problem by establishing an efficient means of using several polynomials so that the x values may be chosen from smaller intervals rather than one large interval. This means that the average polynomial values are smaller than the average of g and have a higher probability of factoring over small primes than the $g(x)$ values in the ordinary quadratic sieve. This then is a way of running the ordinary quadratic sieve in parallel.

◆ **Multipolynomial Quadratic Sieve (MPQS)**

In this algorithm, $n \in \mathbb{N}$ is assumed to be a large composite number. The goal is to split n.

(1) (**Select bounds**): Choose a large smoothness bound B and an $M \in \mathbb{N}$ with $(\sqrt{2n}/M)^{1/4} > B$.

(2) (**Select a factor base**): Choose a set of $L \in \mathbb{N}$ primes as a factor base

(see the discussion on page 43) that is fixed for the algorithm:[5.8]

$$\mathcal{F} = \{p_i : p_i \text{ is prime and } \left(\frac{n}{p_i}\right) = 1 \text{ for } i = 1, 2, \ldots, L\},$$

where the symbol is the Legendre symbol. For $p_i \in \mathcal{F}$ with $q_i = p_i^{a_i} < B$, compute solutions t_{q_i} to the congruences

$$t_{q_i}^2 \equiv n \,(\text{mod } q_i)$$

for $0 < t_{q_i} \leq q_i/2$.

(3) (**Create a quadratic polynomial**): Choose

$$W(x) = a^2 x^2 + 2bx + c,$$

where a, b, c satisfy:

$$a^2 \approx \sqrt{2n}/M, \,^{5.9} \quad b^2 - n = a^2 c, \quad |b| < a^2/2. \tag{5.11}$$

(4) (**Test W(x) for divisibility by factor base elements**): If $q_i \mid W(j)$ for some $j \in [-M, M]$, called a *sieve number*, then $q_i \mid (a^2 j + b)^2 - n$, so

$$j \equiv a^{-2}(\pm t_{q_i} - b)\,(\text{mod } q_i),$$

since $\gcd(a, q_i) = 1$ by Exercise 5.35. We compute $a^{-2}\,(\text{mod } q_i)$ for all such q_i via:

$$a^{-2} \equiv g_{i_1}^{-2} g_{i_2}^{-2} \cdots g_{i_k}^{-2}\,(\text{mod } q_i).$$

Thus, for efficiency, with the calculation of g-primes by the methodology in Exercise 5.35, we also compute and save, for $i = 1, 2, \ldots, r$, all the numbers $g_i^{-2}\,(\text{mod } q_i)$ for each $q_i = p_i^{a_i} < B$ where $p_i \in \mathcal{F}$.

(5) (**Sieving**[5.10]): Define a $(2M + 1)$-tuple:

$$s = (s(-M), s(-M + 1), \ldots, s(j), \ldots, s(M)),$$

which we initialize by setting $s(j) = 0$ for all $j \in [-M, M]$. For each sieve number $j \in [-M, M]$, i.e., those for which some prime power $q_i = p_i^{a_i} \mid W(j)$, reset

$$s(j) = \ln p_i + s(j).\,^{5.11}$$

[5.8]One method of generating a factor base containing many small primes is to multiply n by a suitable small integer m called a *multiplier* and factoring mn rather than n (see [191, p. 391]).

[5.9]The symbol $\approx$ means *approximately equals*, and we will not define it more rigorously since we will see in practice how it works. Moreover, it should be noted that the polynomial $W(x)$ plays the role of $g(x)$ in Exercise 5.20. The reader may also go to Exercise 5.35 for an efficient means of calculating such a, b, c.

[5.10]In general with the MPQS, the amount of time spent on sieving takes more than 85% of the total computing time.

[5.11]An idea attributed to Schroeppel (see [32]) is that since $W(j)$ is a quadratic polynomial, $q_i \mid W(j + q_i \ell)$ for all $\ell \in \mathbb{Z}$, whenever j is a sieve value. Thus we may efficiently calculate the sieve values.

(6) (**Selection of factor candidates**): Define the *report threshold* [5.12] to be

$$RT = \ln\left(\frac{1}{2}M\sqrt{n/2}\right).$$

Select from step (5) all those values j for which $s(j) \approx RT$, test $W(j)$, and save a, b, j, (and thus tacitly c via the choice in Exercise 5.35) only if $W(j)$ factors over $\mathcal{F}$. If the number of $W(j)$ selected is less than $L+2$, go to step (3). Otherwise, go to step (7).

(7) (**Creation of exponent vector**): Since we have $L+2$ sieve values j, we form:

$$W(j) = (-1)^{b_{j_0}} \prod_{i=1}^{L} p_i^{b_{j_i}}, \text{ and } b_{j_i} \leq a_i, \text{ for } j = 1, 2, \ldots, L+2$$

and associate with $W(j)$ the exponent vector

$$v_j = (b_{j_0}, b_{j_1}, \ldots, b_{j_L}) \,(\mathrm{mod}\ 2),$$

so we have a binary $L+1$-tuple for each $j = 1, 2, \ldots, L+2$. Since we have $L+2$ vectors with $L+1$ coordinates, then there is at least one subset

$$\mathcal{J} \subseteq \{1, 2, \ldots, L+2\}^{5.13}$$

such that

$$\sum_{j \in \mathcal{J}} v_j \equiv 0 \,(\mathrm{mod}\ 2),$$

so

$$\prod_{j \in \mathcal{J}} W(j) \equiv z^2 \,(\mathrm{mod}\ n).$$

(8) (**Factor n**): Since $(a^2 x + b)^2 \equiv a^2 W(x) \,(\mathrm{mod}\ n)$, then

$$X^2 \equiv \prod_{j \in \mathcal{J}} (a^2 j + b)^2 \equiv z^2 \prod_{j \in \mathcal{J}} a^2 \equiv Y^2 \,(\mathrm{mod}\ n),$$

so if $1 < \gcd(X - Y, n) < n$, then we have a nontrivial factor of n.

There are many variations that will speed up the MPQS such as those found in [32], which we will not discuss here. As noted in Section 5.3, the ECM should be used in advance of the MPQS to find small prime factors of n, say up to thirty decimal digits. When a number has smallest prime divisor larger than thirty decimal digits, we use the MPQS. One big advantage of the MPQS over

---

[5.12]The report threshold is the average of $\ln|W(j)|$ for $j \in [-M, M]$. When $s(j) \geq RT$, $W(j)$ is a good candidate for factoring over the factor base.

[5.13]We can use Gaussian elimination modulo 2 on the matrix whose columns are v_j to find a set $\mathcal{J}$. See Appendix C on page 222.

the ordinary quadratic sieve is that one can generate many a, b, c values, and switch polynomials when the residues grow too large. In Exercise 5.35, we see that if we have a fixed k and set of r g-primes, the number of polynomials which may be calculated is

$$2^{k-1} \binom{r}{k} = 2^{k-1} \frac{r!}{(r-k)! k!}.$$

As we have seen, residues from different polynomials can be merged later. Thus the MPQS is suitable for parallel processing when each processor has its own polynomials. The current factorization record for (the *three-large-prime* variant of) the MPQS is a 135-digit number (which is actually a factor of the Cunningham number $2^{1606} + 1$, see page 108) accomplished on August 29, 2001 by B. Dodson, A.K. Lenstra, P. Leyland, A. Muffet, and S. Wagstaff. The previous record was the factorization of RSA-129 (see Footnote 3.7 on page 63) factored in 1994 (see [9] and Example 3.10 on page 63). In Section 5.5, we will have a look at another powerhouse for factoring — the number field sieve — which has supplanted the MPQS as the most widely used factoring algorithm.

Exercises

5.35. The following is an efficient means of calculating the a, b, c in the MPQS, and hence of selecting the polynomials $W(x)$. Choose $r, k \in \mathbb{N}$ with $1 < k < r$.[5.14] Generate primes $g_1, g_2, \ldots, g_r$, which are called *g-primes*, satisfying:

(a) $g_i \approx (\sqrt{2n}/M)^{1/(2k)}$,

(b) $(\frac{n}{g_i}) = 1$, and

(c) for each $i = 1, 2, \ldots, r$, $\gcd(g_i, q) = 1$ for all $q \in \mathcal{F}$.

For some choice of k of the g-primes where $1 \le i_1 < i_2 < \cdots < i_k \le r$, let

$$a = g_{i_1} g_{i_2} \cdots g_{i_k}.$$

Now, solve for b_i with $i = 1, 2, \ldots, r$ in:

$$b_i^2 \equiv n \pmod{g_i^2}.$$

Then use the Chinese Remainder Theorem C.13 on page 215, to solve the system of congruences, for a specific choice of signs:

$$b \equiv \pm b_{i_1} \pmod{g_{i_1}^2}; \quad b \equiv \pm b_{i_2} \pmod{g_{i_2}^2}; \quad \cdots \quad b \equiv \pm b_{i_k} \pmod{g_{i_k}^2}.$$

For this solution b, set $c = (b^2 - n)/a^2$.

Show that the conditions in (5.11) are satisfied by this choice of a, b, c. Then given $n = 72452183$, $k = 2$, $r = 3$, $B = 5$, $M = 3$, $\mathcal{F} = \{2, 3\}$, and g-primes $11, 17$, determine a, b, c in this case.

[5.14] In this exercise, we will choose a small values of r for pedagogical purposes but in practice, the MPQS typically uses a value such as $r = 30$, for instance.

5.5 ☞ The Number Field Sieve

I have often admired the mystical way of Pythagorus, and the secret magic of numbers.
    Sir Thomas Browne (1605–1682) English writer and physician

This section is optional and requires some basic knowledge of the theory of algebraic number fields (see [164], for instance). The basic notion behind the quadratic sieve and the MPQS studied in Section 5.4, is that we try to generate many smooth quadratic residues of n close to $\sqrt{n}$, where n is our composite candidate for splitting. This notion was extended by Pollard[5.15] in 1988 who circulated a manuscript, in the factoring community, that was essentially the template for a new integer factoring algorithm. The idea was to use cubic integers (such as $\mathbb{Z}[\sqrt[3]{-2}]$) to factor by trying to generate many smooth cubic residues of n close to $\sqrt[3]{n}$. Later, the idea was extended to the fifth degree (such as $\mathbb{Z}[\sqrt[5]{2}]$, for example) and used to factor the ninth Fermat number, $\mathfrak{F}_9$ (see [138]). Pollard had been motivated in 1986 by the DLP given in [63], where quadratic fields were employed (for an overview of these developments see [165, pp. 207–212]). Pollard then began to look at a more general scenario by considering composite $n \in \mathbb{N}$ that are "close" to being powers, namely,

$$n = r^t - s \text{ for small } r, |s| \in \mathbb{N} \text{ and larger } t \in \mathbb{N}. \tag{5.12}$$

For instance, if $|s| = 1$, such numbers are called *Cunningham numbers*, which have a rich history in factoring (see [46]), and factoring them has come to be known as the *Cunningham project*, which began with the 1925 publication [65] by Cunningham and Woodall.

The special number field sieve (SNFS), so called by the authors of [48], can factor integers of type (5.12) in expected running time $L_n(1/3, (32/9)^{1/3})$ (in the notation of (B.5) on page 206), whereas the general number field sieve (GNFS), which we will study below, has expected running time for *arbitrary* integers n of $L_n(1/3, 1.9229)$. This makes the SNFS asymptotically faster than any known algorithm *for the class of integers in* (5.12), whereas the GNFS is a faster algorithm for arbitrary integers.

The record for the GNFS is factorization of a 158-digit number (which thereby completed the factorization of $2^{953} + 1$, since the 158 digit number was the last unsplit composite factor of it). This was accomplished as a product of a 73-digit prime and an 86-digit prime on January, 18, 2002 by F. Bahr, J. Franke,

---

[5.15]John M. Pollard was born near London, England on October 25, 1941. He received his B.A. in 1963, his M.A. in 1965, and his doctorate in 1978 all from Cambridge University, England. He is a mathematician who worked for British Telecom from 1968 to 1986. Certainly, he is best known for his research in cryptography, and has his name attached to several algorithms, including: *Pollard's rho method, Pollard's p − 1 method,* and his introduction of the *number field sieve,* as well as *lattice sieving.* On January 18, 1999, RSA Data Security Inc. announced that he was a co-recipient of the 1999 RSA award. This award was instituted in 1998 for recognition of those people and organizations that have made "significant, ongoing contributions to security issues and cryptography in the areas of mathematics, public policy, and industry."

and T. Kleinjunj (with primality of the factors verified by Herman te Riele — see http://www.crypto-world.com/announcements/c158.txt). The previous record was factorization of RSA-155, accomplished on August 22, 1999 by Herman te Riele's group (see http://www.crypto-world.com/announcements/RSA155.txt). It is known that the record for the MPQS, cited on page 107, would have taken about one-sixth the time used by the MPQS, if the GNFS had been used. In fact, it took 8000 MIPS-years for the MPQS to factor the 135-digit number, whereas it would have taken the GNFS about 1300 MIPS-years (see Footnote 3.7 on page 63). Similarly, it is known that the previous record of factoring RSA-129 by the MPQS, would have taken a quarter of the time with the use of the GNFS (see [77]). Hence, the following is now the clear front-runner as the most accepted and widely used factoring algorithm.

In order to simplify the following description, we first set the stage. Despite the fact that we are now going to be dealing with a much more powerful and sophisticated algorithm for factoring, the basis underlying idea is still Fermat's difference of squares method. Let's give a simple preview.

The setup and the goal: Given a composite n to split, what we will be setting out to achieve is the following. We select an appropriate monic polynomial $f(x)$, irreducible over $\mathbb{Z}$, where $m \in \mathbb{N}$ with $f(m) \equiv 0 \,(\mathrm{mod}\ n)$, and $\alpha \in \mathbb{C}$ a root of f. This setup allows us to define the natural homomorphism, $\phi : \mathbb{Z}[\alpha] \mapsto \mathbb{Z}/n\mathbb{Z}$, via $\phi(\alpha) = m$ which ensures that, for any $g(x) \in \mathbb{Z}[x]$, we have $\phi(g(\alpha)) \equiv g(m) \,(\mathrm{mod}\ n)$. Thus, we will seek a set $\mathcal{S}$ of polynomials g over $\mathbb{Z}$ such that both $\prod_{g \in \mathcal{S}} g(\alpha) = \beta^2 \in \mathbb{Z}[\alpha]$, and $\prod_{g \in \mathcal{S}} g(m) = y^2 \in \mathbb{Z}$. Then by setting $\phi(\beta) \equiv x \,(\mathrm{mod}\ n)$ we get,

$$x^2 = \phi(\beta)^2 \equiv \phi(\beta^2) \equiv \phi\left(\prod_{g \in \mathcal{S}} g(\alpha)\right) \equiv \prod_{g \in \mathcal{S}} g(m) \equiv y^2 \,(\mathrm{mod}\ n), \qquad (5.13)$$

and we are back to Fermat's method in (4.7) on page 89, where we have a nontrivial factor of n if $x \not\equiv \pm y \,(\mathrm{mod}\ n)$! However, the devil is in the details so here we go.

◆ General Number Field Sieve (GNFS)

We make some initial simplifying assumptions the reasons for which the reader may find in [48]. We assume that a smoothness bound B and the degree d of the polynomial f have been chosen from experimental data.[5.16] Now, we let $m = \lfloor n^{1/d} \rfloor$ and write n in base m,

$$n = m^d + c_{d-1}m^{d-1} + \cdots + c_0, \text{ with } 0 \le c_j \le m - 1 \text{ for } j = 0, 1, \ldots, d. \ (5.14)$$

Now set $f(x) = x^d + c_{d-1}x^{d-1} + \cdots + c_0 \in \mathbb{Z}[x]$, and we have a monic polynomial with $f(m) = n$. However, we wanted f to be irreducible. If it is

---

[5.16]Heuristic complexity arguments determine the choices to be optimal when $B = L_n(1/3, c)$ for $c = (8/9)^{1/3+\epsilon}$, and $d = ((2/c)^{1/2}[\ln n/(\ln \ln n)]^{1/3}) = \ln(L_n(1/3, (2/c)^{1/2}))$. These choices ensure that $B^d \approx n^{2/d}$. Hence, $n > 2^{d^2}$, which is needed to ensure that n is monic in (5.14), a straightforward exercise to verify.

not, then we have no need of the number field sieve, since then $f(x) = g(x)h(x)$ where g and h have unequal positive degrees, so $g(x)h(x) = f(m) = n$, and we have a nontrivial factor of n. Hence, we may assume that f is irreducible (as are most monic polynomials over $\mathbb{Z}$). Thus, we have our polynomial f, B, and d values, and a number field $F = \mathbb{Q}(\alpha)$ of degree d over $\mathbb{Q}$.

In the following, we have to extend our definition of *smooth* given in Definition 5.2. We call $a + b\alpha \in \mathbb{Z}[\alpha]$ *B-smooth* if $|N_F(a + b\alpha)|$ is B-smooth where N_F is the norm map from the field F to $\mathbb{Q}$. Also, for a given prime $p \le B$, set

$$R(p) = \{r \in \mathbb{Z} : 0 \le r \le p - 1 \text{ and } f(r) \equiv 0 \,(\mathrm{mod}\ p)\}.$$

Then whenever (a, b) are coprime, $p \mid N_F(a - b\alpha)$ if and only if $p \mid (a - br)$ for some $r \in R(p)$ with $p \nmid b$. Then r is called the (unique) *signature* of $N(a - b\alpha)$ modulo p. Hence, for each coprime (a, b)-pair, there exist $|R(p)| = \mathfrak{r}$ primes $p \le B$ dividing $N(a - b\alpha)$. We will let these be denoted by $p_1, p_2, \ldots, p_{\mathfrak{r}}$. Then if $a - b\alpha$ is B-smooth, we have:

$$N(a - b\alpha) = (-1)^{a(0)} \prod_{i=1}^{\mathfrak{r}} p^{a(p_i)}, \text{ where } a(0) \in \{0, 1\}.$$

Based on this we can now define exponent vectors. Let

$$v(a - b\alpha) = (a(0), a(p_1), a(p_2), \ldots, a(p_{\mathfrak{r}})).$$

However, based on our goal set above, we want not only $a - b\alpha$ to be B-smooth, but also $a - bm$ to be B-smooth. If the latter is the case, then let $q_{\mathfrak{r}+1}, q_{\mathfrak{r}+2}, \ldots, q_{\mathfrak{s}}$ be all the primes less than or equal to B dividing $a - bm$, and write

$$a - bm = (-1)^{b(0)} \prod_{i=\mathfrak{r}+1}^{\mathfrak{s}} q_i^{b(q_i)},$$

and define, $v(a - bm) = (b(0), b(q_{\mathfrak{r}+1}), \ldots, b(q_{\mathfrak{s}}))$. Finally set

$$v(a, b) \equiv (v(a - b\alpha), v(a - bm)) \,(\mathrm{mod}\ 2).$$

Hence, $v(a, b)$ is a binary vector of length $\mathfrak{r} + \mathfrak{s} + 2$.

For ease of elucidation, we make the simplifying assumption that if we find a set $\mathcal{S} = \{(a, b) \in \mathbb{Z} \times \mathbb{Z} : \gcd(a, b) = 1\}$ such that $\sum_{(a,b) \in \mathcal{S}} v(a, b)$ is the zero vector modulo 2, then both $\prod_{(a,b) \in \mathcal{S}} (a - bm)$ will be a square in $\mathbb{Z}$ and $\prod_{(a,b) \in \mathcal{S}} (a - b\alpha)$ will be a square in $\mathbb{Z}[\alpha]$ (see [189] for the means of dealing with the obstructions when this is not the case). Hence, all we do is to sieve over coprime integer pairs (a, b) with $0 < b \le B$, $|a| \le B$ until the above is achieved. Then we are in the situation (5.13) and we proceed to factor n.

There is also a variant of the SNFS developed by Dan Bernstein [25], called the *multiple-lattice number field sieve*. He demonstrates that the formal relation between the multiple-lattice number field sieve and the SNFS is the same as the formal relation between the MPQS and the quadratic sieve.

Chapter 6

Security of RSA

From the contagion of the world's slow stain, he is secure, and can never mourn, a heart grown cold, a head grown grey in vain.

Percy Bysshe Shelley (1792–1822) English poet

6.1   Implementation Attacks

In Exercise 3.44 on page 76, we demonstrated one glaring misuse of the RSA cipher, namely the use of a *common modulus*, which leads to a common modulus protocol failure.[6.1]   This is *not* a weakness of RSA, but rather an exceptionally *bad implementation* of the cipher. We saw in the aforementioned exercise that Eve could retrieve the plaintext without either knowledge of a decryption exponent or having to factor n. What this demonstrates is that an RSA modulus should *never* be used by more than one entity. It cannot be emphasized forcefully enough that true security for the RSA cryptosystem requires a secure implementation. Without this, any other measures taken, such as using a 1024-bit private key, as suggested in the discussion on page 74, will do nothing to overcome the bad implementation. However, there was one implementation attack, the discovery of which was somewhat troubling.

In 1995, a Stanford undergraduate named Paul Kocher [120] discovered that he could cryptanalyze RSA and recover the decryption exponent by a careful timing of the computation times for a sequence of decryptions. This weakness was a surprising and unexpected discovery since the cryptographic community felt that the nearly two-decades-old, time-tested cryptosystem was well understood. It turns out, as we shall see, that there are means of thwarting the attack, but it was disturbing nevertheless. Perhaps the lesson to be learned is never to be overconfident, and always be vigilant. Even the discovery of more efficient algorithms, such as the number field sieve discussed in Section 5.5, lay waste to the claims made in the heady heydays of the discovery of RSA in 1977. In Martin Gardiner's *Scientific American* article [93] in that year, it was touted that to

---

[6.1]The *common modulus attack*, as it is often called, is attributed first to Simmons [212] and (for a later contribution) to DeLaurentis [67].

factor the RSA-129 challenge number (at that time) with the fastest computers and most efficient algorithms would take several times longer than the life of the known universe. Yet in 1994, after only eight months of computation, if was indeed factored (see Footnote 3.7 on page 63). New mathematical discoveries of efficient algorithms perhaps pose the greatest challenge. Indeed, Gardiner stated in his article that "Rivest and his associates have no proof that at some future time no one will discover a fast algorithm for factoring composites as large as ... [but] they consider the possibility extremely remote."

For the following, the reader will need to engage in solving Exercise 6.1 in order to be familiar with notation and some elementary facts from probability theory. Moreover, we must review the repeated squaring method on page 50 since we will be referring directly to it. For the convenience of the reader and to alter notation to fit the algorithm below, we re-present the repeated squaring method here.

◆ The Repeated Squaring Method Revisited

Given $n \in \mathbb{N}$ and $d = \sum_{j=0}^{k} d_j 2^j$, $d_j \in \{0, 1\}$, the goal is to find $x^d \pmod{n}$. Set $x_0 = x$, $c_0 = 1$, $j = 0$, and execute the following steps.

(1) If $d_j = 1$, then set $c_j \equiv x_j \cdot x \pmod{n}$.

(2) If $d_j = 0$, then set $c_j \equiv x_j \pmod{n}$.

(3) Compute $x_{j+1} \equiv c_j^2 \pmod{n}$.

(4) Reset j to $j + 1$. If $j = k + 1$, then terminate the algorithm with

$$c_k \equiv x^d \pmod{n}.$$

Otherwise, go to step (1).

◆ Timing Attack

We assume that Eve knows the hardware, such as a smart card or computer, being used. Moreover, suppose that, prior to her attack, Eve has measured the time values that it takes the hardware to compute x_i^d in the repeated squaring method for some large number r of ciphertexts x_i. Eve wants to obtain the RSA decryption exponent d, and she knows the RSA modulus n. Since Eve knows that d is odd, she already has $d_0 = 1$. Suppose that she has obtained $d_0, d_1, \ldots, d_{\ell-1}$ for some $\ell \in \mathbb{N}$. Since Eve knows the hardware, she therefore knows the time t_i (for each x_i) that it takes the hardware to compute $c_\ell, \ldots, c_k$ in the above repeated squaring method. She now wants to determine d_ℓ. If $d_\ell = 1$, then the multiplication $x_\ell \cdot x \pmod{n}$ is calculated for each ciphertext x_i, and Eve knows that this takes time q_i, say. However, if $d_\ell = 0$, this multiplication does not take place. Now suppose it takes time s_i for the computer to complete the calculation *after* the multiplication. Then by Exercise 6.1, if $d_\ell = 1$,

$$\mathrm{var}(\{t_i\}_{i=1}^r) = \mathrm{var}(\{q_i\}_{i=1}^r) + \mathrm{var}(\{s_i\}_{i=1}^r) > \mathrm{var}(\{s_i\}_{i=1}^r), \qquad (6.1)$$

and if $d_\ell = 0$, then this fails to hold. Hence, Eve can determine d_ℓ, and similarly d_j for each $j > \ell$. This simple observation allows for Eve to find d without having to factor n.

To summarize: the attack essentially consists of simulating the computation to some point, then building a decision criterion (6.1), with only one correct interpretation possible, depending on the selected value, and finally deciding the bit value by observing whether (6.1) holds or not. This attack is most effective against smart cards and such devices where timing measurements can be obtained precisely.

So how do we defend against Eve's clever intrusion using Kocher's idea? There are two basic methods for defence against Kocher's timing attack. The simplest is to ensure that the modular exponentiation being used always takes a *fixed* amount of time. This may be accomplished by adding a suitable delay factor in each operation. The second method of defence is attributable to Rivest (see Footnote 3.2 on page 61). The method is called *blinding*. Suppose that b is the ciphertext and e is the RSA encryption exponent. Then prior to deciphering b, a random $r \in (\mathbb{Z}/n\mathbb{Z})^*$ is chosen and $b' = b \cdot r^e \pmod{n}$ is computed, followed by $c' \equiv (b')^d \pmod{n}$. Then the computer sets $c = c' \cdot r^{-1} \pmod{n}$. By so doing, the computer is exponentiating a random b' with d that is totally unknown to Eve, who cannot mount the attack as a result.

◆ Other Attacks and Implementation Security Issues

Kocher had another clever idea. It is called *power cryptanalysis*. By a very careful measurement of the computer's *power consumption* during decryption, Eve could recover the secret key. This works since during multi-precision multiplications the computer's power consumption is necessarily higher than it would normally be. Hence, if Eve measures the length of these high consumption episodes she can easily decide when the computer is performing one or two multiplications, and this gives away the bits of d.

We conclude this section with a discussion of some randomness requirements for proper security of RSA not often mentioned in the literature. In [38], it is shown that implementing the RSA cryptosystem as described on page 62, which we will call *plain* RSA (namely, *without* any preprocessing of plaintext message units) is insecure. The authors show that for any RSA public key (n, e), given ciphertext $c \equiv m^e \pmod{n}$, it is possible to recover the plaintext m in the time it takes to compute $2^{m/2+1}$ modular exponentiations. Moreover, the attack they present succeeds with 18% probability over choices of $m \in \{0, 1, \ldots, n-1\}$. (They have similar data for the ElGamal cryptosystem, which we will not discuss here.)

Albeit simple, the attack displayed in [38] is powerful against plain RSA thereby demonstrating that it is fundamentally insecure in that it does not satisfy basic security requirements. One such requirement is *semantic security*, which means that for a given public key (n, e) and ciphertext c, it should be intractable to recover *any* information about the plaintext m (see [100] for details on semantically secure cryptosystems). One example of how plain RSA is semantically insecure is that the Jacobi symbol (m/n) is easily computed from the ciphertext c, since $\left(\frac{c}{n}\right) = \left(\frac{m^e}{n}\right) = \left(\frac{m}{n}\right)^e$.

There are methods for adding *randomness* to the enciphering stage in RSA (see [19] for instance). In order to obtain a *secure* RSA cryptosystem from a *plain* RSA cryptosystem, there should be an application of a preprocessing function to the plaintext *before enciphering*. In [19], there are new standards for "padding" plain RSA so that it is secure against certain chosen ciphertext attacks (see Section 1.3). The attack against plain RSA given in [38] shows that even though an m-bit key is used in plain RSA, the *effective security* is $m/2$ bits. Hence, it is essential that, before encryption, a preprocessing step be implemented that uses the so-called *Optimal Asymmetric Encryption Padding* (OAEP) (introduced in [19]) such as [183], a recent standard from RSA Labs. (Also, see [145], [207] for some cautionary notes, as well as [35], [92] for some new considerations.)

There are also methods required to make hybrid cryptosystems (see Section 3.4) secure. For instance, in [91] it is shown how to ensure a secure hybrid cryptosystem. This is necessary since hybrid cryptosystem usage is insecure in general, *even if both asymmetric and symmetric encryption are themselves secure*. Also, in [186] there is a presentation of how to protect any one-way cryptosystem from chosen-ciphertext attacks.

In the next section, we will be looking at attacks on low exponents for RSA. These attacks break RSA completely in some cases.

Exercises

6.1. Let R be a randomized algorithm (see page 205) that produces $t \in \mathbb{R}$ as output where t is the amount of time it takes for the computer to complete a calculation for a given input. We record the outputs $t_1, t_2, \ldots, t_r$ for given inputs and compute the mean (*average*, or the *expected value*)

$$m = \frac{1}{r}\sum_{j=1}^{r} t_j.$$

The *variance* of the $\{t_j\}$ is defined to be:

$$\mathrm{var}(\{t_j\}_{j=1}^r) = \frac{1}{r}\sum_{j=1}^{r}(t_j - m)^2,$$

and the *standard deviation* is $\sqrt{\mathrm{var}(\{t_j\}_{j=1}^r)}$.[6.2] Prove that for another set of outputs $s_1, s_2, \ldots, s_r$,

$$\mathrm{var}(\{s_j\}_{j=1}^r) + \mathrm{var}(\{t_j\}_{j=1}^r) = \mathrm{var}(\{s_j\}_{j=1}^r + \{t_j\}_{j=1}^r).$$

Calculate the variance and standard deviation of $\{s_j\}_{j=1}^r = \{4, 4, 4, 4\}$ and $\{t_j\}_{j=1}^r = \{5, 10, 50, 100\}$.

[6.2]In statistics and probability these terms are used to determine "population variance" for a sample of a variate having a distribution with a known mean, and the shortened term "variance" is used when there is no confusion with what is called the "sample variance".

6.2 Exponent Attacks

Philosophy is written in that great book which ever lies before our eyes —
I mean the universe ... This book is written in mathematical language ....
        Galileo Galilei (1564–1642) Italian astronomer and physicist

In this section, we look at attacks based upon the choice of low encryption and decryption exponents. We begin with the encryption exponent case.

◆ Low Public RSA Exponent

In order to make encryption more efficient, one can use a small public exponent e. By Exercise 6.2, the smallest possible value, and one commonly used, is $e = 3$, which requires only one modular multiplication and one modular squaring. However, Exercise 6.3 shows us that sending the same message, that fits into a single block, to three different entities allows recovery of the plaintext. This can be avoided by use of another recommended exponent $e = 2^{16} + 1 = 65537$, because $(65537)_{10} = (10000000000000001)_2$, so encryption using the repeated squaring method (revisited in Section 6.1) needs only 16 modular squarings and 1 modular exponentiation; and here to recover the plaintext by the method of Exercise 6.4 would require sending the same message to $2^{16} + 1$ entities, not likely to occur. In any case, the method of attack in Exercise 6.4 can be thwarted by appending a randomly generated bitstring of suitable length to the plaintext message prior to encryption, a practice called *salting the message*, since we alter the "taste" of the message. The term *padding* is also commonly used. Moreover, the random bitstring should be independently generated for each separate encryption. Notice that we are suggesting a preprocessing stage that is analogous to that delineated in Section 6.1 for the implementation attacks discussed therein.

There are more serious attacks on small public RSA exponents that even work against salted messages. One of the most serious is the following introduced by Coppersmith in 1996 (see [61]–[62]). We refer the reader to [61] for a proof of the following since it is beyond the scope of this text. For the interested reader, the following is based upon the so-called LLL lattice based reduction algorithm (see [143]).

Theorem 6.2 (Coppersmith)
If $n \in \mathbb{N}$ is composite, $f(x) = a_d x^d + a_{d-1} x^{d-1} + \cdots + a_0 \in \mathbb{Z}[x]$, $d \in \mathbb{N}$, and there exists an integer x_0 such that $f(x_0) \equiv 0 \,(\mathrm{mod}\, n)$ with $|x_0| < n^{1/d}$, then x_0 can be computed in time that is polynomial in d and $\ln(n)$.

Theorem 6.2 provides an efficient algorithm for finding all roots x_0 of f modulo n when $x_0 < x = n^{1/d}$. As x decreases so does the running time of the algorithm. Notice that n is assumed to be composite since there exist much better algorithms for finding roots modulo a *prime* than the result in Theorem 6.2. In view of Coppersmith's attack, there is motivation to use large random public encryption keys. In [140], the authors state that "Values such as 3 and 17 can no longer be recommended, but commonly used values such as $2^{16} + 1 = 65,537$ still seem to be fine. If one prefers to stay on the safe side one may select an odd 32-bit or 64-bit public exponent at random." Moreover, the authors of [41] provide evidence that breaking low exponent RSA *cannot* be equivalent to factoring. Even though we still do not have a proof that breaking RSA is equivalent to factoring the RSA modulus, this should give us pause.

One important application of Theorem 6.2 is a generalization of the notion given in Exercise 6.4 and first introduced by Hastad [107]. Although the attack in the exercise can seemingly be thwarted by padding the message, Hastad showed that certain types of padding are insecure. Now we present a stronger version of Hastad's method proved in [34].

Theorem 6.3 (Strong Hastad Broadcast Attack)
    Let $n_1, \ldots, n_r \in \mathbb{N}$ be pairwise relatively prime with $n_1 \leq n_j$ for all $j = 1, \ldots, r$, and let $f_j(x) \in (\mathbb{Z}/n_j\mathbb{Z})[x]$ with the maximum degree of f_j (for $j = 1, \ldots, r$) being ℓ. If there exists a unique $m < n_1$ such that $f_j(m) \equiv 0 \pmod{n_j}$ for all $j = 1, \ldots, r$ and $r > \ell$, then m can be efficiently calculated.

Example 6.4 *Suppose that Alice is communicating with entities $B_1, B_2, \ldots, B_r$ and she has RSA encryption exponent $e < r$. She wishes to send the same message m to each of them. First she pads m for each recipient B_j as*

$$f_j(m) = 2^t j + m$$

where t is the bitlength of m. Then she sends

$$f_j^e(m) \equiv (2^t j + m)^e \pmod{n_j} \text{ for } j = 1, 2, \ldots, r.$$

If Eve intercepts more than e of these she can retrieve m via Theorem 6.3.

Example 6.4 is an instance where the messages are *linearly* related, and this is a special case of Hastad's demonstration that, in fact, any fixed polynomial applied as padding is insecure. Therefore, a proper means of defending against the broadcast attack is to pad with a *randomized* polynomial, not a *fixed* one.

There are also *naive* random padding algorithms that can be cryptanalyzed via another attack by Coppersmith. Let's discuss how this is done. To understand the basic scenario, we enlist Alice, Bob, and Mallory once again. Alice pads a message m that she wants to send to Bob, then enciphers it and transmits it. However, malicious Mallory intercepts the message and prevents it from reaching Bob (a man-in-the-middle attack, see page 27). Yet when Bob does not respond to her message, she decides to randomly pad m again, encrypts it and sends it to Bob. Now Mallory intercepts the second message and has two different encipherments of m using different random pads. The following theorem describes how Mallory can recover m.

Theorem 6.5 (Coppersmith's Short Pad Attack)

Let n be an N-bit RSA modulus with public enciphering key e, and set $\ell = \lfloor N/e^2 \rfloor$. Suppose that $m \in (\mathbb{Z}/n\mathbb{Z})^*$ is a plaintext message unit of bitlength at most $N - \ell$, and set $m_1 = 2^\ell m + r_1$ and $m_2 = 2^\ell m + r_2$ with $r_1, r_2 \in \mathbb{Z}$, $r_1 \neq r_2$, $0 \leq r_1, r_2 < 2^\ell$. Then with knowledge of n, e, m_1, m_2 (but not r_1 or r_2) m can be efficiently recovered.

Proof. See [60]. $\qquad\square$

Remark 6.6 *When $e = 3$, the attack can be mounted if Alice sends messages with pads of maximum bitlength less than $1/9$-th that of the message length, since in that case, if the message length is M, $r_1, r_2 < 2^{M/9} < 2^{N/9} = 2^\ell$. It can also be seen that the above attack could not be mounted if $e = 65337$ is chosen.*

Let's have a look at an illustration of the attack and how it is manifested.

Example 6.7 *Suppose we are given RSA enciphering exponent $e = 3$, RSA modulus n, a plaintext message m and a pad r. If we set $M = 2^\ell m$, we may consider the simple situation where Alice sends:*

$$M^3 \equiv c_1 \,(\mathrm{mod}\ n) \ \text{ and } \ (M + r)^3 \equiv c_2 \,(\mathrm{mod}\ n),$$

(since if $m_1 = M + r_1$ and $m_2 = M + r_2$, then we may rewrite $m_2 = m_1 + r_2 - r_1$). By using resultants[6.3] (about which we need not concern ourselves here except to note that the following can be done), Eve can eliminate M as a variable and be left with

$$r^9 + 3(c_1 - c_2)r^6 + 3(c_2^2 + 7c_1 c_2 + c_1^2)r^3 + (c_1 - c_2) \equiv 0 \,(\mathrm{mod}\ n),$$

so as long as the pad r satisfies $|r| \leq n^{1/9}$, she can recover r by using Theorem 6.2. But how does this allow her to recover m? Essentially the method is due to Coppersmith's generalization of a result by Franklin and Reiter (see [62]). To see how this is done, we calculate $g = \gcd(x^3 - c_1, (x + r)^3 - c_2)$ in $\mathbb{Z}/n\mathbb{Z}[x]$ using the Euclidean algorithm.[6.4] It is clear that $x - m$ divides g. Moreover, since $\gcd(3, \phi(n)) = 1$ (given that $e = 3$ is the RSA public exponent) then $x^3 - c_1$ cannot factor into 3 linear terms in $\mathbb{Z}/n\mathbb{Z}[x]$, so $x^3 - c_1 = (x - m)q(x)$ where $q(x)$ is an irreducible quadratic. Hence, since it can be shown that g is linear, $g = x - m$ and we have recovered M from which we get m. (See [90] for $e = 3$.)

The following attack involves a low public key e *and* knowledge of some of the bits of d. Suppose that Eve somehow recovers a fraction of the bits of d. Then via the following theorem, she can recover *all* of d, thereby rendering a total break of the RSA cryptosystem.

[6.3]For the reader interested in the details surrounding resultants and the matrix theoretic connections, see [131, p. 360].

[6.4]For the most attentive reader, we note that although $\mathbb{Z}/n\mathbb{Z}[x]$ is not a Euclidean ring, if the Euclidean algorithm fails then we get a nontrivial factor of n, so Eve wins out in any case.

Theorem 6.8 (Coppersmith's Partial Key Exposure Attack)

If $n = pq$ is an RSA modulus of bitlength ℓ, then given either the $\lfloor \ell/4 \rfloor$ least significant bits of p or the $\lfloor \ell/4 \rfloor$ most significant bits of p, n can be factored efficiently.

Proof. See [61].   □

Another way of stating Theorem 6.8, is that for a given $n = pq$ and an estimate x for p such that $|x - p| \leq n^{1/4}$, then p and q can be computed in polynomial time (actually it can be shown to be computable in time that is polynomial in ℓ).[6.5] This attack generalizes a key result given in [37], wherein they show that if $e < \sqrt{n}$, then one can recover d from knowledge of a fraction of its bits. The result in [37] is often called the *partial key exposure attack*, but we retain the label given for Theorem 6.8 above since the result in [37] follows from Coppersmith's result (see Exercise 6.11). This attack is actually a devastating attack upon a variant of RSA given by Vanstone and Zucchereto in [231] where high order bits of p and q are prescribed in advance. The lesson here is clear: Safeguard *all* of the bits of d. In other words, keep the entirety of d secure.

◆ Low Secret RSA Exponent

Now we turn our attention to small *secret* RSA exponents. Again, as with the reason for choosing small public exponents, we want increased efficiency in the decryption process, so we choose small secret exponents. For instance, given a 1024-bit RSA modulus, the decryption process can have efficiency increased ten-fold with the choice of small d. However, Weiner [233] developed an attack that yields a total break[6.6] of the RSA cryptosystem.

Theorem 6.9 (Weiner's Attack)

If $n = pq$ where p and q are primes such that $q < p < 2q$ and $d < n^{1/4}/3$, then given a public key e with $ed \equiv 1 \pmod{\phi(n)}$, d can be efficiently calculated.

The manner in which Weiner proved this was to use the classical theory of simple continued fractions (see page 223). The exact connection is given as follows. If $ed = 1 + k\phi(n)$ for some $k \in \mathbb{N}$, then under the hypothesis of Theorem 6.9, k/d is a convergent in the simple continued fraction expansion of e/n. Therefore, if the decryption exponent is small enough (as defined in Theorem 6.9) a cryptanalyst can compute d via the elementary task of computing a few convergents in the *publicly known* value e/n. Hence, we must conclude that Weiner's attack leads to a total break of the RSA cryptosystem if small decryption exponents are used. Therefore, use of small secret exponents to gain efficiency has the result of a total loss of security. Boneh and Durfee [36] improved Weiner's method by showing that the RSA cryptosystem is insecure

---

[6.5]The reader may be interested to know that in 1907, D.N. Lehmer [134] proved the following. If $n = pq$ and $|p - q| < 2n^{1/4}$, then n can be factored efficiently.

[6.6]By a *total break*, we mean that a cryptanalyst can recover d, hence retrieve all plaintext from ciphertext.

for any $d < n^{0.292}$. They admit that their attack cannot be stated as a theorem since they were unable to prove that it always succeeds but, on the other hand, they could not find a single example where the attack failed.

Exercises

6.2. Show that $e = 1, 2$ as an RSA encryption exponent yields an insecure RSA cipher.

6.3. Given RSA public enciphering modulus $e = 3$, show how a plaintext message m can be recovered if it is enciphered and sent to three different entities having pairwise relatively prime moduli n_i with $m < n_i$ for each for $i = 1, 2, 3$.

   (*Hint: Use the Chinese Remainder Theorem.*)

6.4. Generalize Exercise 6.3 by showing that a plaintext m can be recovered if e is the RSA enciphering exponent and m is sent to $k \geq e$ recipients with pairwise relatively prime RSA moduli n_i such that $m < n_i$ for $i = 1, 2, \ldots, k$.

   In Exercises 6.5–6.10, find the RSA decryption exponent d for each given RSA modulus n and RSA encryption exponent e.

6.5. $n = 3359 \cdot 1759$ and $e = 5$.

6.6. $n = 1979 \cdot 2281$ and $e = 7$.

6.7. $n = 4217 \cdot 4919$ and $e = 3$.

6.8. $n = 6373 \cdot 6761$ and $e = 7$.

6.9. $n = 7487 \cdot 7559$ and $e = 3$.

6.10. $n = 8443 \cdot 8861$ and $e = 11$.

☆ 6.11. Suppose that we have an RSA modulus n and RSA encryption and decryption exponents e and d. Then we know that there exists an integer m such that

$$ed - m(n - p - q + 1) = 1.$$

Assume that Eve knows $n/4$ of the least significant bits of d and that she can find solutions to the quadratic congruence

$$mnp^2 - (mn + m + 1)p + mn \equiv 0 \,(\mathrm{mod}\ 2^{n/4})$$

by having to check no more than

$$e \log_2 e$$

possible values for p. Show how Eve can efficiently factor n.

6.3   Strong Moduli

Even if strength fail, boldness at least will deserve praise: in great endeavors even to have had the will is enough.

                                               Propertius (50–16 B.C.) Roman Poet

We have already seen how poor implementation and bad choices can lead to an insecure RSA cryptosystem. Now we look at proper choices for RSA moduli.

Exercise 3.12 on page 59 shows how primes p and q being close together (in the sense described therein) results in a means of factoring the modulus $n = pq$ using Fermat's difference of squares method. Thus, one clearly bad choice is such an RSA modulus.

To avoid the above, we can select p and q (not close together — see the discussion following Gordon's algorithm below) such that $(p-1)/2$ and $(q-1)/2$ are primes. Such p and q are called *safe primes*. Although it is assumed that there are infinitely many safe primes, we have no proof of this suspicion.

◆ **Algorithm for Generating (Probable) Safe Primes**

Let b be the input bitlength of the required prime. Execute the following steps.

(1) Select a $(b-1)$-bit odd random $n \in \mathbb{N}$ and a smoothness bound B (determined experimentally).

(2) Trial divide n by primes $p \leq B$. If n is divisible by any such p, go to step (1). Otherwise, go to step (3).

(3) Use the Miller-Selfridge-Rabin test on page 87 to test n for primality. If it declares that 'n is probably prime', then go to step (4). Otherwise, go to step (1).

(4) Compute $2n + 1 = q$ and use the Miller-Selfridge-Rabin test on q. If it declares q to be a probable prime, terminate the algorithm with q as a 'probable safe prime'. Otherwise go to step (1).

Since long-term security of RSA moduli require products of primes having 512 bits, we would initiate the above algorithm with $b = 512$. However, there are primes that have even more constraints to ensure security of the RSA modulus. They are given as follows.

Definition 6.10 (Strong Primes)

A prime p is called a strong prime *if each of the following hold.*

(1) $p - 1$ *has a large prime factor q.*

(2) $p + 1$ *has a large prime factor r.*

(3) $q - 1$ *has a large prime factor s.*

The following algorithm was initiated in [102].

◆ Gordon's Algorithm for Generating (Probable) Strong Primes

(1) Generate two large (probable) primes $r \neq s$ of roughly equal bitlength using the Miller-Selfridge-Rabin test on page 87.

(2) Select the first prime in the sequence $\{2js + 1\}_{j \in \mathbb{N}}$, and let

$$q = 2js + 1$$

be that prime.

(3) Compute $p_0 \equiv r^{q-1} - q^{r-1} \pmod{rq}$.

(4) Find the first prime in the sequence $\{p_0 + 2iqr\}_{i \in \mathbb{N}}$, and let

$$p = p_0 + 2iqr$$

be that prime, which is a strong prime (see Exercise 6.16).

If one wants p to be a safe prime as well as a strong prime, this can be done by the methods in [241], but the algorithm therein is not as efficient as Gordon's algorithm above. We see that an RSA modulus consisting of two strong primes is not susceptible to the $p - 1$ or $p + 1$ factoring methods discussed in Section 5.2 on page 96. In general, having strong primes in the modulus makes it more difficult to factor, but also harder to find such primes. Also, the reader should be aware that the definition of strong prime is not consistent throughout the literature. There are some more minimal versions than the one we gave in Definition 6.10. Nevertheless, we may be fairly certain that under our definition, if the primes are chosen to be strong and large enough, the RSA cryptosystem is not susceptible to a direct factorization attack. Even the number field sieve that we studied in Section 5.5 on page 108 will not threaten the RSA cryptosystem if we choose the primes large enough. The new device, called *Twinkle*,[6.7] designed by Shamir to run the number field on general purpose computers (by a factor of as much as 1000 times faster) will not qualitatively change this since the factor of 1000 can be retrieved by choosing a larger modulus. That is the essence of the discussion, since even those who hold the opinion that having strong primes in the modulus make that modulus no harder to factor than one with "weaker" primes, agree that choosing random primes "large enough" will thwart direct factoring techniques.

In [203], Shamir proposed an new variant of RSA which is called *unbalanced RSA*. In this version, we may choose the RSA modulus $n = pq$ such that p is much larger than q and we lose no efficiency. For instance, p could be a 5000-bit

---

[6.7]This is an acronym for T*he* W*eizmann* IN*stitute* K*ey* L*ocating* E*ngine*, which is an electro-optical sieving device that is capable of executing sieve-based factoring algorithms, such as the GNFS, up to three orders of magnitude faster than on a conventional computer. (http://www.datastreamconsulting.com/RSAdocs/bulletn12.pdf)

prime and q could be a 500-bit prime. Such primes put the RSA modulus far away from any known factoring attack. To decrypt

$$c \equiv m^e \,(\text{mod } n),$$

we compute

$$m' \equiv c^{d'} \,(\text{mod } q) \text{ where } d' \equiv d \,(\text{mod } q - 1).$$

Then for $0 \leq m < q$, we must have $m' = m$. Hence, decryption in the unbalanced version of RSA involves one exponentiation modulo a 500-bit prime, so we have the same efficiency as that obtained with regular RSA having a 500-bit modulus.

We conclude this section with a discussion of an attack on RSA which will answer a question in the mind of the most attentive reader: If the reason for (1)–(2) in Definition 6.10 is to thwart the $p \pm 1$ factoring methods, what is the reason for (3)? The reader may prepare for the following by solving Exercise 6.18, which is a special case of the following, and we will need that exercise in its description.

◆ Cycling Attack on RSA

Let $n = pq$ be an RSA modulus and let e be the encryption exponent, so that $c \equiv m^e \,(\text{mod } n)$ is the enciphering function. Suppose that $t \in \mathbb{N}$ is the smallest value such that

$$g = \gcd(c^{e^t} - c, n) > 1.$$

If *exactly one* of p or q divides $(c^{e^t} - c)$, say, p, then $g = p$ and we have factored n. If *both* p and q divide $(c^{e^t} - c)$, then $g = n$ and we are in the situation in Exercise 6.18, namely, we can retrieve m. Such an attack on RSA is called a *cycling attack*.

It is most often the case that the factorization of n will occur with a cycling attack than that we find m. Therefore, the cycling attack is typically viewed as an algorithm for factoring n. Since we have already seen that this is infeasible for a properly chosen RSA modulus, then this does not present a serious threat. However, it does explain why (3) is part of Definition 6.10, namely, to thwart the cycling attack. Although it is true that a randomly chosen pair of large primes for an RSA modulus will result in only a negligible chance of the cycling attack succeeding, we should still choose strong primes since they are at least as secure as random primes and it takes only an insignificant amount of additional time to compute them, as compared to using random primes. Thus, there is no good reason or significant savings in *not* using them. The bottom line in all this is that, as we have seen throughout the text, we cannot be certain of what mathematical algorithms will be developed in the future, so choosing strong primes may help to thwart those attacks, and is one more deposit in our bank account for security of RSA.

Exercises

6.12. Given $n \in \mathbb{N}$, prove that the following are equivalent.

(a) $a^n \equiv a \pmod{n}$ for all $a \in \mathbb{Z}$.

(b) $a^{n-1} \equiv 1 \pmod{n}$ for all $a \in \mathbb{Z}$ such that $\gcd(a, n) = 1$.

(c) n is squarefree and $(p-1) \mid (n-1)$ for all primes p dividing n.

(*These equivalent conditions are known as* Korselt's criterion *discovered in 1899 (but implicit in Gauss's* [95, Article 92, pp. 60–61]). *Compare with Definition 4.3 on page 82.*)

In what follows, let $\lambda(n)$ be the Carmichael function defined in Exercise 5.2 on page 95.

6.13. Prove that if n is squarefree, then

$$a^{\lambda(n)+1} \equiv a \pmod{n},$$

for all $a \in \mathbb{Z}$. Give a counterexample to the latter when n is not squarefree.

☆ 6.14. Given $n \in \mathbb{N}$, $a \in \mathbb{Z}$, prove that the following are equivalent.

(a) There exist $s, t \in \mathbb{N}$ such that $a^{t\lambda(n)+s} \equiv a^s \pmod{n}$.

(b) $\gcd(a^{s+1}, n) \mid a^s$.

6.15. Let $n = pq$ be an RSA modulus, (e, d) a public/private key pair, and define a function f by

$$f(x) \equiv x^e \pmod{n}, \text{ where } ed \equiv 1 \pmod{\lambda(n)}.$$

Prove that f is a one-way function with d as trapdoor.

6.16. Prove that the prime p output by Gordon's algorithm is indeed a strong prime.

6.17. Suppose that we obtain two strong primes p and q using Gordon's algorithm on page 121. Also, we set

$$n = pq, \qquad n' = (p-1)(q-1)/4,$$

and choose a random integer e such that

$$\gcd(e, n') = 1, \quad (p-1) \nmid (e-1), \quad (q-1) \nmid (e-1).$$

Show how to set up an RSA cryptosystem with public enciphering key e and RSA modulus n.

6.18. Given an RSA modulus n and public enciphering exponent e, we encrypt via the function $f(m) \equiv m^e \equiv c \pmod{n}$. Show that there exists an $\ell \in \mathbb{N}$ such that $c^{e^\ell} \equiv c \pmod{n}$, and prove that knowledge of ℓ can be used to retrieve m.

6.4　Generation of Random Primes

Truth, as light, manifests itself and darkness, thus truth is the standard of itself and of error.

Baruch Spinoza (1632–1677) Dutch philosopher

Given the discussion at the end of Section 6.3, it is clearly important for the security of RSA that we be able to generate large random primes efficiently. This section is devoted to a discussion of methods for accomplishing this task. For the first of our generation algorithms, we will use one of the probabilistic primality testing algorithms discussed in Chapter 4.

◆ Large (Probable) Prime Generation

We let b be the input bitlength of the desired prime and let B be the input smoothness bound (empirically determined). Execute the following steps.

(1)　Randomly generate an odd b-bit integer n.

(2)　Use trial division to test for divisibility of n by all odd primes no bigger than B. If n is so divisible, go to step (1). Otherwise go to step (3).

(3)　Use the Miller-Selfridge-Rabin (MSR) probabilistic primality test on page 87 to test n for primality. If it is declared to be a probable prime, then output n as such. Otherwise, go to step (1).

There is a variant of the above test that injects a step after the MSR test by invoking an elliptic curve primality *proving* algorithm (see Footnote 5.7 on page 103 and [165, Theorem 6.30, p. 240]) namely, a deterministic primality test in which case the output is called a *provable prime*. However, we have not studied these elliptic curve primality proving algorithms herein, so we are going to now describe a prime generating algorithm that produces provable primes based upon the contents of Exercise 4.1 on page 83 — Pocklington's theorem from 1914 (see [184]).

We first restate Pocklington's theorem here in a different format and notation.

Theorem 6.11 (Pocklington's Theorem)
　Suppose that

$$n - 1 = 2AB \text{ where } A, B \in \mathbb{N} \text{ with } B > 2A.$$

If there exists a prime divisor p of B and an integer $m > 1$ such that both

$$m^{n-1} \equiv 1 \,(\mathrm{mod}\ n),$$

and

$$\gcd(m^{(n-1)/p} - 1, n) = 1,$$

then n is a provable prime.

◆ **Maurer's Large (Provable) Prime Generation (Brief Version)**

The goal is to output a random large provable prime of a given required bitlength.

(1) Randomly generate an odd integer B (for instance see Exercise 6.23).

(2) Select a random integer $A < B$.

(3) Test $n = 2AB + 1$ using Pocklington's Theorem 6.11. If n is prime, then go to step (4). Otherwise go to step (2).

(4) If n is of the required bitlength, then output n as a provable prime. Otherwise, set $n = B$ and go to step (2).

The full version of the above may be found in [149], [151]. The running time of Maurer's test is only slightly larger than that for the probable prime test described before it. Of course, the larger the value of the required bitlength, the faster the former will run over the latter, but the latter has *provable* security attached to it that the former does not have.

The decision to use either of the above algorithms reduces to the determination of the size of the prime to be generated, how often the algorithm is implemented, the computing facilities available, and the degree of security sought. Today, with increasingly fast processors available, especially special-purpose hardware for RSA (see [40]), the small amount of extra time needed to output a provable vs. a probable prime is insignificant in terms of taking advantage of the power a provable prime ensures.

We now give an illustration (deliberately contrived) of Maurer's algorithm.

Example 6.12 *Suppose that we want to generate a 35-bit random (provable) prime. We randomly generate* $B = 168$ *and* $A = 3$, *so* $n = 2AB + 1 = 1009$. *By choosing* $m = 2$ *and* $p = 3$ *we get* $2^{1008} \equiv 1 \,(\mathrm{mod}\ n)$ *and* $\gcd(2^{1008/3} - 1, n) = 1$, *so* $n = 1009$ *is a provable prime. However, this is only a 10-bit prime since* $1009 = (1111110001)_2$. *Thus we set* $n = B$ *and go back to step (2). We randomly select* $A = 146$, *so* $n = 2AB + 1 = 2 \cdot 146 \cdot 1009 + 1 = 294629$ *and apply Theorem 6.11 with* $m = 3$ *and* $p = 2$. *Then*

$$3^{n-1} \equiv 1 \,(\mathrm{mod}\ n) \text{ and } \gcd(3^{(n-1)/p} - 1, n) = 1,$$

so $n = 294629$ *is a provable prime. However,* $n = (1000111111011100101)_2$ *is a 19-bit prime, so we set* $B = 294629$ *and go to step (2). We randomly select* $A = 50415$ *and set* $n = 2AB + 1 = 2 \cdot 50415 \cdot 294629 + 1 = 29707442071$. *For* $m = 2$, *and* $p = 294629$ *we get,*

$$2^{n-1} \equiv 1 \,(\mathrm{mod}\ n) \text{ and } \gcd(2^{(n-1)/p} - 1, n) = 1,$$

so $n = 29707442071$ *is a provable prime. Moreover,*

$$n = (110111010101011001110010111110010111)_2,$$

a 35-bit prime, so we are done. See Exercises 6.19–6.22.

Maurer's algorithm makes repeated use of Pocklington's Theorem by reselecting values of A and testing for successively larger prime values until the desired bitlength is achieved and we have a provable prime of the required size.

There is yet another method similar to the above that uses Pocklington's Theorem to generate large primes given by Ribenboim in [194], which the reader may also find in [245]. Ribenboim gives evidence (in the absence of a proof) that for producing primes of a given size the algorithm will run in polynomial time. In any case, in public-key cryptography, we require efficient algorithms for generating large, random primes as an indispensable tool.

Exercises

In Exercises 6.19–6.22, take the given values of A, B, m, p, and execute Maurer's algorithm until the provable prime of given bitlength b is achieved.

6.19. $A = 3$, $B = 51$, $m = 2$, $p = 17$, and $b = 9$.

6.20. $A = 7$, $B = 75$, $m = 2$, $p = 5$, and $b = 14$.

6.21. $A = 11$, $B = 105$, $m = 2$, $p = 5$, and $b = 12$.

6.22. $A = 11$, $B = 211$, $m = 2$, $p = 211$, and $b = 13$.

6.23. Let $n = pq$ be an RSA modulus. Randomly select $a \in \mathbb{N}$ such that $\gcd(a, \phi(n)) = 1$ and choose a *seed* $s_0 \in \mathbb{N}$ where $1 \leq s_0 \leq n - 1$. Recursively define
$$s_j \equiv s_{j-1}^a \,(\mathrm{mod}\; n), \quad 1 \leq j < \ell,$$
where ℓ is the least integer such that $s_{\ell+1} = s_j$ for some natural number $j \leq \ell$. Then
$$f(s_0) = (s_1, s_2, \ldots, s_\ell)$$
is called the *RSA pseudo-random number generator*. The value ℓ is called the *period length* of f and a is called the *exponent*.

   (a) Prove that ℓ *must* exist.

   (b) Find f when $a = 3$, $s_0 = 2$, $b = 0$ and $n = 561$.

   (c) find ℓ in (b).

(*Bit generators such as the one above are generally called* pseudorandom bit-generators (*PRBG*). *The RSA-PRBG is deemed to be cryptographically secure since it too is based upon the intractability of the integer factoring problem. A predecessor to it is the linear congruential generator, those of the form* $s_j \equiv as_{j-1} + b \,(\mathrm{mod}\; n)$ (*see* [165]), *discovered by D.H. Lehmer in 1949 (see Footnote 4.3).*)

Chapter 7

Authentication

Whoever, in discussion, adduces authority uses not intellect but rather memory.

Leonardo da Vinci (1452–1519) Italian painter and designer

7.1 Identification, Impersonation, and Signatures

In this chapter, we will be concerned with *authentication*, meaning verification of the identity and data origin of a legitimate entity in a protocol by another (legitimate) entity. In this section, we will discuss *impersonation* (the assumption of the identity of a legitimate entity by an adversary) as well as identification and an introduction to digital signatures. We have already had a taste of a type of impersonation attack when we discussed the man-in-the-middle attack on page 27. The following is a description of such an attack on a general public-key cryptosystem.

Suppose that Alice wishes to send a message m to Bob using a public-key cryptosystem such as RSA. His public enciphering key is e, say. If Mallory, impersonating Bob, sends Alice his public key e' and Alice assumes this is Bob's public key, she will send $m^{e'}$. Mallory intercepts and using his private key d' computes $(m^{e'})^{d'} = m$. Then he enciphers m with Bob's public key and sends m^e to Bob. Neither Alice nor Bob knows that they have been duped by Mallory. This is illustrated as follows.

Diagram 7.1 (Impersonation Attack on Public-Key Cryptosystems)

$$\boxed{Alice} \quad \overset{e'}{\underset{E_{e'}(m)=c'}{\overset{\longleftarrow}{\longrightarrow}}} \quad \boxed{\begin{array}{c} Mallory \\ \mathcal{D}_{d'}(c')=m \end{array}} \quad \overset{E_e(m)=c}{\longrightarrow} \quad \boxed{Bob}$$

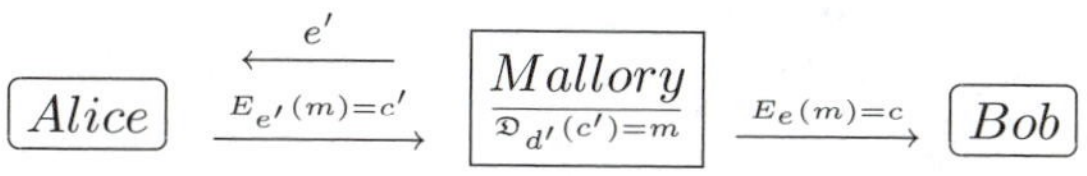

This illustrates why we need a means of ensuring that Alice and Bob can verify that they are actually talking to one another and not Mallory. One means of thwarting the above attack is for Alice to obtain a random number r from Bob and sign it with her private key d that Bob can verify. This is illustrated as follows.

Diagram 7.2 (Challenge-Response Protocol)

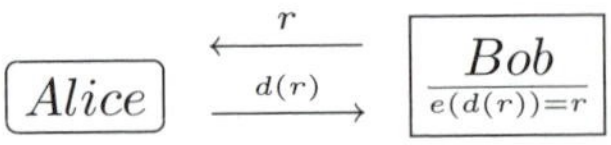

$$\boxed{Alice} \quad \begin{array}{c} \xleftarrow{\quad r \quad} \\ \xrightarrow{\quad d(r) \quad} \end{array} \quad \boxed{\begin{array}{c} Bob \\ \hline e(d(r))=r \end{array}}$$

However, the caveat is that Bob must obtain Alice's *authentic* public key. Otherwise, Mallory can trick Bob as he did Alice in Diagram 7.1. However, if we bring Trent into the picture, this can be avoided. One means for doing this was described at the bottom of page 73.

We need more protocols for identification. We have seen one means for legitimate entities to identify themselves on page 34 where we discussed the Fiege-Fiat-Shamir Identification Protocol. In order to present an alternative, and more desirable, protocol to the latter, we need to engage in a discussion about "signatures".

We are all aware that when we affix our signature to a legal document, credit card, or a personal letter, we are identifying ourselves by signing. Of course, we also know that this is subject to forgery (impersonation) by criminal elements. There are verification procedures such as a salesperson verifying the signature on the credit card slip by comparing it with that on the credit card itself. What we see here is a two-stage process of *signing* and *verifying*. The same is true for *digital signatures* meaning digital data strings that associate a given message with its sender. There is a more formal setup.

Definition 7.3 (Digital Signature Schemes)

Let $\mathcal{M}$ be a message space, *$\mathcal{K}$ a* keyspace, *and $\mathcal{S}$ a set of bitstrings of fixed length, called a* signature space. *For $k \in \mathcal{K}$, a digital signature algorithm $\mathrm{sig}_k : \mathcal{M} \mapsto \mathcal{S}$ is a method for producing a digital signature. A digital verification algorithm $\mathrm{ver}_k : \mathcal{M} \times \mathcal{S} \mapsto \{0,1\} = \mathbb{F}_2$ is a method for verifying that a digital signature, $\mathrm{sig}_k(m) = c$, is authentic (where 1 represents "true", when $\mathrm{sig}_k(m) = c$, and 0 represents "false", when $\mathrm{sig}_k(m) \neq c$). A digital signature scheme is comprised of a digital signature algorithm and a digital verification algorithm.*

Typically digital signature schemes must meet two criteria to be secure. If Alice signs a message m with $\mathrm{sig}_k(m)$, it must be computationally infeasible for an adversary to retrieve the pair $(m, \mathrm{sig}_k(m))$, called the *unforgeable property*. Secondly, if Bob receives $\mathrm{sig}_k(m) = c$ from Alice, then Bob must be able to verify that this is Alice's signature using $\mathrm{ver}_k(c)$, called the *authentic property*. It is also desirable that a digital signature scheme satisfy two more properties. The first is that after being transmitted, neither Bob nor Mallory can alter m, called the *not alterable property*; and Bob must be able to instantly detect if a m is being resent, called the *not reusable property*.

We also need the following notion before presenting the next identification scheme. A *certificate* is a quantity of information that has been signed by a trusted authority, such as Trent. Types of certificates include *identification* certificates that contain an entity's identifying information such as birth certificate, passport information, and perhaps list of public keys, for instance.

Now we are in a position to describe an identification scheme that is an alternative to the Fiege-Fiat-Shamir protocol. The security of the following is based upon the intractability of the discrete log problem (DLP) and requires the services of Trent. The following was introduced in 1991 [199].

◆ Schnorr Identification Protocol

Setup Stage: Trent selects each of the following:

(1) a large prime p such that the DLP in $\mathbb{F}_p^*$ is intractable (say, $p \geq 2^{1024}$).

(2) a large prime divisor q of $p - 1$ (say, $q \geq 2^{160}$).

(3) $\alpha \in \mathbb{F}_p^*$ such that $\operatorname{ord}_p(\alpha) = q$ (say, $\alpha = \beta^{(p-1)/q}$ where β is a primitive root modulo p).

(4) a parameter t such that $q > 2^t$ (usually $t \geq 40$).

(5) a secure signature scheme embodying a secret digital signing algorithm $\operatorname{sig}_{T(k)}$ and a public digital verifying algorithm $\operatorname{ver}_{T(k)}$ for verification of Trent's signatures. (Typically $\operatorname{sig}_{T(k)}$ involves a cryptographic hash function for security (see page 55), but we will omit this here for increased clarity of presentation.)

Then Trent creates a certificate for Alice as follows:

(6) Trent establishes a bitstring containing information I_A that identifies Alice. Then Alice selects a private random nonnegative exponent $e \leq q - 1$ and she computes $v \equiv \alpha^{-e} \pmod{p}$, which she sends to Trent. Upon receipt, Trent generates a signature $s = \operatorname{sig}_{T(k)}(I_A, v)$. Then he sends the certificate $C(A) = (I_A, v, s)$ to Alice.

Identification Protocol: Alice wishes to identify herself to Bob, who must verify her identity.

(7) Alice selects a random nonnegative integer $k \leq q - 1$ and computes

$$\gamma \equiv \alpha^k \pmod{p}.$$

(Here, k is called the *commitment*.)

(8) Alice sends her certificate $C(A)$ and γ, called the *witness*, to Bob.

(9) Bob computes $\operatorname{ver}_{T(k)}((I_A, v, s)) = 1$, thereby verifying Trent's signature. Then Bob selects a random natural number $r \leq 2^t$, which he sends to Alice. (Here r is called the *challenge*.)

(10) Alice computes $y \equiv k + er \pmod{q}$, which she sends to Bob. (Here y is called the *response.*)

(11) Bob computes $\delta \equiv \alpha^y v^r \pmod{p}$, and if $\delta \equiv \gamma \pmod{p}$, he accepts Alice's identity. Otherwise, he rejects it.

Steps (8)–(10) represent an example of a *three-pass protocol*. Notice that Alice proves *knowledge of e* (without revealing e) by her response y to the challenge r in step (10). Hence, this is an example of a *proof of knowledge* about which we will say more following the example.

For illustrative purposes, the following uses much smaller values than those suggested in the above algorithm, as do Exercises 7.2–7.5.

Example 7.4 *In this illustration of the above protocol, we will assume that Trent has created a certificate for Alice and that it has been verified by Bob that indeed we have Trent's signature. We proceed with the rest of the protocol. Suppose that we have $p = 4937$, $q = 617$ and $t = 9$. We know that $\alpha = 1624$ has order 617 in $\mathbb{F}_p^*$, since 3 is a primitive root modulo p and $\alpha \equiv 3^{(p-1)/q} \pmod{p}$. If Alice's private exponent is $e = 55$, then she computes*

$$v \equiv \alpha^{-e} \equiv 1624^{-55} \equiv 2967 \pmod{p}.$$

This completes the setup stage. Now if Alice selects $k = 29$, she computes $\gamma \equiv \alpha^k \equiv 1624^{29} \equiv 4585 \pmod{4937}$. If Bob chooses $r = 105$ and sends it to Alice, she computes

$$y \equiv k + er \equiv 29 + 55 \cdot 105 \equiv 251 \pmod{617},$$

which she sends to Bob who computes,

$$\delta \equiv \alpha^y v^r \equiv 1624^{251} \cdot 2967^{105} \equiv 4585 \equiv \gamma \pmod{4937},$$

so Bob accepts Alice's identity as valid.

We now discuss some features of Schnorr's protocol. First, how secure is it? By Exercise 7.6, if Mallory has a non-negligible probability of successfully executing Schnorr's protocol, then he must (essentially) "know" Alice's private exponent e. This is a property called *soundness*, i.e., this demonstrates that Schnorr's protocol is a *proof of knowledge of e*. Obtaining e would be similar to getting the PIN for a bank card since, with knowledge of it, an adversary can convince a receiver of the identity of the sender. However, unlike a PIN, e is never revealed. Alice merely *proves* her knowledge of it. Once Alice proves her identity in this fashion, and Bob accepts her proof in step (11), the protocol is said to have the *completeness* property. However, soundness and completeness are insufficient to guarantee security. For example, Alice could just reveal her private exponent e to Mallory, who could then impersonate her, and the protocol would still have the soundness and completeness properties. Thus, Alice must ensure that no information about e is leaked.

If Mallory does not have knowledge of e, and he tries to impersonate Alice, then he is in the position in step (10) of having to compute y, which is a function of e, in response to Bob's challenge r. However, computing e from v involves solving an instance of the DLP, which is assumed to be intractable. With all this being said, it has not been proved that Schnorr's protocol is secure.

When compared with the Fiege-Fiat-Shamir protocol, Schnorr's protocol is much more efficient. The reason for this is that the most computationally intensive operation is the modular exponentiation in step (7). However, by design, this may be computed offline. In step (10) there is one modular addition and one modular multiplication, so the online computations are very moderate. The computations for Fiege-Fiat-Shamir protocol are significantly greater as we saw in Section 2.1. The Schnorr algorithm was designed with this computational efficiency in mind for such applications as smart cards with low computing power. In general, the Schnorr protocol is quite suitable when Alice has restricted computing power. Notice as well that more computational efficiency is gained by using a subgroup of order $q \mid (p-1)$, which lowers the number of bits needed for transmission. We also mentioned the three-pass protocol involved in steps (8)–(10). This was a built-in design of the protocol to reduce bandwidth,[7.1] especially in comparison to the Fiege-Fiat-Shamir protocol.

The above discussion shows that Schnorr's protocol is efficient and presumed secure based on the intractability of the DLP, but nobody has proved that the protocol is secure. The following modification of the protocol, however, has been proved to be secure under the assumption of the intractability of a *particular* discrete log. This first appeared in 1993 [176].

◆ Okamoto Identification Protocol

Setup Stage: Trent selects each of the following:

(1) a large prime p and a large prime divisor q of $p-1$.

(2) $\alpha_1, \alpha_2 \in \mathbb{F}_p^*$ such that $\mathrm{ord}_p(\alpha_1) = q = \mathrm{ord}_p(\alpha_2)$.

(3) $e = \log_{\alpha_1}(\alpha_2)$ is kept secret from **all** participants, except of course Trent, and we assume that it is intractable to compute e.

Then Trent executes steps (4)–(5) of Schnorr's protocol, and creates a certificate for Alice as follows.

(4) As in step (6) of Schnorr's protocol, I_A is established, and Alice secretly selects nonnegative integers $e_1, e_2 \leq q-1$. She computes

$$v \equiv \alpha_1^{-e_1} \alpha_2^{-e_2} \pmod{p},$$

[7.1] Bandwidth is the width of the range of frequencies that an electronic signal occupies on the given transmission medium. In other words, it is the speed of data on a given transmission path. Bandwidth is measured in *megabits*, where a megabit is 10^6 bits. *Megabits per second* is denoted by *Mbps*. There have been alternative, albeit less accepted, interpretations of the meaning of a megabit, but the preceding is now considered to be the standard.

which she sends to Trent, who generates $s = \text{sig}_{T(k)}(I_A, v)$. The certificate

$$C(A) = (I_A, v, s)$$

is then sent to Alice.

Identification Protocol:

(5) Alice selects random nonnegative integers $k_1, k_2 \leq q - 1$ and computes

$$\gamma \equiv \alpha_1^{k_1} \alpha_2^{k_2} \pmod{p}.$$

(6) Alice sends her certificate $C(A)$ and γ to Bob.

(7) Bob computes $\text{ver}_{T(k)}((I_A, v, s))$, and if it is 1 he accepts the signature. Otherwise he rejects it and terminates the protocol. Upon acceptance, Bob selects a random natural number $r \leq 2^t$, which he sends to Alice.

(8) Alice computes $y_1 \equiv k_1 + e_1 r \pmod{q}$, and $y_2 \equiv k_2 + e_2 r \pmod{q}$, and sends y_1, y_2 to Bob.

(9) Bob computes $\delta \equiv \alpha_1^{y_1} \alpha_2^{y_2} v^r \pmod{p}$, and if $\delta \equiv \gamma \pmod{p}$, he accepts Alice's identity. Otherwise, he rejects it.

For pedagogical purposes, we will use unrealistically small values in the following example, as do we in Exercises 7.9–7.12.

Example 7.5 *Let $p = 7487$, $q = 197$, $t = 7$, $\alpha_1 = 64$, $\alpha_2 = 99$. Note that both α_1 and α_2 have order q in $\mathbb{F}_p^*$ since 5 is a primitive root modulo p and*

$$\alpha_1 = 5^{(p-1)/q} \text{ while } \alpha_2 = 5^{3(p-1)/q} \text{ with } 3 \nmid (p-1).$$

If Alice secretly chooses $e_1 = 101$ and $e_2 = 51$, then she computes

$$v \equiv \alpha_1^{-e_1} \alpha_2^{-e_2} \equiv 64^{-101} \cdot 99^{-51} \equiv 1119 \pmod{7487}.$$

Then Trent creates $C(A) = (I_A, v, s)$ for her, where $s = \text{sig}_{T(k)}(I_A, v)$. This completes the setup stage. Alice now chooses $k_1 = 180$ and $k_2 = 8$ at random. She computes

$$\gamma \equiv 64^{180} \cdot 99^8 \equiv 5843 \pmod{7487}.$$

Then she sends γ and $C(A)$ to Bob. Bob selects the challenge $r = 5$ and sends it to Alice. She then computes both

$$y_1 \equiv k_1 + e_1 r \equiv 180 + 101 \cdot 5 \equiv 94 \pmod{197},$$

$$y_2 \equiv k_2 + e_2 r \equiv 8 + 51 \cdot 5 \equiv 66 \pmod{197},$$

and sends them to Bob. He computes

$$\delta \equiv 64^{94} \cdot 99^{66} \cdot 1119^5 \equiv 5843 \equiv \gamma \pmod{7487},$$

so he accepts Alice's proof of identity.

The principal difference between the Schnorr and Okamoto protocols is that the latter is based upon the intractability of computing the discrete log $\log_{\alpha_1}(\alpha_2) = e$. It can be shown that, without collaboration with Alice in the protocol, Mallory cannot get any information about her private exponents e_1 and e_2, so he is faced with the intractable problem of computing e, a discrete log problem. This is a proof of the security of Okamoto's protocol under the assumption of intractability of computing e. With all this being said, Schnorr's protocol is still faster, given the additional computations involved in the Okamoto protocol. Thus, although there is no proof of the security of the Schnorr protocol, it still has not been successfully cryptanalyzed. In other words, no weaknesses have been found. Thus, for efficiency, especially in the use of smart cards, Schnorr's protocol may be chosen in practice.

We close this section with a brief discussion of some attacks on identification protocols. This supplements our discussion of attacks in Section 1.3. We have already discussed the impersonation attack in this section, so we exclude it from what follows.

● Classification of Attacks on Identification Protocols

The following describes only three of the most prominent attacks.

◆ Chosen-Text Attack

This is an attack on a challenge-response protocol where Mallory chooses challenges according to some plan designed to recover information about Alice's private key. For instance, in Schnorr's protocol, Alice enciphers the challenge r with y, so this attack involves chosen-plaintext (see page 26). A method for thwarting this type of attack is to embed in each challenge-response a random number.

◆ Forced Delay

This type of attack involves Mallory intercepting a message and relaying it later. This is a kind of man-in-the-middle attack (see page 27). Defence against this type of attack may include the use of random numbers in conjunction with short response time-outs.

◆ Replay (or Playback) Attack

This kind of attack involves the use of information from a previous execution of the protocol in order to attempt to deceive the verifier. Defence against this attack might contain the use of challenge-response methods, as well as the use of *nonces*, which are numbers used exactly one time for a given protocol. Often in challenge-response protocols, the term is used to refer to a random number with differing criteria for the randomness (see Remark 1.13 on page 8).

There are other types of attacks and other types of identification protocols of course. However, we have presented the above as a reasonable introduction to the topic and as a motivator for the next section since identification protocols are ideally suited for conversion into signature schemes.

Exercises

7.1. Show how the final step in the Schnorr Identification Protocol establishes Alice's identity by verifying that $\delta \equiv \gamma \pmod{p}$.

In Exercises 7.2–7.5, use the Schnorr Identification Protocol to decide whether or not Bob should accept Alice's proof of identity. You are given $(p, q, \alpha, t, e, k, r, \gamma, v)$, and you may assume that Trent's signature has been verified in step (9). Thus, you need only execute the final steps (10)–(11).

7.2. $(p, q, \alpha, t, e, k, r, \gamma, v) = (2729, 31, 2484, 4, 25, 6, 16, 532, 532)$.

7.3. $(p, q, \alpha, t, e, k, r, \gamma, v) = (3119, 1559, 49, 10, 151, 1001, 512, 501, 460)$.

7.4. $(p, q, \alpha, t, e, k, r, \gamma, v) = (3623, 1811, 25, 10, 998, 1007, 256, 979, 2850)$.

7.5. $(p, q, \alpha, t, e, k, r, \gamma, v) = (7481, 17, 3668, 4, 8, 4, 3, 4104, 4508)$.

☆ 7.6. Assume that if Mallory knows a value of γ in Schnorr's Identification Protocol, and with this value has a probability of at least 2^{1-t} of successfully impersonating Alice. Prove that Mallory can compute Alice's private exponent e in polynomial time. (*What this says is that anyone who can successfully impersonate Alice must essentially "know" the private key e, i.e., be able to compute it in polynomial time.*)

7.7. Suppose that, in the Schnorr Identification Protocol, Mallory attempts to impersonate Alice by forging a certificate $C'(A) = (I_A, v', s')$, where $v' \neq v$. Demonstrate how Bob will detect the forgery, assuming a protocol has been set up properly to ensure security.

7.8. Suppose that Mallory obtains Alice's legitimate certificate $C(A)$ in the Schnorr identification protocol, but he does not know e. Show how a secure protocol will thwart Mallory from successfully impersonating Alice.

In Exercises 7.9–7.12, use the Okamoto Identification Protocol to decide whether or not Bob should accept Alice's proof of identity. Use the parameters given, and you may assume that Trent's signature has been verified in step (7). Thus, you need only execute the final steps (8)–(9).

7.9. $(p, q, \alpha_1, \alpha_2, t, e_1, e_2, k_1, k_2, r, \gamma, v) =$
$$(2729, 11, 443, 2541, 3, 25, 27, 10, 9, 8, 2490, 1768).$$

7.10. $(p, q, \alpha_1, \alpha_2, t, e_1, e_2, k_1, k_2, r, \gamma, v) =$
$$(3119, 1559, 49, 2246, 10, 151, 279, 1001, 901, 512, 2172, 939).$$

7.11. $(p, q, \alpha_1, \alpha_2, t, e_1, e_2, k_1, k_2, r, \gamma, v) =$
$$(3623, 1811, 25, 1133, 10, 998, 5, 1007, 506, 256, 3000, 771).$$

7.12. $(p, q, \alpha_1, \alpha_2, t, e_1, e_2, k_1, k_2, r, \gamma, v) =$
$$(7481, 17, 3668, 3311, 4, 8, 10, 4, 8, 3, 4500, 3311).$$

7.2 Digital Signature Schemes

Man, proud man, drest in a little brief authority, most ignorant of what he's most assured...

William Shakespeare

One of the primary applications of RSA is in the use of digital signatures, to which we were given a brief introduction in Section 7.1. Therein we saw that a digital signature is a mechanism for binding information to an entity, and includes a verification procedure. The first signature scheme that we discuss is one with *message recovery*, which means that the message being sent is not required as input to the verification algorithm. In this case, the original message is recovered from the signature itself.

◆ **RSA Signature Scheme**

Setup Stage: Alice wishes to send a message $m \in \mathcal{M} = \mathbb{Z}/n\mathbb{Z} = \mathcal{C}$ to Bob. She selects an RSA modulus $n = pq$ and an RSA key pair (e, d) obtained via the RSA key generation algorithm given on page 61. The keyspace is

$$\mathcal{K} = \{k = (n, p, q, e, d) : ed \equiv 1 \,(\mathrm{mod}\ \phi(n))\},$$

where n, e are public and p, q, d are private.

Signing Stage: Alice's private digital signature sig_k is given by

$$\mathrm{sig}_k(m) \equiv m^d \equiv c \,(\mathrm{mod}\ n),$$

and ver_k is her public verification algorithm. She sends (m, c) to Bob.

Verification Stage: Bob obtains Alice's public (e, ver_k), and computes $\mathrm{ver}_k(m, c)$ which is 1 precisely when $m \equiv c^e \,(\mathrm{mod}\ n)$, in which case he accepts the signature, and rejects it otherwise.

Example 7.6 *If Alice generates* $n = pq = 701 \cdot 911 = 638611$, $e = 251$, $d = 553251$, *then* $\phi(n) = 637000$. *If she wishes to send* $m = 11911$ *to Bob, she computes* $\mathrm{sig}_k(m) \equiv 11911^d \equiv 341076 \,(\mathrm{mod}\ n)$ *and sends* $(m, c) = (11911, 341076)$ *to Bob. After getting Alice's public data* (e, ver_k), *he computes* $\mathrm{ver}_k(m, c) = 1$ *since* $c^e \equiv 341076^{251} \equiv 11911 \equiv m \,(\mathrm{mod}\ 638611)$. *(See Exercises 7.13–7.16.)*

As with the RSA cryptosystem itself, an adversary who is able to factor the modulus n can compute $\phi(n)$, then use the extended Euclidean algorithm to get d from $\phi(n)$ and e (see Exercises 6.5–6.10 on page 119). Hence, factoring results in a total break of the scheme. Therefore, p and q must be chosen to make factoring n computationally infeasible (see Chapter 6). In the following discussion, we assume that the RSA signature scheme has been properly implemented, ensuring that Alice's private key d is secure.

Note that in the RSA digital signature scheme, anyone, not just Bob, can verify Alice's signature since e is made public. However, only Alice can sign the message since $\mathrm{sig}_k = d$ is private. This also ensures that Alice cannot deny later that she sent the message, since nobody else could have computed m^d. This is called *non-repudiation*. Another safeguard is to ensure that a digital signature is not reused. One means of doing this, with RSA for instance, is to affix a *timestamp*. Thus, instead of just the message m, Alice would have a message with a timestamp t, so the original message would be $M = (m, t)$.

An analogue of the RSA signature scheme is Alice's signing a postcard and sending it to Bob. There is a variant of the RSA scheme that involves a combination of signing and public-key encryption. If Alice were to write a letter on paper, she would sign it and put it in an envelope. An analogue of this is to digitally sign the message, then encrypt it.[7.2] Thus, after the above signing stage, she would add an *encryption stage* where she enciphers with Bob's public exponent e_B, so $(m, c)^{e_B}$ is sent. Then Bob uses his private RSA exponent d_B to calculate $((m, c)^{e_B})^{d_B} \equiv (m, c) \pmod{n}$, and he uses Alice's public RSA exponent to compute $c^e \equiv m \pmod{n}$. This further encryption of the entire message with Bob's public key ensures confidentiality. This variant of the RSA signature scheme can be applied to any signature scheme with message recovery, namely, by hashing the message and signing the hash. What results is a *signature scheme with appendix*, which means that the message is required as input for the verification algorithm.

If we choose a small public exponent (see page 115) the verification is considerably faster than the signing. Thus, the RSA signature scheme is well-suited to circumstances where signature verification is the primary operation used. It becomes even more efficient if we enlist Trent to create a certificate of identification for Alice, which he has to do only once. Then verification may take place numerous times by Bob and other entities with whom Alice has communication. It can be shown that for messages no longer than half the RSA modulus, the RSA signature scheme with message recovery is most efficient, whereas if message blocking is required, then the most bandwidth-efficient (see Footnote 7.1 on page 131) method is the RSA signature scheme with appendix.

Another important signature scheme was developed by ElGamal in 1985 [79]–[80]. This is motivated by our discussion of the ElGamal cryptosystem in Section 3.3. The following scheme is more specific to digital signatures than the RSA scheme which, as we have seen, can be used as both a public-key cryptosystem and a signature scheme. Moreover, the following scheme is a randomized algorithm, whereas the RSA signature scheme is deterministic.

Typically a hashing of the message before signing is executed in what follows, but we will eliminate this for increased clarity and simplicity of presentation.

◆ ElGamal Signature Scheme

Setup Stage: Alice wants to sign and send a message to Bob for verification. First Alice engages in ElGamal key generation as described on page 67, so her public key is (p, α, y), α being a primitive root modulo a large random prime p (with intractable DLP in $\mathbb{F}_p$) and her private key is a, where $y \equiv \alpha^a \pmod{p}$. The message to be signed is $m \in \mathbb{F}_p^*$.

[7.2]However, in certain circumstances the "sign then encrypt" scheme has been shown to be insecure by Krawczyk [128].

Signing Stage: Alice performs each of the following.

(1) Select a random $r \in (\mathbb{Z}/(p-1)\mathbb{Z})^*$.

(2) Compute $\beta \equiv \alpha^r \pmod{p}$.

(3) Compute $\gamma \equiv (m - a\beta)r^{-1} \pmod{p-1}$.

(4) For $k = (p, \alpha, a, y)$ the signed message $\text{sig}_k(m, r) = (\beta, \gamma)$ is sent to Bob.

Verification Stage: Bob does each of the following.

(5) Using Alice's public key (p, α, y) verify that $\beta \in \mathbb{F}_p^*$ and reject if not.

(6) Compute $\delta \equiv y^\beta \beta^\gamma \pmod{p}$.

(7) Compute $\sigma \equiv \alpha^m \pmod{p}$.

(8) $\text{ver}_k(m, (\beta, \gamma)) = 1$ if and only if $\sigma \equiv \delta \pmod{p}$. Otherwise reject.

By Exercise 7.18, the signature verification is valid. The following is an illustration with small values for pedagogical purposes.

Example 7.7 *Let $p = 3361$, with primitive root $\alpha = 22$. Alice selects $a = 111$ as her private key and computes $\alpha^a \equiv 22^{111} \equiv 174 \pmod{3361}$. Thus, her public key is $(p, \alpha, y) = (3361, 22, 174)$. If $m = 119$, and she chooses $r = 331$, then she computes $\beta \equiv 22^{331} \equiv 1218 \pmod{3361}$. Then she computes*

$$\gamma \equiv (m - a\beta)r^{-1} \equiv (119 - 111 \cdot 1218) \cdot 331^{-1} \equiv 2891 \pmod{3360},$$

and sends $\text{sig}_k(119, 333) = (\beta, \gamma) = (1218, 2891)$ to Bob. First Bob verifies that $\beta \in (\mathbb{Z}/p\mathbb{Z})^*$, then computes,

$$\delta \equiv y^\beta \beta^\gamma \equiv 174^{1218} 1218^{2891} \equiv 74 \equiv 22^{119} \equiv \alpha^m \pmod{3361},$$

so Bob accepts the signature as valid. (See Exercises 7.19–7.22.)

How secure is the ElGamal scheme? Suppose that Mallory tries to forge Alice's signature on m by choosing a random $r_1 \in (\mathbb{Z}/(p-1)\mathbb{Z})^*$ and computing $\beta' \equiv \alpha^{r_1} \pmod{p}$. Mallory is now in the position of having to compute $\gamma' \equiv (m - a\beta')r_1^{-1} \pmod{p-1}$. However, if the DLP in $\mathbb{F}_p$ is intractable, then this computation is infeasible so only a guess at the value of γ' is possible with a probability of success being $1/p$. For large p, this is insignificant.

As noted prior to the description of the ElGamal scheme, we purposely did not hash the message. However, one must hash the message or else Mallory can forge a signature on a random message as described in Exercise 7.23. Nevertheless, even this production of a valid forged signature does not allow Mallory to forge a signature of his own choosing, without first solving the discrete log problem. Hence, this type of forgery is not a serious threat to the security of the

ElGamal scheme. However, if step (5) is not enforced, then Mallory can forge certain signatures of his own choosing if he has a previous legitimate message signed by Alice, as demonstrated in Exercise 7.24. For attacks that break the scheme based upon "poor use", see Exercises 7.25–7.26.

In 1991, a variant of the ElGamal scheme was derived by Schnorr from his identification protocol, studied in Section 7.1, and is described in the same paper [199] where he describes identification. The original idea for developing signature schemes from identification protocols is in Fiat and Shamir [88].

◆ Schnorr Signature Scheme

Setup Stage: The goal is for Alice to sign a binary message $m \in \mathcal{B}$ (bit-strings of arbitrary length) to be verified by Bob. As in the setup stage for the Schnorr identification scheme, p is a large prime such that the DLP in $\mathbb{F}_p^*$ is intractable, q is a large prime divisor of $p - 1$, and $\alpha \in \mathbb{F}_p^*$ such that $\text{ord}_p(\alpha) = q$. Also, here we will need a cryptographic hash function[7.3] $h : \mathcal{B} \mapsto \mathbb{F}_q$. Alice's private key is the natural number $e \leq q - 1$ and her public key is (p, q, α, y), where $y \equiv \alpha^{-e} \pmod{p}$. Set $k = (p, q, \alpha, y, a)$ for sig_k and ver_k defined below.

Signing Stage: Alice performs the following.

(1) Select a random $r \in \mathbb{N}$ such that $r \leq q - 1$.

(2) Compute $\beta \equiv \alpha^r \pmod{p}$.

(3) Compute $h \equiv h(m) \pmod{q}$.

(4) Compute $\gamma \equiv r + eh \pmod{q}$ and send $\text{sig}_k(m, r) = (\beta, h, \gamma)$ to Bob.

Verification Stage: Bob executes the following steps.

(5) Obtain Alice's public key (p, q, α, y).

(6) Compute $\delta \equiv \alpha^\gamma y^h \pmod{p}$.

(7) $\text{ver}_k(\beta, h, \gamma) = 1$ if and only if $\delta \equiv \beta \pmod{p}$, in which case he accepts the signature, and rejects otherwise.

By Exercise 7.27, step (7) does indeed verify Alice's valid signature. As usual, the following is a small illustration in terms of the values.

Example 7.8 *Let* $p = 1319$, $\alpha = 169$, $q = 659$, *and* $m = 1110001111$. *(Note that α has order q since $\alpha = 13^2$ where 13 is a primitive root modulo p and $p - 1 = 2q$.)* *If Alice's private key is* $e = 5$, *then she computes*

$$y \equiv \alpha^{-e} \equiv 169^{-5} \equiv 883 \pmod{1319}.$$

[7.3] A cryptographic hash function that depends on a private key, called a *key-dependent one-way hash function*, is often called a *message authentication code* (MAC). From this perspective MACs could be called *signatures* since the security provided by the private key will ensure a low probability of being altered or forged.

She selects $r = 7$, computes $\beta \equiv \alpha^r \equiv 169^7 \equiv 355 \,(\text{mod } 1319)$, $h(m) = 119$, and $\gamma \equiv r + eh \equiv 7 + 5 \cdot 119 \equiv 602 \,(\text{mod } 659)$. Then the signature $\text{sig}_k(m, r) = (\beta, h, \gamma) = (355, 119, 602)$ is sent to Bob. After obtaining Alice's public key $(p, q, \alpha, y) = (1319, 659, 169, 883)$, he computes

$$\delta \equiv \alpha^\gamma y^h \equiv 169^{602} \cdot 883^{119} \equiv 355 \equiv \beta \,(\text{mod } 1319),$$

so he accepts the signature. (See Exercises 7.28–7.31.)

Features of the Schnorr scheme include the fact that the most expensive of the computations, the exponentiation modulo p, may be done in a preprocessing stage offline. Also, use of the subgroup of order q allows for smaller signatures (at the same level of security) as those generated by the ElGamal scheme or by the RSA scheme.

We close this section with a discussion of the first *Digital Signature Algorithm* (DSA) recognized by any government. In August of 1991, the U.S. government's *National Institute of Standards and Technology* (NIST) proposed a new DSA, and in May of 1994, it became the U.S. *Federal Information Processing Standard* (FIPS 186).[7.4] DSA is similar to both the ElGamal and Schnorr schemes studied above, and it is a digital signature scheme with appendix. However, in 1997, NIST solicited comments on augmenting FIPS 186 with other digital signature schemes, most notably the RSA and elliptic curve schemes.[7.5] On December 15, 1998 NIST announced the approval of FIPS 186-1 as an interim final standard for a new *Digital Signature Standard* (DSS), and this included an RSA signature scheme, but not an elliptic curve scheme. On February 15, 2000, NIST announced the approval of FIPS 186-2, superseding FIPS 186-1, and including three FIPS-approved algorithms: (1) DSA; (2) RSA (as specified in ANSI X9.31); and (3) ECDSA. For the sake of simplicity of presentation, we will provide only the original DSA here.

One may think of the DSA as playing a role analogous to that played by DES (see page 22). The governmental plans for DSA included electronic mail, cash transactions, data exchange, software distribution, and data storage, among others.

◆ Digital Signature Algorithm (DSA)

Setup Stage: Alice selects a prime q with $2^{159} < q < 2^{160}$. Then for some nonnegative integer $t \leq 8$, she chooses a prime p with $2^{511+64t} < p < 2^{512+64t}$, such that q divides $p - 1$. She chooses an $\alpha \in \mathbb{F}_p^*$ with $\text{ord}_p(\alpha) = q$. A

[7.4]See http://www.nist.gov/.

[7.5]The *Elliptic Curve Digital Signature Algorithm* (ECDSA) is the elliptic curve analogue of the original DSA. In 1999, it was accepted as an *American National Standards Institute* (ANSI) standard: ANSI X9.62, and as a NIST standard in 2000. Its security is based upon the intractability of the elliptic curve DLP for which there is no known subexponential time algorithm (see Footnote 5.7 on page 103). Since elliptic curves are already an optional topic in this text, we will not describe the ECDSA herein. The reader interested in the details may go to http://csrc.nist.gov/publications/fips/ and download the publication FIPS 186-2.

cryptographic hash function $h : \mathbb{F}_q^* \mapsto \mathcal{B}_{160}$ (bitstrings of length 160) is selected. She chooses a private key $e \in \mathbb{N}$ such that $e < q$ and computes $\beta \equiv \alpha^e \pmod{p}$. Her public key is (p, q, α, β) and her private key is e.[7.6]

Signing Stage: Alice performs the following in order to sign a message $m \in \mathbb{F}_q^*$. In what follows, we will assume that any powers of α or β have been reduced modulo p before being used in any congruence modulo q.

(1) Select a random $r \in \mathbb{N}$ such that $r \leq q - 1$.

(2) Compute $\gamma \equiv \alpha^r \pmod{q}$.

(3) Compute $\sigma \equiv r^{-1}(h(m) + e\gamma) \pmod{q}$.

(4) Alice sends m and $\mathrm{sig}_k(m, r) = (\gamma, \sigma)$ to Bob.

Verification Stage: Bob executes the following steps.

(5) Obtain Alice's public data (p, q, α, β).

(6) Compute $\delta_1 \equiv \sigma^{-1} h(m) \pmod{q}$ and $\delta_2 \equiv \sigma^{-1} \gamma \pmod{q}$.

(7) Compute $\delta \equiv \alpha^{\delta_1} \beta^{\delta_2} \pmod{q}$.

(8) $\mathrm{ver}_k(m, (\gamma, \sigma)) = 1$ if and only if $\delta \equiv \gamma \pmod{q}$, in which case Bob accepts, and rejects otherwise. (*By Exercise 7.32, Alice's signature is indeed verified by the verification algorithm in step* (8).)

An advantage of DSA is that in a precomputation stage, the exponentiation of α can be done offline and need not be part of the signature generation. Another positive feature is that DSA has relatively short signatures of 320 bits so the signing can be done efficiently. The security of DSA is based on the DLP which is a reason for working modulo q. For instance, the Silver-Pohlig-Hellman algorithm for computing discrete logs (discussed on page 39) is useless as an attack for sufficiently large prime divisors q of $p - 1$. Hence, DSA essentially relies only on mod q information. Also, note that there are more modular exponentiations (the computationally costly ones) in the ElGamal scheme than in DSA, which is therefore faster.

Some disadvantages of DSA include the fact that it cannot be used for key exchange. Moreover, the modulus at a mere 512 bits can be a drawback for security, so the prime p should actually be chosen such that $2^{1023} < p < 2^{1024}$ for long-term security.

One important feature of DSA that links to Chapter Four is that the Miller-Selfridge-Rabin probabilistic primality test (discussed on page 87) is actually embedded as part on DSA (see FIPS PUB 186, Appendix 2, which can be downloaded at http://www.itl.nist.gov/fipspubs/fip186.htm).

[7.6]The prime q is selected first (and fixed at 160 bits according to FIPS 186), then a search is performed for a prime p (which can be any multiple of 64 between 512 and 1024 bits inclusive) such that q divides $(p - 1)$. The algorithm for generating DSA primes is given in [89].

There are also signatures called *one-time signatures*, which are signature schemes constructed from one-way functions. Only one message can be signed, since otherwise (as with a one-time pad) signatures can be forged. We have sufficient data on signatures to proceed, so we will not study these here. The reader interested in one-time signatures may consult [156].

Exercises

In Exercises 7.13–7.16, use the RSA signature scheme to verify or reject the signature $\text{sig}_k(m)$ *given* m, *the exponent* e, *and RSA modulus* n.

7.13. $(n, \text{sig}_k(m), m, e) = (20497, 15971, 911, 133)$.

7.14. $(n, \text{sig}_k(m), m, e) = (713387, 554189, 96, 511)$.

7.15. $(n, \text{sig}_k(m), m, e) = (583201, 317905, 1111, 611)$.

7.16. $(n, \text{sig}_k(m), m, e) = (1757947, 226213, 9666, 323)$.

7.17. Suppose that we want a variant of the RSA signature scheme that allows Alice to sign a message *without knowing what the message says*. Here is how it is done. After the setup stage in the RSA signature scheme on page 135, we add a *blind message stage* wherein Bob selects a random integer:

$$z \in (\mathbb{Z}/n\mathbb{Z})^*,$$

called the *blinding factor*, and computes

$$s \equiv z^e m \pmod{n},$$

called *blinding* and sends s, called a *blinded message* to Alice. She then computes

$$t \equiv s^d \pmod{n},$$

called a *blind signature*, and sends t to Bob. He computes $tz^{-1} \pmod{n}$, called *unblinding*. Prove that

$$s^d z^{-1} \equiv m^d \pmod{n}.$$

Why is it a bad idea for Alice to sign in this fashion (without additional safeguards)?

(Such blind signature schemes are due to Chaum. See [51]–[53].)

7.18. Prove that the signature verification in step (8) of the ElGamal signature scheme actually does verify Alice's signature.

In Exercises 7.19–7.22, use the ElGamal signature scheme to verify or reject the signature $\text{sig}_k(m, r) = (\beta, \gamma)$ *for the given* (p, α, y).

7.19. $\text{sig}_k(m, r) = \text{sig}_k(207, 11) = (11, 227) = (\beta, \gamma); (p, \alpha, y) = (277, 5, 11)$.

7.20. $\text{sig}_k(m, r) = \text{sig}_k(119, 5) = (236, 183) = (\beta, \gamma); \ (p, \alpha, y) = (409, 21, 190).$

7.21. $\text{sig}_k(m, r) = \text{sig}_k(191, 5) = (330, 37) = (\beta, \gamma); \ (p, \alpha, y) = (769, 11, 711).$

7.22. $\text{sig}_k(m, r) = \text{sig}_k(911, 11) = (968, 765) = (\beta, \gamma); \ (p, \alpha, y) = (1021, 10, 326).$

7.23. In the ElGamal scheme Mallory can forge a signature as follows. Suppose that Mallory selects

$$r_1, r_2 \in (\mathbb{Z}/(p-1)\mathbb{Z})^*.$$

He then computes

$$\beta_1 \equiv \alpha^{r_1} y^{r_2} \pmod{p}$$

and

$$\gamma_1 \equiv -\beta_1 r_2^{-1} \pmod{p-1}.$$

Prove that (β_1, γ_1) is a valid signature for the message

$$m_1 \equiv \gamma_1 r_1 \pmod{p-1}.$$

7.24. Mallory can forge certain signatures in the ElGamal scheme if he has intercepted a previous legitimate signature (β, γ) by Alice for a message m, as follows. Suppose that Mallory is lucky and $m^{-1} \pmod{p-1}$ exists, and Mallory chooses a message m_1 to forge. Mallory computes both congruences $t \equiv m_1 m^{-1} \pmod{p-1}$ and $\gamma_1 \equiv t\gamma \pmod{p-1}$. By the Chinese remainder theorem, he can also compute a solution $x = \beta_1$ to the congruences: $x \equiv \beta t \pmod{p-1}$ and $x \equiv \beta \pmod{p}$. Prove that (β_1, γ_1) is a valid signature for m_1, if step (5) in the ElGamal scheme is ignored.

7.25. Suppose that Alice is careless in her use of the ElGamal scheme and she leaks knowledge of the value of r. Show how Mallory can compute her private key a from this knowledge, resulting in a total break of the system.

7.26. Suppose that Alice decides to use the same value of r for signing two different messages in the ElGamal scheme. Show how Mallory can use this knowledge to find r and break the system (see Exercise 7.25).

7.27. Prove that step (7) of Schnorr's signature scheme verifies Alice's valid signature.

In Exercises 7.28–7.31, use the Schnorr signature scheme to verify or reject the signature $\text{sig}_k(m, r) = (\beta, \gamma)$ for the given values.

7.28. $(p, q, \alpha, y) = (1489, 31, 132, 1291), \ (\beta, h, \gamma) = (1401, 1011, 9).$

7.29. $(p, q, \alpha, y) = (1657, 23, 1220, 1456), \ (\beta, h, \gamma) = (913, 101, 6).$

7.30. $(p, q, \alpha, y) = (1871, 17, 81, 693), \ (\beta, h, \gamma) = (1273, 11, 10).$

7.31. $(p, q, \alpha, y) = (3191, 29, 1079, 427), \ (\beta, h, \gamma) = (1217, 21, 25).$

7.32. Prove that step (8) in the DSA signature scheme actually verifies Alice's (valid) signature.

7.3 Digital Cash and Electronic Commerce

Ah, take the cash and let the credit go, nor heed the rumble of a distant drum!

Edward Fitzgerald (1809–1883) English scholar and poet

Increasingly in the modern world, doing business by computer — electronic commerce — is becoming a more prevalent part of our everyday lives. One such means is the use of *digital cash systems* which consist of a set of protocols for transferring money in its various forms, most often for the purchase of goods and services. One may think of digital cash as an emulation of the underlying "real" money via digital data. However, unlike "hard" cash, digital cash is easily copied since essentially electronic files are exchanged. Moreover, we don't want someone such as Mallory to hack into our transaction and steal our credit card number or banking data and wipe out our savings, for instance. Therefore, we need to ensure that safeguards are built into our electronic transactions. What characteristics should a safe electronic commerce scheme possess? The first involves the title of this chapter and the contents of Section 7.2.

◆ **Authenticity**: Entities engaged in e-commerce must be protected from impersonation in terms of both their identities and digital signatures.

◆ **Integrity**: Any data exchanged during a transaction by legitimate entities cannot be altered by an adversary.

◆ **Offline Payments**: The protocol involving transfer of digital cash for purchases is done offline, which means that a third party (such as a bank or credit card company) are not required in the transfer between payer and payee.

◆ **Security**: Information such as credit card numbers and bank card PINs are protected and digital cash cannot be copied and reused.

◆ **Untraceability**: The privacy of the legitimate entities engaged in e-commerce transactions is protected and there is no means of tracing entities via their transactions.

In [177], Okamoto and Ohta gave a list of properties of an *ideal* cash system which includes the above. Essentially what makes digital cash, and therefore e-commerce, possible are public-key cryptography and digital signatures. An example of how digital cash works is that banks, for instance, affix their digital signatures to a digital equivalent of a money order using their private key, which a vendor or customer can verify using the bank's public key. We are now going to look at a couple of widely used digital cash schemes. Before getting into the details of the first system called *ECash*,[7.7] we need to become familiar with some terminology from e-commerce.

[7.7]ECash technology, which is used by numerous banks worldwide, was originated by *Digi-Cash*, a company started in April 1990. However, on February 19, 2002, it was announced that *Infospace Incorporated*, a provider of wireless and internet software and applications services "acquired substantially all of the technology and intellectual property of *ECash Technologies Incorporated*". See http://www.digicash.com.

◆ **Mint** The *Mint* represents any electronic bank that is connected to the Internet and plays an essential but unobtrusive role. The Mint has an RSA modulus n, a private key d that it keeps secure, and a public key e.

◆ **ECash Coins** A coin in ECash is a pair of integers $(m, m^d) \pmod{n}$ where m is the unique identification number of the coin. The value $m^d \pmod{n}$ is the *Mint's signature* or *special digital stamp* on the coin. Coins will typically have different denominations, but for simplicity we will assume that the denomination is the same for each of them. Moreover, to ensure a coin is used only once, the Mint records all identification numbers m in its *Used Coin Database*. If m has been so recorded and someone tries to "spend" the coin, the Mint would inform that the coin is a worthless copy.

◆ **Blinding** This is the technique illustrated in Exercise 7.17 on page 141, wherein Alice blindly signs a message sent by Bob. In the ECash scheme, Alice will be replaced by a bank, for instance, and Bob will be replaced by a bank customer. However, as noted in the exercise, there has to be some built-in safeguards. The cut-and-choose protocol below is one means for ensuring this.

◆ **Money Order** This consists of digital data that contains a given entity's identifying data together with the identification number m of an ECash coin, and its denomination. For instance, for Alice, a $100 money order might be $(\$100, m \pmod{n}, I_A)$ where I_A is a digital data string uniquely identifying her. Thus, m is a "blank" coin awaiting the banks signature m^d to validate it.

◆ **Cut-and-Choose Protocol** The classic cut-and-choose protocol, for dividing anything equitably, is described as follows. Alice cuts the thing in half, Bob chooses a half for himself, and leaves the other half for Alice. For instance, if they both want a piece of an apple, this ensures that Alice will be as fair as possible in her cutting, since Bob chooses first. With the ECash scheme, this is implemented as follows. The customer prepares 100 money orders, say, and each has a different blinding factor in its blinding protocol. The Mint, upon receipt, chooses 99 of the money orders and requests that the customer unblinds them. If the coin and personal identification data are all correct, the Mint signs the 100-th money order and sends it back to the customer. Otherwise, it will not sign the remaining message. The details will be provided below.

◆ **Secret Splitting**[7.8] This is any protocol that takes a message, and divides it into pieces each of which is meaningless in itself, but when pieced back together yields the original message. For instance, given the assistance of Trent, Alice and Bob can split a message m via the following protocol.

(1) Trent generates a random bitstring b with bitlength equal to that of m and creates $b \oplus m = r$, where $\oplus$ is addition modulo 2.

(2) Trent gives b to Alice and r to Bob, with b and r having no meaning unto themselves individually.

[7.8]For the reader interested in more detail and applications of this topic, also called *secret sharing*, we devote the entirety of Section 8.1 to it.

(3) Alice and Bob can piece together the information to retrieve the original message via $b \oplus r = m$.

● ECash™ Scheme

Suppose that Bob wants to withdraw \$100.00 from his account at the Mint and use it to spend at a Vendor to purchase merchandise electronically. The following protocol is enacted.

(1) Bob generates 100 sets of unique digital data strings $\mathcal{S}_j = \{I_{j_k}\}_{k=1}^{100}$ for $1 \leq j \leq 100$ such that *each* I_{j_k} uniquely identifies him with information that the Mint wants to see to ensure his authenticity.

(2) He then engages in a secret splitting protocol (see above) so that, for each $j = 1, \ldots, 100$, the digital data string is split into two pieces $I_{j_k} = \{L_{j_k}, R_{j_k}\}$ with $I_\ell \oplus I_r$ identifying Bob if and only if $\ell = r$, where $\ell, r \in \{j_k\}_{k=1}^{100}$.

(3) Bob prepares 100 money orders for \$100 each:

$$M_j = (\$100, m_j, \{L_{j_k}, R_{j_k}\}_{k=1}^{100}) \quad (1 \leq j \leq 100),$$

where m_j is a randomly generated number (by Bob's computer) that is an ECash coin's identification number with $m_j \neq m_i$ for $j \neq i$.

(4) Bob executes a blinding protocol for each of the money orders by selecting a random $z_j \pmod{n}$ for $1 \leq j \leq 100$ and sends the 100 blinded money orders $(\$100, z_j^e m_j, \{L_{j_k}, R_{j_k}\}_{k=1}^{100})$ to the Mint.

(5) Using a cut-and-choose protocol, the Mint opens 99 of the money orders and checks that the amounts are all the same, \$100, that $m_j \neq m_i$ for $j \neq i$, and that each $L_{j_k} \oplus R_{j_k}$ is a valid identity string. If the Mint sees no evidence of fraud, it (blindly) signs the remaining money order, M_{100} say, and sends the validated money order

$$(\$100, z_{100}^e m_{100}, (z_{100}^e m_{100})^d, \{L_{100_k}, R_{100_k}\}_{k=1}^{100})$$

to Bob, withdrawing \$100 from his account. Otherwise, it does not, and informs of the problem.

(6) Bob unblinds to get the ECash coin (m_{100}, m_{100}^d), which he can now spend with a Vendor via the money order M_{100}.

(7) The Vendor verifies the Mint's signature by computing $(m_{100}^d)^e = m_{100}$. Then the Vendor gives Bob a random 100-bit binary string $(b_1 b_2 \ldots b_{100})$, and requests that Bob reveal L_{100_k} if $b_k = 1$, and R_{100_k} if $b_k = 0$ for each of $k = 1, 2, \ldots, 100$, which Bob does.

(8) The Vendor sends the money order to the Mint for verification from its database.

(9) The Mint checks its used coin database to ensure that m_{100} is not there. If it is not, then the Mint deposits \$100 into the Vendor's account, and records m_{100} in its used coin database along with the identity string selected by Bob via the binary string in step (7). The Vendor then sends the goods to Bob along with a receipt.

(10) If m_{100} is in the used coin database, the Mint rejects the money order. Then it compares the identity string on the bogus money order with the stored identity string attached to m_{100}. If they are the same, then the Mint knows the Vendor duplicated the money order. If they differ, then the Mint knows that the entity who gave it to the Vendor must have copied it. Given that the coin (m_{100}, m_{100}^d) was spent with another Vendor, then that Vendor gave Bob a different binary string. The Mint compares the differing strings until it finds a position where the bits differ, say the i-th position. This is where one Vendor asked Bob to open L_i and the other asked Bob to open R_i. Thus, then Mint forms $L_i \oplus R_i$, revealing Bob's identity.

The above scheme ensures anonymity for Bob, as a legitimate user. When he spends the coin, the Mint must honour it since the Mint's signature is on it. However, since it is unable to recognize the specific coin, given that it was blinded when signed, the Mint does not know who made the payment. However, if Bob is not a legitimate user and tries to spend the coin twice, called *double spending*, then the Mint can detect him in step (10). The attentive reader will have noticed that we assumed that the binary strings were different in step (10) if Bob is illegitimate. This is not 100% certain but the probability that they *are* the same is 1 in 2^{100}, which is *extremely* unlikely. Thus, there is no traceability of legitimate entities via their transactions, which means that the ECash scheme satisifes the property *untraceability*, which means that we have *anonymous digital cash*. Moreover, step (10) tells us that the coins cannot be copied and reused. Since the Mint keeps its signature d secure, as well as identity data, then the ECash scheme has the *security* property. The scheme also satisifes the *integrity* property since the scheme is based upon the security of RSA, which we have seen to be valid when properly implemented. Step (10) also verifies that we have the property of *authenticity* since Bob is protected from impersonation. Of the properties discussed at the outset of this section, this leaves only the offline property to discuss. Since an illegitimate user can be identified in step (10), then it is not necessary to check the coins immediately since a cheater would be identified later. Thus, the offline property exists for the scheme, since the Vendor does not have to check at the time of payment (online), but rather can do so later (offline).

We have not discussed such things as what happens if there is a system crash, or hard disk crash, in the middle of a payment attempt over the Internet, for instance. However, it turns out that there is a special recovery protocol executed between Bob and the Mint that allows all the coins that have been withdrawn by Bob to be reconstructed. These reconstructed coins can be redeemed at the

Mint (but only those coins not already in its used coin database). This is called the *recovery* property, which the ECash scheme possesses. It turns out that recovery of ECash coins can be accomplished over the Internet with the click of a button.

There is also a method for setting up an ECash account at the Mint that we have not addressed. First the Mint gives Bob a PIN P and an account number N. At home Bob installs the Mint's secure software on his computer and generates his own RSA public/private key pair (e_B, d_B). Then he registers his account with the Mint by sending $(N^e, e_B^e, P^e) \pmod{n}$. The private key d_B can be password-encrypted and stored on Bob's hard disc. Transactions with the Mint can now begin.

Another aspect that our simplified version of the ECash scheme did not address is the issue of different coin denominations. The ECash scheme uses a different RSA public exponent for each denomination, but the same RSA modulus n for each of them. Then the above ECash scheme is executed in parallel for as many iterations necessary to withdraw the required amount.

Here is a summary ECash scheme.

● ECash Withdrawal Scheme in Brief

(1) Bob's computer generates a random coin identification numbers corresponding to an amount of ECash required.

(2) Bob blinds the identification numbers and sends it and his identity data to the Mint.

(3) The Mint, upon verification of validity, encrypts the identification numbers with its signature to form valid coins, debits Bob's account, and sends the coins to Bob.

(4) Bob unblinds the signed coins.

(5) The coins are stored in Bob's computer.

● ECash Spending Scheme (Offline) in Brief

(1) Bob's computer collects the total number of ECash coins required.

(2) Bob sends the coins to the Vendor.

(3) The Vendor sends the coins to the Mint, and sends the goods to Bob.

(4) The Mint verifies the validity of the coins and credits the Vendor's account.

The ECash scheme is an *anonymous cash* scheme since the use of (legitimate) ECash is untraceable as is the case with real paper cash. The other type of digital cash is *identified digital cash* which is similar to the use of a credit card that allows the bank to track the money as it moves through the system.

However, users desire the untraceabilty property, so digital cash systems aim for anonymity. The roots of ECash can be found in the works of Chaum (see [51] and [54], for example), who invented the notion of digital coins and the basic protocols for digital cash. There are numerous alternative approaches to the ECash scheme that we will not address (see [87] and [200], for instance). There is one more digital cash scheme that is partly based upon Schnorr's signature scheme, studied in Section 7.2, and which the inventor argues to be the model most suited for the Internet. We close this section with a discussion of it. However, due to the relative complexity of the scheme, this is to be considered optional material.

The following scheme is due to Stefan Brands (see [44]–[45]). We need to set the stage for the scheme with some terminology. First we employ a new cast of characters. We will let Alice play the role now (to give Bob a rest), replace the Mint by the Bank, and the Vendor by the Merchant. For simplicity of presentation, we assume that there is only one denomination of coin, which will be, in this scheme, a six-tuple of integers. The details are given below.

● ☞ Brands' Digital Cash Scheme

Setup Stage: The Bank performs the following steps.

(1) Choose a large prime p such that $(p-1)/2 = q$ is also prime, and select α to be the square of a primitive root modulo p. Also, we assume that the DLP in $(\mathbb{Z}/p\mathbb{Z})^*$ is intractible.

(2) Choose two random $x_1, x_2 \in (\mathbb{Z}/q\mathbb{Z})^*$, compute $g_1 \equiv \alpha^{x_1} \pmod{p}$ and $g_2 \equiv \alpha^{x_2} \pmod{p}$, then discard x_1, x_2. (Note that by (1), $g_1 \equiv g_2 \pmod{p}$ if and only if $x_1 \equiv x_2 \pmod{q}$.) Make (α, g_1, g_2) public.

(3) Select a random secret $x \in (\mathbb{Z}/q\mathbb{Z})^*$ and compute,

$$h \equiv \alpha^x \pmod{p}, \quad h_1 \equiv g_1^x \pmod{p}, \quad \text{and} \quad h_2 \equiv g_2^x \pmod{p}.$$

Then (h, h_1, h_2) is the Bank's public key and x is the Bank's private key.

(4) Choose two public cryptographic hash functions,

$$H_1 : ((\mathbb{Z}/p\mathbb{Z})^*)^5 \mapsto (\mathbb{Z}/q\mathbb{Z})^* \quad \text{and} \quad H_2 : ((\mathbb{Z}/p\mathbb{Z})^*)^4 \mapsto (\mathbb{Z}/q\mathbb{Z})^*.$$

(5) The Merchant registers identification number M with the Bank.

Opening Alice's Account:

(1) Alice generates $e_1, e_2 \in (\mathbb{Z}/q\mathbb{Z})^*$ at random and computes

$$A \equiv g_1^{e_1} g_2^{e_2} \not\equiv 1 \pmod{p},$$

which she sends to the bank.

(2) The Bank stores (A, I_A, N_A) in its database where I_A is a digital data string uniquely identifying Alice and N_A is her account number.

Identification Protocol:[7.9] When Alice wishes to withdraw coins from her account, she must first identify herself to the Bank's satisfaction.

(1) Alice generates $f_1, f_2 \in (\mathbb{Z}/q\mathbb{Z})^*$, at random, computes $f \equiv g_1^{f_1} g_2^{f_2} \pmod{p}$, and sends f to the Bank.

(2) The Bank generates a random $k \in (\mathbb{Z}/q\mathbb{Z})^*$ (the challenge), and sends it to Alice.

(3) Alice computes $\ell_1 \equiv f_1 + ke_1 \pmod{q}$ and $\ell_2 \equiv f_2 + ke_2 \pmod{q}$ (the responses) and sends (ℓ_1, ℓ_2) to the Bank.

(4) The Bank accepts her response if and only if $fA^k \equiv g_1^{\ell_1} g_2^{\ell_2} \pmod{p}$.

   (*By Exercise 7.33, step* (4) *does indeed identify Alice uniquely.*)

(5) If the Bank accepts her response in step (4), it sends her an identification number $y_1 = A^x$.

   (*By completing step* (5), *Alice proves that she owns A. She does this by a proof of knowledge of (e_1, e_2).*)

Coin Withdrawal Protocol: For simplicity, we assume that Alice wants to withdraw only one coin, a six-tuple of integers (X, Y, Y_1, Y_2, Y_3, Z), which we will now see how to construct.

(1) The Bank chooses a random $w \in (\mathbb{Z}/q\mathbb{Z})^*$, computes $y_2 \equiv \alpha^w \pmod{p}$, $y_3 \equiv A^w \pmod{p}$, and sends (y_2, y_3) to Alice.

(2) Alice selects three random integers $z_1 \in (\mathbb{Z}/q\mathbb{Z})^*$ and $z_2, z_3, \in \mathbb{Z}/q\mathbb{Z}$. She computes the following where all congruences are modulo p,

$$y_1' \equiv A^{z_1}, \quad Y_1 \equiv y_1^{z_1}, \quad Y_2 \equiv y_2^{z_2} \alpha^{z_3} \quad \text{and} \quad Y_3 \equiv y_3^{z_1 z_2} A^{z_1 z_3}.$$

Now she computes $s_1, s_2, t_1, t_2, u_1, u_2 \in (\mathbb{Z}/q\mathbb{Z})^*$ such that:

$$e_1 z_1 \equiv s_1 + s_2 \pmod{q}, \ e_2 z_1 \equiv t_1 + t_2 \pmod{q}, \ z_1 \equiv u_1 + u_2 \pmod{q}.$$

Then she calculates

$$X \equiv g_1^{s_1} g_2^{t_1} A^{u_1} \pmod{p} \text{ and } Y \equiv g_1^{s_2} g_2^{t_2} A^{u_2} \pmod{p}.$$

(*Note that by Exercise 7.34, $XY \equiv y_1' \pmod{p}$, which is Alice's blinded identity.*)

[7.9]In the Brands scheme this step is often called the *representation problem step*. It turns out that the Brands scheme is built on the Schnorr signature scheme and the representation problem which is given as follows. In a group of prime order G with generators $(g_1, g_2, \ldots, g_s)$ for $s \geq 2$, $g_j \in G$, and a given $h \in G$, find a representation such that $h = \prod_{j=1}^{s} g_j^{b_j}$ for $b_j \geq 0$. The reader will note that this is related to a discrete log problem and so is difficult without knowledge of the b_j.

(3) Alice computes a challenge

$$c_1 = H_1(y_1', Y_1, Y_2, Y_3, X),$$

and blinds it with $c \equiv c_1 z_2^{-1} \pmod{q}$, which she sends to the Bank.

(4) The Bank sends a response $r \equiv xc + w \pmod{q}$ to Alice, and debits her account. Alice accepts r if and only if

$$\alpha^r \equiv h^c y_2 \pmod{p} \text{ and } A^r \equiv y_1^c y_3 \pmod{p}.$$

(*See Exercise 7.35.*)

(5) Alice computes $Z \equiv r z_2 + z_3 \pmod{q}$. Her coin is

$$C = (X, Y, Y_1, Y_2, Y_3, Z),$$

which she can now spend.

(*Essentially* (Y_1, Y_2, Y_3, Z) *is the Banks's signature on* (X, Y)*, so we write* $(X, Y, \mathrm{sig}(X, Y))$ *for* C *in what follows for simplicity.*)

Spending Protocol: Alice wishes to purchase some goods from the Merchant.

(1) She sends the Merchant her coin $(X, Y, \mathrm{sig}(X, Y))$.

(2) The Merchant verifies that $XY \neq 1$ (see Exercise 7.37), then sends a challenge
$$c = H_2(X, Y, M, T_M)$$
to Alice, where T_M is a timestamp with the date and time on it.

(3) Alice computes the responses,

$$r_1 = s_1 + s_2 c \pmod{q}; \quad r_2 \equiv t_1 + t_2 c \pmod{q}; \text{ and } \quad r_3 \equiv u_1 + u_2 c \pmod{q}$$

which she sends to the Merchant.

(4) The Merchant verifies that $g_1^{r_1} g_2^{r_2} A^{r_3} \equiv XY^c \pmod{p}$ holds and if so accepts the payment. (*See Exercise 7.36.*)

(5) The Merchant sends $(X, Y, \mathrm{sig}(X, Y), T_M, c, r_1, r_2)$ to the Bank.

(6) The Bank verifies the signature $\mathrm{sig}(X, Y)$, that no double spending has occurred, and that c and r_1, r_2 are valid challenge response protocols. If all holds true, the Bank pays the Merchant.

Deposit Protocol:

(1) The merchant sends $(X, Y, \mathrm{sig}(X, Y), T_M, c, r_1, r_2)$ to the Bank.

(2) The Bank checks that $\text{sig}(X, Y)$ is valid, that the coin has not already been spent, and that the Merchant's challenge and Alice's responses r_1, r_2 are valid. If all of this holds true, the Bank pays the Merchant.

As with the ECash scheme, Brands' scheme requires the customer to reveal enough information without revealing identity. However, if Alice tries to double-spend, Exercise 7.38 tells us that she will be identified and charged with fraud. However, if she is legitimate, then her identity is not revealed. Thus, Brands' scheme provides anonymity to legitimate entities since Alice never has to provide identification, as is the case with paper money. As with the ECash scheme, Brands' scheme also ensures untraceability of legitimate entities. However, as shown in Exercise 7.38, the Bank can identify a double-spender. Brands' scheme possesses authenticity since the scheme is secure against impersonation due to the fact that it is based upon the intractability of the DLP.

One of the major advantages of Brands' method is that it does not use any cut-and-choose protocol or secret splitting, which are time-costly. Thus, with Brands' scheme, the Bank does not have to engage in such protocols. Moreover, since Brands' scheme is based upon the DLP, then the integer factoring problem does not come into play as it does with the use of an RSA modulus, used in the ECash scheme. Now we have a look at the parameters involved in Brands' method.

Since g_1, g_2 are made public, and $A \equiv g_1^{e_1} g_2^{e_2} \pmod{p}$, then g_1, g_2 must be chosen large enough to make it computationally infeasible for an adversary to compute a representation of Alice's account. Nevertheless, the Bank must be able to accommodate all its customers with the pairs (e_1, e_2), so the Bank has to ensure that g_1, g_2 are not chosen so large as to prevent this. The exponents e_1, e_2 are in $(\mathbb{Z}/q\mathbb{Z})^*$ and Brands suggests that q should have 140 bits while e_1, e_2 should be around 70 bits. With such large e_j, Alice would be better served if her e_j were stored on a smart card to save both her personal memory (it's hard to remember a 70-bit integer) and her computer memory where Mallory might be able to find them and impersonate her. As usual, the system is only as secure as the implementation and security of the private/secret keys.

Although Brands' scheme is relatively complicated mathematically, most of the work is required to preserve both anonymity and to prevent double-spending. Given the above advantages, the consensus is that Brands' scheme is preferable to the ECash scheme in most implementations. There exists a variant of Brands' scheme that can be used on smart cards (see [87]) and this is generally regarded as the best implementation. In this variant, one can store a *counter* and amounts can be withdrawn from the card[7.10] up to the limit set by the counter. This is an alternative to the digital coins method presented in the ECash and Brands' schemes. The *CAFE* (Conditional Access for Europe) project (see [33]) uses a

---

[7.10]*Stored-value cards* (a type of *smart card* — see Section 9.4 on page 198) have an embedded microchip that can be preprogrammed with a specific monetary value or counter. For spending, the card is swiped through a reader that debits the amount of the purchase from the counter on the card and credits the vendor's account. Telephone callings cards provide an example.

combination of these methods.[7.11] A smart card, for instance, with a counter can be used to withdraw but at the same time one "virtual coin" is considered to be spent, although these virtual coins have no value in and of themselves. If a card has reached its limit and is reloaded, the virtual coins are reinstated with unspent status at the same time. As we have seen in the above two schemes, the use of digital coins ensures that only the Bank can create them, so adversaries are forced to copy existing coins created by the Bank, and a used coin database detects attempts to spend these illegal copies. Thus, the best way to ensure security is the use of digital coins.

We have not discussed the need for a means of standardizing the exchange of financial information in electronic transactions. If credit card companies, for example, are to exchange information with banks and other financial institutions, they must be assured of security, privacy, inegrity, as well as authenticity. This was accomplished in May 1997 when Visa and Master Card completed SET™ (Secure Electronic Transmission), that provides an infrastructure for certifying public keys, with the goal of protecting payment cards used in Internet e-commerce against fraud. We will study SET and related features in Chapter 8, where public-key infrastructure (PKI) will be discussed along with related concepts. SET is a PKI application that is vital since the use of digital signatures, for instance, in e-commerce is realistic only if there is an infrastructure for certifying public keys. Digital signatures opened the door to digital cash, and thereby e-commerce, which would be impossible without them.

Exercises

7.33. Prove that step (4) of the identification protocol in the Brand scheme identifies Alice uniquely.

7.34. Prove that in step (2) of the withdrawal protocol in Brands' scheme

$$XY \equiv y_1' \pmod{p},$$

where y_1' is her blinded identity.

7.35. Explain why Alice accepts the Bank's response in step (4) of the withdrawal protocol in Brands' scheme only under the conditions given therein.

7.36. Verify that step (4) of the Spending protocol in the Brands' scheme holds for valid responses by Alice.

7.37. In step (2) of the spending protocol of Brands' scheme, show that if Alice is legitimate, then $XY \not\equiv 1 \pmod{p}$.

☆ 7.38. Show that if Alice tries to double-spend in Brands' digital cash scheme, then she will be identified by the Bank (and therefore charged with fraud).

[7.11]The CAFE project was born in December of 1992, but died in 1996 due to an inability to come up with a workable system that was acceptable. The goal was to use *electronic wallets* that could be used for digital payments, including guaranteed security and anonymity.

Chapter 8

Key Management

If the doors of perception were cleansed everything would appear to man as it is, infinite.

William Blake (1757–1827) English Poet

A cryptographic scheme is only as strong as the security of its keys. This chapter is devoted to key management — the secure generation, distribution, and storage of keys. These are aspects of public key infrastructure — protocols, services, and standards — used in concert as an edifice to support secure public-key cryptography, which we will discuss in Section 8.3. However, we already have been introduced to a means of key recovery for digital cash schemes discussed in Section 7.3. Thus, we begin with a section that looks at more general such schemes to be used in public-key cryptosystems.

8.1 Secret Sharing

We became acquainted with a secret sharing scheme in Section 7.3, where we discussed *secret splitting* between two entities on page 144. This is the simplest form of secret sharing, and can easily be generalized to $N \in \mathbb{N}$ entities by simply breaking up the message into N distinct pieces m_j so that $m = m_1 \oplus m_2 \oplus \cdots \oplus m_N$. We now look at some other secret sharing schemes.

The prime motivations for secret sharing of cryptographic keys are to ensure schemes for key recovery (of lost keys) and to address the problem of authorization of key use. Secret sharing assures that the key can be used only when certain groups of authorized participants are present and agree. This protects the key from misuse and allows for security of backup copies of keys.

The origins of secret sharing may be traced to the 1979 (independent) work of Blakely [27] and Shamir [202]. We begin with a secret sharing scheme due to Shamir.

Suppose that we have a bank vault and no single individual can be trusted with the combination. One way is to split the combination among three entities, any two of whom can open the vault. This notion can be formalized.

Definition 8.1 (Threshold Schemes)[8.1]

Let $t, w \in \mathbb{N}$ such that $t \leq w$. A (t, w)-threshold scheme is a method of sharing a message m among a set $\mathcal{S}$ of w entities such that any subset of t of them can recover m, but no smaller subset can do so. If $\mathcal{P}_j \in \mathcal{S}$ knows m_j, then m_j is called a share *(sometimes called a* shadow*) for the participant $\mathcal{P}_j$.*

The secret splitting discussed in Section 7.3 is an example of a $(2, 2)$ threshold scheme. The following also goes by the name of the *Lagrange Interpolation Scheme*. In fact, the reader must be familiar with the Lagrange interpolation formula, Theorem C.40, given in Appendix C on page 222.

◆ **Shamir's Threshold Scheme**

Trent distributes shares of m to $w \in \mathbb{N}$ participants of whom any $t \leq w$ of them will be able to recover m.

Setup Stage: Trent performs the following.

(1) Choose a prime $p > \max(m, w)$, where p is public, and set $m_0 = m \in \mathbb{Z}/p\mathbb{Z}$.

(2) Select $t - 1$ random integers c_j for $j = 1, 2, \ldots, t - 1$ and set

$$p(x) \equiv m + \sum_{j=1}^{t-1} c_j x^j \equiv \sum_{j=0}^{t-1} c_j x^j \pmod{p},$$

where $c_0 = m$.

(3) Compute $p(x_k) \equiv m_k \pmod{p}$ for distinct integers $x_k \leq p - 1$ and securely distribute the share (x_k, m_k) to participant $\mathcal{P}_k$ for $1 \leq k \leq w$.

Pooling Shares: Without loss of generality, suppose a group of t participants $\mathcal{P}_k$ for $1 \leq k \leq t$ get together and plug their shares into the Lagrange interpolation formula:

$$f(x) = \sum_{k=1}^{t} m_k \prod_{\substack{1 \leq \ell \leq t \\ \ell \neq k}} \frac{x - x_\ell}{x_k - x_\ell} = \sum_{k=1}^{t} m_k K_k(x),$$

where

$$K_k(x) = \prod_{\substack{1 \leq \ell \leq t \\ \ell \neq k}} \frac{x - x_\ell}{x_k - x_\ell}.$$

Then by Exercise 8.1,

$$f(x_k) \equiv m_k \pmod{p}, \text{ for } 1 \leq k \leq t$$

[8.1]It can be shown that for any $t \leq w$ and $t > 1$, there exists a (t, w)-threshold scheme (see [127]).

and by Exercise 8.2,

$$p(0) \equiv f(0) \equiv \sum_{k=1}^{t} m_k K_k(0) \equiv m \,(\mathrm{mod}\ p).$$

Example 8.2 *Let $p = 3361$, $m = 3001$, $w = 5$, and $t = 3$ in the Shamir threshold scheme. Select*

$$p(x) = 111x^2 + 256x + 3001.$$

Then compute $p(x_i)$ for $x_1 = 1, x_2 = 2, x_3 = 3, x_4 = 4, x_5 = 5$, since we only need ensure that the x_i are distinct and less than p. We get

$$(x_1, m_1) = (1, 7), (x_2, m_2) = (2, 596), (x_3, m_3) = (3, 1407),$$

$$(x_4, m_4) = (4, 2440), \text{ and } (5, m_5) = (5, 334).$$

we select $i = 1, 3, 5$ for pooling among the three participants and calculate the Lagrange interpolation formula to be,

$$f(x) = -\frac{2473}{8} x^2 + \frac{3873}{2} x - \frac{12963}{8},$$

Since

$$8^{-1} \equiv 2941 \,(\mathrm{mod}\ p), \qquad 2^{-1} \equiv 1681 \,(\mathrm{mod}\ p),$$

then

$$-2473 \cdot 2941 \equiv 111 \,(\mathrm{mod}\ p), \qquad 3873 \cdot 1681 \equiv 256 \,(\mathrm{mod}\ p),$$

and

$$-2941 \cdot 12963 \equiv 3001 \,(\mathrm{mod}\ p).$$

Thus, the participants consider $f(x) \equiv 111x^2 + 256x + 3001 \,(\mathrm{mod}\ p)$, and compute $f(0) \equiv 3001 \equiv m \,(\mathrm{mod}\ p)$ to recover the message. See Exercises 8.3–8.7.

In Shamir's scheme, no fewer than t participants can recover m (a property that makes it an example of what is sometimes called a *perfect threshold scheme*). Thus, knowing only $t - 1$ shares does not give an advantage to a participant over an adversary who does not know any of the shares, other than the size of m, which does not help to recover it. Furthermore, the security of Shamir's scheme does not rely upon the assumed intractability of such problems as the DLP or the IFP. Thus, the scheme is as secure as a one-time pad in the sense that an exhaustive search of all possible shares will reveal to an adversary that *any* message m could be the secret.

There are variations on Shamir's scheme. For instance, suppose that the CEO of a large corporation wants to control the majority of the shares in a scheme for secret sharing the combination to the company vault. Suppose that $t = 7$ shares are required and the CEO has 5 shares while other participants have only 1 share. Therefore, the CEO gets together with two underlings to recover the combination, but without participation by the CEO, it takes 7 participants to recover it.

Another variation on Shamir's scheme is illustrated by the following scenario. Suppose that two corporations A and B hold their securities in the same vault. They wish to create a scheme where 2 participants from corporation A and 3 participants from corporation B hold shares. Here is how they do it. Take a linear polynomial p_1 and a quadratic polynomial p_2, and form their product. Then give $w_1 \geq 2$ employees of corporation A a share $p_1(x_i)$ for $1 \leq i \leq w_1$, and give $w_2 \geq 3$ employees from corporation B a share $p_2(y_i)$ for $1 \leq i \leq w_2$. Then any two participants from corporation A can get together and recover p_1 but not p_2, and any 3 participants from corporation B can get together and recover p_2 but not p_1. Participants from both corporations A and B must act in concert to recover the full combination determined by the product $p_1 p_2$ acting on the individual shares.

Another secret sharing scheme [7] developed in 1983 which uses the Chinese Remainder Theorem, is the following.

◆ **Asmuth-Bloom Threshold Scheme**

This is a (t, w)-threshold scheme for sharing a secret m.

Setup Stage: Each of the following is executed.

(1) Choose a large prime $p > m$, where p is public.

(2) Select pairwise relatively prime values $a_j \in (\mathbb{Z}/p\mathbb{Z})^*$ for $j = 1, 2, \ldots, w$ such that $a_1 < a_2 < \cdots < a_w$.

Share Creation and Distribution: Each of the following is executed.

(1) Select a random value $r \in \mathbb{N}$ such that $m + rp < \prod_{j=1}^{t} a_j$. (One must be careful to select the values so that (the rare event) $m + rp < a_j$ does *not* occur. Otherwise, participant j has the secret.)

(3) Securely distribute shares $s_j \equiv m + rp \,(\text{mod } a_j)$ for $j = 1, 2, \ldots, w \geq t$, together with a_j and p.

Pooling Shares: Any t participants can get together to recover m using the Chinese Remainder Theorem, but less than t cannot.

Example 8.3 *Let $p = 2111$, $t = 3$, $w = 4$, $m = 291$, and select $(a_1, a_2, a_3, a_4) = (1193, 1213, 1217, 1223)$. Then randomly choose $r = 834264$. Since*

$$m + rp = 1761131595 < a_1 \cdot a_2 \cdot a_3 = 1761131653,$$

then we compute the shares $s_1 \equiv 1135 \,(\text{mod } 1193)$; $s_2 \equiv 1155 \,(\text{mod } 1213)$; $s_3 \equiv 1159 \,(\text{mod } 1217)$, and securely distribute them. The participants gather and use the Chinese Remainder Theorem to solve the three congruences to recover $m + rp$, and the secret message via, $1761131595 \equiv 291 \,(\text{mod } 2111)$. See Exercises 8.8– 8.12

We close this section with a secret sharing scheme based upon vector spaces and matrices (see Appendix C for a reminder of the basic notions). The following scheme [27] was introduced in 1979. This is not a (t, w)-threshold scheme.

◆ **Blakely's Secret Sharing Vector Scheme**
The secret message is m_1 to be reconstructed by $t > 2$ participants.

Setup Stage: The following are executed.

(1) Choose a large prime $p > m_1$, where p is made public, and select $m_2, m_3, \ldots, m_t \in \mathbb{F}_p$ at random. Then $m = (m_1, m_2, \ldots, m_t)$ is a point in the t-dimensional vector space $\mathbb{F}_p^t$.

(2) For each $j = 1, 2, \ldots, t$, select $n_1^{(j)}, \ldots, n_{t-1}^{(j)} \in \mathbb{F}_p$ at random and set

$$c_j \equiv m_t - \sum_{i=1}^{t-1} n_i^{(j)} m_i \pmod{p}.$$

(3) Each of the t participants is given the equation for a hyperplane in $\mathbb{F}_p$ as follows,

$$\ell_j \equiv c_j + \sum_{i=1}^{t-1} n_i^{(j)} x_i \pmod{p},$$

for $j = 1, 2, \ldots, t$, where the intersection of the t hyperplanes must be the point m.

Pooling Stage: The participants gather to reconstruct the secret message as follows. They form the equations,

$$\sum_{i=1}^{t-1} n_i^{(j)} x_i - \ell_j \equiv -c_j \pmod{p},$$

for $j = 1, 2, \ldots, t$. In matrix terminology, this translates into the following

$$AX \equiv \begin{pmatrix} n_1^{(1)} & n_2^{(1)} & \cdots n_{t-1}^{(1)} & -1 \\ n_1^{(2)} & n_2^{(2)} & \cdots n_{t-1}^{(2)} & -1 \\ \vdots & \vdots & \vdots & \vdots \\ n_1^{(t)} & n_2^{(t)} & \cdots n_{t-1}^{(t)} & -1 \end{pmatrix} \begin{pmatrix} x_1 \\ x_2 \\ \vdots \\ x_t \end{pmatrix} \equiv \begin{pmatrix} -c_1 \\ -c_2 \\ \vdots \\ -c_t \end{pmatrix} \pmod{p}. \quad (8.4)$$

It follows from Theorem C.38 on page 221 that, if $\det(A) \neq 0$, then there is the unique solution $X = (m_1, \ldots, m_t)$, so the secret m_1 is recovered.

Example 8.5 *Let $p = 409$, $m_1 = 96$ and $t = 3$. If we randomly select $m_2 = 109$ and $m_3 = 208$, then $m = (96, 109, 208)$. We select the following at random:*

$$n_1^{(1)} = 2; \quad n_2^{(1)} = 51; \quad n_1^{(2)} = 25; \quad n_2^{(2)} = 111; \quad n_1^{(3)} = 105; \quad n_2^{(3)} = 308.$$

Then the values in step (2) are:

$$c_1 \equiv m_3 - n_1^{(1)}m_1 - n_2^{(1)}m_2 \equiv 208 - 2 \cdot 96 - 51 \cdot 109 \equiv 183 \,(\mathrm{mod}\ 409),$$

$$c_2 \equiv m_3 - n_1^{(2)}m_1 - n_2^{(2)}m_2 \equiv 208 - 25 \cdot 96 - 111 \cdot 109 \equiv 24 \,(\mathrm{mod}\ 409),$$

$$c_3 \equiv m_3 - n_1^{(3)}m_1 - n_2^{(3)}m_2 \equiv 208 - 105 \cdot 96 - 308 \cdot 109 \equiv 319 \,(\mathrm{mod}\ 409).$$

Thus, the hyperplanes distributed to the participants are:

$$\ell_1 \equiv 183 + 2x_1 + 51x_2 \,(\mathrm{mod}\ 409),$$

$$\ell_2 \equiv 24 + 25x_1 + 111x_2 \,(\mathrm{mod}\ 409),$$

$$\ell_3 \equiv 319 + 105x_1 + 308x_2 \,(\mathrm{mod}\ 409).$$

Plugging these values into (8.4), we get,

$$AX = \begin{pmatrix} 2 & 51 & -1 \\ 25 & 111 & -1 \\ 105 & 308 & -1 \end{pmatrix} \begin{pmatrix} x_1 \\ x_2 \\ x_3 \end{pmatrix} = \begin{pmatrix} -183 \\ -24 \\ -319 \end{pmatrix} = C.$$

then solving for $\det(A) = 269$, *and using Cramer's rule, we get*

$$(x_1, x_2, x_3) = (-49023/269, 19505/269, 945936/269).$$

However, $269^{-1} \equiv 260 \,(\mathrm{mod}\ 409)$ *and*

$$-49023 \cdot 260 \equiv 96; \quad 19505 \cdot 260 \equiv 109; \quad 945936 \cdot 260 \equiv 208,$$

all congruences modulo 409. Thus, $(m_1, m_2, m_3) = (96, 109, 208)$, *and the secret message* $m_1 = 96$ *is retrieved. See Exercises 8.13–8.16.*

It is not guaranteed that $\det(A) \neq 0$ in (8.4), but if we choose p large enough, then it is highly probable that A is indeed invertible. Shamir's method is essentially a special case of the Blakely method since Shamir's method effectively deals with a Vandermonde matrix for A, the determinant of which is zero if and only if some $x_k \equiv x_i \,(\mathrm{mod}\ p)$, but we chose these values to be distinct in step (3) of the algorithm, (see Exercise 8.17). This gives Shamir's method an advantage over Blakely's method. Moreover, Shamir's method clearly requires each participant to have less information in their respective shares.

There are numerous other secret sharing schemes, and related topics not covered herein. The interested reader may consult [17], [28], [50], [64], [108], [129], [156], [198], [208], and [214]. The bottom line is that the motivation for secret sharing is secure key management, about which we will learn more as this chapter unfolds.

Exercises

8.1. In Shamir's threshold scheme, prove: $f(x_k) \equiv m_k \pmod{p}$ for $1 \leq k \leq t$.

8.2. Prove that $p(x) \equiv f(x) \pmod{p}$ in the Shamir threshold scheme.

8.3. Given $g(x) = 444x^2 + 68x + 2856$, with $g(1) = 7$, $g(2) = 1407$, and $g(3) = 334$, which are the same values produced by $f(x)$ in Example 8.2, where $g(x)$ is taken modulo $p = 3361$. Explain why this does not contradict Lagrange's Theorem C.40.

In Exercises 8.4–8.7, use the Shamir threshold scheme with $(t, w) = (3, 5)$ to recover the message $m = p(0)$ with the given values of $p(x)$ for $x_1 = 1$, $x_3 = 3$, and $x_5 = 5$ as done in Example 8.2 modulo the given prime p.

8.4. $(p(x), p) = (561x^2 + 273x + 111, 2707)$.

8.5. $(p(x), p) = (225x^2 + 56x + 207, 4231)$.

8.6. $(p(x), p) = (714x^2 + 541x + 147, 5417)$.

8.7. $(p(x), p) = (69x^2 + 19x + 999, 6991)$.

8.8. Show that in Example 8.3, if we choose $r = 834265$, then we cannot recover the message with the Amuth-Bloom scheme. Why?

In Exercises 8.9–8.12, use the Asmuth-Bloom scheme with $(t, w) = (3, 4)$ to recover the message m with the given values of p, r, a_i for $i = 1, 2, 3$.

8.9. $(p, m, a_1, a_2, a_3, r) = (5039, 519, 5, 7, 7103, 10)$.

8.10. $(p, m, a_1, a_2, a_3, r) = (6379, 679, 5, 706, 7001, 101)$.

8.11. $(p, m, a_1, a_2, a_3, r) = (7103, 3071, 10, 11, 7001, 100)$.

8.12. $(p, m, a_1, a_2, a_3, r) = (8179, 1703, 51, 91, 7156, 515)$.

In Exercises 8.13–8.16, use the Blakely secret sharing scheme with $t = 3$ with each of the given values to show how the secret message m_1 can be recovered using the same technique as in Example 8.5. In each of the following, we are given: $(m_1, m_2, m_3, n_1^{(1)}, n_2^{(1)}, n_1^{(2)}, n_2^{(2)}, n_1^{(3)}, n_2^{(3)}, p)$.

8.13. $(59, 409, 90, 15, 52, 11, 123, 308, 400, 503)$.

8.14. $(77, 88, 99, 85, 95, 103, 204, 305, 406, 607)$.

8.15. $(107, 1, 718, 297, 306, 419, 537, 698, 709, 719)$.

8.16. $(333, 737, 900, 157, 206, 315, 473, 585, 696, 937)$.

8.17. Show that the modular equations $m_k \equiv m + \sum_{j=1}^{t} c_j x_k^j \pmod{p}$ in Shamir's threshold scheme give rise to a matrix equation modulo p, where the matrix having the values x_k^j is the Vandermonde matrix whose determinant is 0 modulo p exactly when some $x_k \equiv x_i \pmod{p}$ for $k \neq i$. (See page 222.) (*Hint: Use (C.37) on page 221.*)

8.2   Key Establishment

The management of a balance of power is a permanent undertaking, not an exertion that has a foreseeable end.

Henry Kissinger (1923–) American politician

A *Key establishment protocol* is a protocol using cryptography to securely establish a shared secret (symmetric) key. One would expect that public-key cryptography would not need such a protocol since public enciphering keys are stored in public databases. However, recall the discussion on page 75, wherein we looked at the more practical use of a hybrid cryptosystem. The reason that they are used in practice is that public-key methods are much slower than those using symmetric keys. Thus, RSA, for example, could be used to transfer symmetric keys, which could then be used for the bulk of the data communication.

Key establishment can manifest itself in two distinct ways. One is called *key distribution* (also called *key transfer*, and *key transport*) where one entity generates a symmetric key and sends it to other entities. For instance, Alice could use Bob's public key to encipher the symmetric key (see page 128). The second method for key establishment is *key agreement* which is a protocol where entities act in concert by contributing to the generation of a symmetric key. For example, the Diffie-Hellman key-exchange protocol discussed on page 49 is such a protocol. Note that the (less accurate) term *key exchange* is often used in the literature since two entities perform an *exchange* of information, the result of which is their *agreement* on a shared key. Key management here is crucial since, in practice, adversaries will likely attack the key management scheme before the cryptographic scheme. Thus, the ease of key management with public-key cryptography may be seen as the primary advantage over symmetric-key cryptography. For instance, in a network having a large number of users, such as that described on page 73, public-key cryptography has a huge advantage in key management.

One of the weaknesses of the aforementioned Diffie-Hellman protocol is that it is susceptible to the man-in-the-middle attack (see pages 27 and 127). One solution to this weakness is the use of digital signatures (see Section 7.1). For example, there is a variant of Diffie-Hellman that employs entity authentication and mutual explicit key authentication, all with entity anonymity guaranteed against eavesdroppers. An instance of such an *authenticated key agreement protocol* (meaning that the key agreement protocol itself authenticates the participants identities) is given as follows as a variant of the Diffie-Hellman exchange. This is due to Diffie, Van Oorschot, and Weiner [75], which evolved from earlier work (see [73, p. 568]).

The following is considered to be a *three-pass variant* of the Diffie-Hellman protocol. Compare this with the discussion of Shamir's three-pass protocol (embedded in steps (3)–(5) of the Massey-Omura cryptosystem on page 71), which itself can be used as a key agreement scheme.

◆ **Station-to-Station Protocol (STS)**[8.2]

Alice and Bob exchange three messages with the goal of authenticated agreement on a symmetric key k.

Background Assumptions:

(a) Alice and Bob possess identification certificates,

$$C(A) = (I_A, \text{ver}_A, \text{sig}_A(I_A, \text{ver}_A)), \quad C(B) = (I_B, \text{ver}_B, \text{sig}_B(I_B, \text{ver}_B)),$$

respectively, where $(\text{sig}_A, \text{ver}_A)$ and $(\text{sig}_B, \text{ver}_B)$ are their respective signing and verification algorithms (see Section 7.1). Also, the signatures in $C(A)$ and $C(B)$ are computed by Trent.

(b) Trent has a signature scheme including a public verification algorithm ver_T.

(c) There is a publicly known prime p and primitive root α modulo p.

(d) E is a symmetric enciphering algorithm (for which a key k will be developed below, and its use with the key denoted by E_k).

Protocol Steps:

(1) Alice selects a random secret $e_A \in (\mathbb{F}_p)^*$ such that $e_A \leq p-2$ and computes $\alpha^{e_A} \pmod{p}$, which she sends to Bob.

(2) Bob chooses a random secret $e_B \in (\mathbb{F}_p)^*$ with $e_B \leq p-2$, calculates

$$k \equiv (\alpha^{e_A})^{e_B} \pmod{p}, \text{ and } s_B = \text{sig}_B(\alpha^{e_B}, \alpha^{e_A}),$$

then sends $(C(B), \alpha^{e_B} \pmod{p}, E_k(s_B))$ to Alice.

(3) Alice computes $k \equiv (\alpha^{e_B})^{e_A} \pmod{p}$, and uses E_k^{-1} to decrypt and obtain s_B, which she verifies using ver_B, and she verifies $C(B)$ using ver_T.

(4) Upon successful completion of step (3), Alice computes

$$s_A = \text{sig}_A(\alpha^{e_A}, \alpha^{e_B}),$$

and sends $(C(A), \alpha^{e_A} \pmod{p}, E_k(s_A))$ to Bob.

(5) Bob uses E_k^{-1} to decipher and get s_A which he verifies using ver_A, and uses ver_T to verify $C(A)$. (A successful completion of this step ensures Alice and Bob have a shared key k.)

The STS protocol establishes a key k, mutually confirmed by Alice and Bob, whose identities have been verified to each other, but not to Eve. Thus, we indeed have an authenticated key agreement protocol. Now Alice and Bob can use k to encrypt all subsequent messages between them.

[8.2]For a relatively new partial attack on STS, and means to defend against it, see: http//:www.cacr.math.uwaterloo.ca/~ajmeneze/publications/sts.ps

Suppose that we want to dispense with the need for (explicit) certificates, such as $C(A)$ and $C(B)$. Then we need a means whereby Alice and Bob can *implicitly* authenticate each other's identity. This is accomplished in the following scheme [97], developed by Girault in 1991, although he cites work established by others as his motivation.

◆ **Girault's Self-Certifying Key Agreement Scheme**

Alice and Bob exchange messages with a goal of implicitly certified agreement on a public key k.

Setup Stage: Trent performs each of the following.

(1) Select primes p and q, form the RSA modulus $n = pq$, and choose an element α of maximum order in $(\mathbb{Z}/n\mathbb{Z})^*$ (see Exercise 8.18). The value of n is public, but its factorization known only to Trent. Also, the DLP in $\mathbb{Z}/n\mathbb{Z}$ must be intractable.

(2) Select an RSA public/private key pair (e, d).

(3) Create identity strings I_A and I_B for Alice and Bob, respectively.

Protocol Steps:

(4) Alice and Bob individually choose private exponents d_A, and d_B, compute $c_A \equiv \alpha^{d_A} \pmod{n}$, $c_B \equiv \alpha^{d_B} \pmod{n}$, and send (c_A, d_A), (c_B, d_B), respectively, to Trent.

(5) Trent verifies that $c_A \equiv \alpha^{d_A} \pmod{n}$ and $c_B \equiv \alpha^{d_B} \pmod{n}$, and if valid, calculates

$$p_A \equiv (c_A - I_A)^d \pmod{n}, \quad p_B \equiv (c_B - I_B)^d \pmod{n},$$

sends p_A to Alice, and p_B to Bob.

(6) Alice randomly chooses $e_A \in (\mathbb{Z}/n\mathbb{Z})^*$, computes $s_A \equiv \alpha^{e_A} \pmod{n}$, and sends (I_A, p_A, s_A) to Bob.

(7) Bob selects $e_B \in (\mathbb{Z}/n\mathbb{Z})^*$ at random, computes $s_B \equiv \alpha^{e_B} \pmod{n}$, and sends (I_B, p_B, s_B) to Alice.

(8) Alice computes

$$k \equiv s_B^{d_A} (p_B^e + I_B)^{e_A} \pmod{n},$$

and Bob computes

$$k \equiv s_A^{d_B} (p_A^e + I_A)^{e_B} \pmod{n}.$$

(See Exercise 8.19.)

Example 8.6 *Suppose that $p = 919$, $q = 839$, so $n = 771041$ is the RSA modulus for the Girault protocol. Trent also chooses $(e, d) = (53, 420929)$ as the RSA key pair, and $\alpha = 77$ as the element of maximum order $384642 = \mathrm{lcm}(838, 918)$ in $(\mathbb{Z}/n\mathbb{Z})^*$. The identification strings chosen by Trent are $I_A = 19$ and $I_B = 39$. Alice selects $d_A = 5$ and computes $c_A \equiv \alpha^{d_A} \equiv 430247 \,(\mathrm{mod}\ n)$. Bob chooses $d_B = 89$ and computes $c_B \equiv \alpha^{d_B} \equiv 264373 \,(\mathrm{mod}\ n)$. Then c_A and c_B are sent to Trent. Trent computes both*

$$p_A \equiv (c_A - I_A)^d \equiv (430247 - 19)^{420929} \equiv 907 \,(\mathrm{mod}\ n),$$

and

$$p_B \equiv (c_B - I_B)^d \equiv (264373 - 39)^{420929} \equiv 682111 \,(\mathrm{mod}\ n),$$

which are sent to Alice and Bob, respectively. Alice chooses $e_A = 15$, computes

$$s_A \equiv \alpha^{e_A} \equiv 77^{15} \equiv 297843 \,(\mathrm{mod}\ n),$$

and sends $(I_A, p_A, s_A) = (19, 907, 297843)$ to Bob. Bob selects $e_B = 83$, computes

$$s_B \equiv \alpha^{e_B} \equiv 77^{83} \equiv 312102 \,(\mathrm{mod}\ n),$$

and sends $(I_B, p_B, s_B) = (39, 682111, 312102)$ to Alice. Alice computes,

$$k \equiv s_B^{d_A}(p_B^e + I_B)^{e_A} \equiv 312102^5 (682111^{53} + 39)^{15} \equiv 517619 \,(\mathrm{mod}\ n),$$

and Bob computes,

$$k \equiv s_A^{d_B}(p_A^e + I_A)^{e_B} \equiv 297843^{89}(907^{53} + 19)^{83} \equiv 517619 \,(\mathrm{mod}\ n),$$

so they have agreed on $k = 517619 \,(\mathrm{mod}\ n)$ via implicit key authentication. (See Exercises 8.21–8.24.)

In step (5) of the Girault scheme, Trent's calculation of p_A and p_B essentially replaces the certificates, since p_A and p_B are the *self-certifying public keys*, given that, for instance, $p_A^e + I_A \equiv c_A \equiv \alpha^{e_A} \,(\mathrm{mod}\ n)$, so from this information anyone can verify Alice's public key α^{e_A}. However, since (I_A, p_A, s_A), for instance, is not signed by Trent, then Mallory could forge this information. Yet, if Mallory takes I_A and produces a forgery s'_A, there is no way for him to compute d'_A corresponding to s'_A, provided that the DLP is intractable in $\mathbb{Z}/n\mathbb{Z}$. Hence, without d'_A, a key construction cannot be accomplished by Mallory who is impersonating Alice. Thus, Trent cannot be duped by Mallory who must be convinced that Alice knows d_A. This is the reason that Alice must send d_A to Trent. He could quite easily compute p_A from c_A without knowing d_A, but Alice must convince him that she knows d_A. Of course, there exist other means for Alice to convince Trent of this knowledge without giving away d_A. For instance, see the discussion surrounding Schnorr's identification scheme at the bottom of page 130. If Trent does not verify knowledge of d_A in some fashion, however, there can be dire consequences (see Exercise 8.20).

There are numerous other key agreement protocols. The above are presented as an introduction to the topic with illustrations sufficient for us to carry on with our investigations. The reader interested in more recent protocols may consult [115], wherein the authors present an authenticated key agreement protocol provably secure against the man-in-the-middle attack. Also see [208] and [234].

We now turn to another aspect of key management, namely, key distribution. It may be argued that secure key distribution is the most difficult aspect of key management. We have already seen a preliminary aspect of key distribution when we discussed *key predistribution* on page 73. By key predistribution we will mean a key agreement protocol where Trent acts as a *key server* who generates and distributes enciphered *session keys* to a network of users. Here we will think of a *network* as a system of computers that transfers data among the entities called *users*. Thus any pair of users can decipher and use a shared key, unknown to all other users except, of course, Trent.

If there are n users on a network using symmetric-key techniques, and there is no key server, then there must be $n(n-1)/2$ centrally stored keys, or approximately n^2. This is called the n^2-*key distribution problem*. Of course, as n gets very big, this becomes unacceptable. Hence, public-key techniques (or a key server) are needed to solve the problem since, as we noted on page 73, only $n-1$ keys are needed in that case.

We now discuss one of the best-known key predistribution schemes in what follows. This was introduced in [31] as a simplification of Blom's original method [29].

◆ Blom's (Simplified) Key Predistribution Scheme

The goal is for the key server to distribute a symmetric key to each user over a secure channel.

Basic Assumptions: We suppose that there is a network of $n \in \mathbb{N}$ users, and that keys are taken from $\mathbb{F}_p$ where $p \geq n$ is a public prime. Each user, such as Alice, has a unique public key $k_A \in \mathbb{F}_p$.

Protocol Steps:

(1) Trent chooses three random values $r_1, r_2, r_3 \in \mathbb{F}_p$, not necessarily distinct, and forms,

$$p(x, y) \equiv r_1 + r_2(x + y) + r_3 xy \pmod{p}.$$

(2) For each user, such as Bob, Trent computes,

$$f_B(x) \equiv r_1 + r_2 k_B + (r_2 + r_3 k_B)x \pmod{p},$$

which he sends to Bob over a secure channel.

(3) Now Alice and Bob may communicate over the network using the shared symmetric key,

$$k \equiv r_1 + r_2(k_A + k_B) + r_3 k_A k_B \pmod{p},$$

where Alice computes k via,

$$k_{A,B} \equiv f_A(k_B) \equiv r_1 + r_2 k_A + (r_2 + r_3 k_A) k_B \,(\text{mod } p),$$

and Bob computes it via,

$$k_{B,A} \equiv f_B(k_A) \equiv r_1 + r_2 k_B + (r_2 + r_3 k_B) k_A \,(\text{mod } p),$$

since $k \equiv k_{A,B} \equiv k_{B,A} \,(\text{mod } p)$.

Example 8.7 *Let $p = 569$, and $n = 3$, with Alice, Bob, and Cathy on the network, where $k_A = 356$, $k_B = 215$, and $k_C = 501$. Assume that Trent chooses $r_1 = 5$, $r_2 = 96$, and $r_3 = 416$, so $p(x,y) \equiv 5 + 96(x + y) + 416xy \,(\text{mod } p)$. Then the individual polynomials sent to Alice, Bob, and Cathy are,*

$$f_A(x) \equiv 41 + 252x; \quad f_B(x) \equiv 161 + 203x; \quad f_C(x) \equiv 305 + 258x.$$

the three session keys are,

$$k_{A,B} \equiv 166 \,(\text{mod } p); \quad k_{A,C} \equiv 544 \,(\text{mod } p); \quad k_{B,C} \equiv 13 \,(\text{mod } p),$$

for communications between Alice and Bob; Alice and Cathy, and Bob and Cathy, respectively.

By Exercise 8.30, Blom's simplified scheme is unconditionally secure against an attack by any individual user, but is vulnerable to a total break by more than one user acting in concert (see Exercise 8.31). However, the scheme can easily be made secure against any $n \in \mathbb{N}$ users acting in concert by altering the choice by Trent in step (1) of the polynomial to be,

$$p(x,y) \equiv \sum_{i=0}^{n} \sum_{j=0}^{n} r_{i,j} x^i y^j \,(\text{mod } p), \tag{8.8}$$

for randomly chosen $r_{i,j} \in \mathbb{F}_p$ with $r_{i,j} \equiv r_{j,i} \,(\text{mod } p)$ for all such i,j. The general setup (8.8) is an aspect of the full Blom protocol.

On page 73, we made a brief reference of Kerberos, which is actually a key predistribution/authentication scheme that we now discuss in detail.

The development of Kerberos began in 1989, and originated from a larger endeavor at MIT, called *Project Athena*, the purpose of which was secure communication across a public network for student access of their files. Kerberos is the authentication protocol aspect of Project Athena, and is based upon a client-server-verifier model described as follows. First, a *client*, Carol, is a user (which might in reality be dedicated software or a person) with some goal to achieve, and this could be as simple as sending e-mail or as complex as installation of a system component. A *server* (and verifier), Victor, provides services to clients. This might involve anything from e-commerce to allowing access to personal files. In the Kerberos model, Trent is a trusted authority, called the *Kerberos authentication server*. The property of producing a new session key each time a pair of users wants to communicate is called *key freshness*.

◆ Kerberos Authentication/Session Key Distribution Protocol — Simplified

Carol interacts with Trent and Victor with the goal of authenticating her identity to Victor, and establishing a session key, k, which she can use to communicate with him.

Basic Assumptions: We are given a symmetric-key algorithm E (such as AES — see page 22). Trent selects a random key k, a timestamp t, and a validity period L, called a *lifetime*. Carol and Trent share a secret symmetric key $k_{C,T}$, and Victor and Trent share one, $k_{V,T}$. Also, I_C, I_V, and I_T are identity strings for Carol, Victor, and Trent, respectively.

Protocol Activities:

(1) Carol sends her request for a session key to use with Victor, together with her identity string I_C to Trent.

(2) Trent computes $m_C = E_{k_{C,T}}(k, I_V, t, L)$ where I_V is Victor's identity string that he has taken from a database. He also computes $m_V = E_{k_{V,T}}(k, I_C, t, L)$, called a *ticket* for Victor, and sends both m_C and m_V to Carol.

(3) Carol uses $E_{k_{C,T}}^{-1}$ to retrieve k, I_V, t, and L from m_C. She verifies that t and L are valid, and that I_V is the identity of Victor. She then creates a fresh timestamp t_C, computes

$$m'_V = E_k(I_C, t_C),$$

called the *authenticator*, which she sends to Victor along with m_V that she received from Trent.

(4) Victor uses $E_{k_{V,T}}^{-1}$ to get k from the ticket, m_V. Then he uses E_k^{-1} to decrypt the authenticator m'_V. He checks that the two copies of I_C from the ticket and authenticator match. He checks that t_C is within the *expiration window*, which can be accomplished by Victor subtracting t_C from the current time which must be within some mutually accepted fixed time interval. Then he checks that his current time is within the lifetime L specified by the ticket. If these three facts hold, he declares Carol to be authentic, and he computes $m'_C = E_k(t_C)$, and sends it to her.

(5) Carol applies E_k^{-1} to m'_C, and checks that t_C matches the value in m_C from step (3). If it does, she declares Victor to be authentic and now has a session key k to communicate with him.

The role of the timestamp t and the lifetime L is to thwart Mallory from storing old messages for retransmission at a later time (a replay attack — see page 133). If any of the checks against t in the above protocol fail, then the protocol terminates since a *stale* timestamp has been discovered. Checks that are valid are those that show the current time to be in the range t to $t + L$. The lifetime L also has the advantage of allowing Carol to re-use Victor's ticket without contacting Trent, so steps (1)–(2) can be eliminated over the lifetime of the ticket. However, each time Carol reuses the ticket, she must create a new authenticator with a fresh timestamp, but the same session key k.

The expiration window can be any agreed fixed amount that takes into consideration such things as clock skew, message transit time, and processing time. This could be measured in seconds or milliseconds, depending on the situation. However, the clocks have to be "roughly synchronized" for all users in the network, and they must be secured to prevent modifications. We say *roughly* synchronized since perfect synchronization is a tremendously difficult task. This is one of the drawbacks of Kerberos, since the current time is used to determine the validity of the session key k. There are solutions that involve synchronized distributed clocks using network protocols, which themselves must be secure.

Going back in time, the basic protocol for Kerberos comes from the following key predistribution protocol introduced in [173] where timestamps are not used. These were later proposed by Denning and Sacco [69].

◆ Needham-Schroeder Key Predistribution Scheme

Alice and Bob communicate with Trent for the purpose of mutual identity and shared key authentication.

Basic Assumptions: E is a symmetric-key algorithm and n_A, n_B are nonces (see page 133) chosen by Alice and Bob, respectively. Trent selects a session key k for Alice and Bob to share.

Setup Stage: Alice and Trent use a shared symmetric key $k_{A,T}$, and Bob and Trent use a shared symmetric key $k_{B,T}$. Also, I_A and I_B are identity strings for Alice and Bob, respectively.

Protocol Steps:

(1) Alice sends (I_A, n_A) to Trent.

(2) Trent sends $E_{k_{A,T}}(n_A, I_B, k, E_{k_{B,T}}(k, I_A))$ to Alice.

(3) Alice uses $E_{k_{A,T}}^{-1}$ to decrypt k and confirms that n_A is the same as in step (1), and verifies I_B. Then she sends $E_{k_{B,T}}(k, I_A)$ to Bob.

(4) Bob deciphers k and sends $E_k(n_B)$ to Alice.

(5) Alice decrypts with E_k^{-1}, generates $n_B - 1$ and sends $E_k(n_B - 1)$ to Bob.

(6) Bob deciphers with E_k^{-1} and verifies $n_B - 1$.

The above protocol differs from Kerberos in that it has no lifetime parameter L or timestamp t, but rather uses nonces. However, step (3) is essentially the same as the ticket in the Kerberos protocol, but with Needham-Schroeder, the ticket is double-encrypted in Step (2). Yet, Bob has no means of verifying the freshness of k. Thus, Mallory can resend the message in step (3), which Bob receives in step (4) and sends the message to Alice, which Mallory intercepts. If Mallory can get access to k, he can send the message in step (5), impersonating Alice, and Bob verifies as in step (6) that it is Alice, when in fact he is now dealing with Mallory.

Needham and Schroeder addressed this problem in [174], which is the same as the Otway-Rees protocol [179]. The above is presented largely for historical reasons since it is not secure. However, they also have a public-key protocol that results in authentication and key agreement on a distributed symmetric key. This was also introduced in [173]. However, almost twenty years after its publication, Lowe [144] discovered an attack that compromises authenticity of the messages. Thus, we present the modification made by Lowe to address this weakness.

◆ Needham-Schroeder (Modified) Public-Key Scheme

Alice and Bob communicate with the goal of mutual identity and key authentication as well as agreement on a symmetric key.

Background Assumptions: P is a public-key encryption algorithm (such as RSA) and $P_A(m_1, m_2)$, for instance, will denote the use of Alice's public key to send messages m_1, m_2. Also, k_A, k_B are Alice and Bob's respective chosen symmetric session keys. Moreover, Alice and Bob are assured of having each other's authentic public key. The strings I_A and I_B identify Alice and Bob.

Protocol Steps:

(1) Alice sends Bob $P_B(k_A, I_A)$.

(2) Bob decrypts with his private key, and verifies Alice's identity I_A. If valid, he sends $P_A(k_A, k_B, I_B)$ to Alice.

(3) Alice uses her private key to decipher and checks that the copy of k_A agrees with what she sent in step (1). Also, she verifies Bob's identity I_B. If these are valid, she sends $P_B(k_B)$ to Bob.

(4) Bob uses his private key to decrypt and verifies that the copy of k_B agrees with what he sent in step (2). If all is valid, then they can use a one-way function f to compute $f(k_A, k_B)$, which they may use as a mutually agreed and authenticated session key for them to communicate.

The difference between the above (modified) protocol and that exhibited by Needham and Schroeder in 1978, is that Lowe added Bob's identification string I_B in step (2). Without this, Alice could unknowingly initiate a run with Mallory who receives $P_M(k_A, I_A)$, which he decrypts using his private key. If he sends $P_B(k_A, I_A)$ to Bob, then Bob verifies Alice's identity I_A, not knowing that he is communicating with Mallory. Bob then sends $P_A(k_A, k_B)$ to Alice, which Mallory allows. She has only to verify that k_A is as she sent it in step

(1), and it is, so she sends $P_M(k_B)$, which Mallory intercepts, and decrypts. He then sends $P_B(k_B)$ to Bob. Bob verifies and establishes the session key. Hence, Mallory can successfully negotiate a man-in-the-middle attack without Alice or Bob knowing they have been duped. The addition of Bob's identity string in step (2) thwarts Mallory in this attack since Bob, by sending $P_A(k_A, k_B, I_B)$ is essentially issuing a challenge, namely, extract k_B from the latter and return $P_B(k_B)$ to Bob. Only Alice can meet this challenge since she has kept her private key secure. If Bob receives $P_B(k_B)$, he knows that only Alice could have received and transformed it. If Mallory tries his little trick above, Alice will immediately know it since she will see that I_B does not correspond to I_M, and Mallory is foiled.

The aforementioned attack on the Needham-Schroeder scheme escaped detection for nearly two decades before Lowe's fix. This demonstrates that due diligence is necessary. Any protocol must be free from offering unwanted openings for an attacker. For instance, in the original protocol above, Alice unwittingly offers Mallory the opportunity to decipher $P_M(k_A)$ from which he can learn k_A. The Lowe fix allows Alice to check I_B against I_M at which point she knows that I_B did not come from Mallory and she can stop the protocol on the spot. Also, a protocol must be provably correct in the sense that there is some authentication test method that it must pass. Work in this direction is being done, for instance, by J. Guttman (see [106]).

Other than the Diffie-Hellman protocol, we have also seen other key agreement schemes based on asymmetric techniques. For example, the ElGamal cryptosystem discussed on page 67 contains the essence of the *ElGamal key-agreement protocol* (called ElGamal Key Generation on page 67). Once Bob has generated his public key (p, α, α^a), Alice may obtain a copy of it. She then chooses a random natural number $n \leq p-1$ and sends $\alpha^n \pmod{p}$ to Bob. Then the shared key is $k \equiv \alpha^{ax} \pmod{p}$, computed by Alice as $k \equiv (\alpha^a)^x \pmod{p}$, and by Bob as $k \equiv (\alpha^x)^a \pmod{p}$. This protocol is sometimes called the *one-pass ElGamal key agreement protocol*, which neither authenticates the identity of individual entities nor guarantees key confirmation (meaning the assurance both that the key is fresh and the entity with the public key is also in possession of the corresponding private key). One way of ensuring this is through the use of public-key certificates, about which we will learn in Section 8.3.

We conclude this section with an example of a hybrid key agreement and authentication protocol. This was introduced [22] in 1992. We will describe it with the Diffie-Hellman implementation, although RSA or ElGamal could also be used, for instance. The Diffie-Hellman implementation is one of the simplest.

◆ Encrypted Key Exchange (EKE) — Diffie-Hellman Implemented

Alice and Bob share a common password k (a secret data item used to authenticate an entity, also called *passcode*, and when expanded into a longer phrase *passphrase*) and E, a symmetric-key algorithm. This protocol establishes both identity authentication and a shared key.

Background Assumptions: From the Diffie-Hellman protocol on page 49, we have a fixed prime p and generator α of $\mathbb{F}_p^*$ publicly known for all users in the network employing this protocol. The string I_A identifies Alice.

Protocol Steps:

(1) (Initialization) Alice randomly selects a number $R_A \in (\mathbb{Z}/p\mathbb{Z})^*$, computes $\alpha^{R_A} \pmod{p}$, and sends $(I_A, \alpha^{R_A} \pmod{p})$ to Bob.

(2) (Challenge) Bob chooses a random $R_B \in (\mathbb{Z}/p\mathbb{Z})^*$ and calculates

$$K \equiv (\alpha^{R_A})^{R_B} \pmod{p}.$$

Then he generates a random string R_B', computes

$$E_k(\alpha^{R_B} \pmod{p}, E_K(R_B')),$$

and sends it to Alice.

(3) (Response/Challenge) Alice uses E_k to decipher $\alpha^{R_B} \pmod{p}$. Then she computes $K \equiv (\alpha^{R_B})^{R_A} \pmod{p}$, and uses E_K to decrypt R_B'. She then generates a random string R_A', computes $E_K(R_A', R_B')$, and sends it to Bob.

(4) (Verification/Response) Bob decrypts R_A', R_B' using E_K, and checks that R_B' is the same as that sent in step (2). If so, he sends $E_K(R_A')$ to Alice.

(5) (Verification) Alice uses E_K to decipher, and if R_A' is the same as the value she sent in step (3), the protocol is complete.

The challenge-response portion of the protocol in steps (2)–(4) is designed to convince Bob that Alice has knowledge of K, while the portion in steps (3)–(5) are designed to convince Alice that Bob knows K. The reader may go to the Kerberos protocol and verify that steps (2)–(5) of that protocol serve the same function for Carol and Victor with the timestamp playing the role of K.

One of the strengths of EKE is against password-guessing attacks, since knowledge of $E_K(R_B')$ does not allow for easy checking of guesses for k. Each choice for k yields a candidate value for α^{R_B}, which in turn yields a candidate value for K. But R_B' was randomly chosen so there is no method for verifying if this guess is correct. Moreover, a passive attacker such as Eve, even with knowledge of k, will still have to solve the DHP in order to determine K. The reason is that she will know $\alpha^{R_A} \pmod{p}$ and $E_k(\alpha^{R_B} \pmod{p}, E_K(R_B'))$, so she has α, $\alpha^{R_A} \pmod{p}$, and $\alpha^{R_B} \pmod{p}$, but not R_A or R_B. Thus, this hybrid protocol, EKE, draws on the strengths of both symmetric and public-key cryptography by working in concert to strengthen both of them. The inventors of EKE, Bellovin and Merritt have a patent on their protocol [23].

With all the above being said, EKE still relies on a shared secret, so if a password is somehow captured, it can be used for impersonation. Since the inception of EKE, authentication protocols have proliferated. In 1996, for example, Jablon [113] developed the *Simple Password Exponential Key Exchange* (SPEKE™), (pronounced "speak") which is an improvement on EKE. (He did so as the founder of *Integrity Sciences*, now part of *Phoenix Technologies*, where he is currently CTO of the Platform Security Division.) The SPEKE protocol has two stages. It first uses a variant of the Diffie-Hellman key exchange protocol to establish a shared key K. The second stage is a zero-knowledge challenge-response protocol for Alice and Bob to mutually authenticate their respective knowledge of K. As with EKE, the end result is mutual identity authentication and shared key agreement. Both EKE and SPEKE allow use of a small password to provide authentication and key agreement over an unsecured channel. Moreover, it is immune to offline dictionary attacks (for example, such attacks might involve an adversary using a list of probable passwords, hashing the list, and comparing them with the list of encrypted passwords in an attempt to find a match). SPEKE has a number of applications from cell phones to smart cards (see http://www.IntegritySciences.com/uses.html).

Password-based protocols are subject to *password sniffing*, which is an attack in which an adversary listens to data traffic that includes secret passwords in order to capture them for use at a later time. Thus, eavesdropping on a TCP/IP[8.3] network can be easily accomplished against protocols that transmit passwords in the clear. Moreover, password protocols require the passwords to be stored on the host, usually hashed, and if revealed would compromise security. To address these concerns, a new protocol was developed at Stanford University in 1997, called *Secure Remote Protocol* (SRP), which differs from EKE, SPEKE in that instead of relying on shared secrets, or password/password equivalents stored on a server, SRP mandates that the server store a salt value and a verifier. Hence, without password storage, SRP is more secure than password schemes, and it performs a secure key exchange in the authentication process.

SRP solves the long-standing problem of having both ease-of-use and security. It relies on a type of Diffie-Hellman exchange. However, authentication to a server is rarely done. Also, it has the additional advantage of *forward secrecy*, which is the property that the loss of a verification secret does not compromise earlier encrypted sessions. SRP is freely available to residents of North America in the form of SRP-enabled Telnet[8.4] and FTP[8.5] software — see:

   http://www-cs-students.stanford.edu/~tjw/srp/download.html.

[8.3] This is the acronym for the protocols packaged in a *network protocol stack* for *Internet protocols*. By a *network protocol stack* we mean a software package that supports networking functions for operations software. *IP* is the acronym for *Internet protocol*, which is the protocol that transports specific bundles between *hosts* — those individual computer systems on a network that communicate with other such systems. The *Internet* itself is the globally interconnected network using the set of Internet protocols.

[8.4] This is the Internet protocol that supports remote connections between computers.

[8.5] This is the acronym for of *File Transfer Protocol*, the protocol used on the Internet for sending files, which are either uploaded from an individual computer to the FTP server, or downloaded from the FTP server to an individual computer. Thus, FTP is a two-way protocol. This is distinct from *Hyper Text Transfer Protocol*, or HTTP, which is a protocol used to transfer files from an Internet server onto a browser in order to view a page that is on the Internet. HTTP is a one-way system — the contents of a page from the server are downloaded onto the computer's browser for viewing, but files are not transferred to the computer's memory.

Exercises

8.18. Prove that if $n = pq$ is an RSA modulus, then the maximum order of an element in $(\mathbb{Z}/n\mathbb{Z})^*$ is $\operatorname{lcm}(p-1, q-1)$.

8.19. Prove that the two values in step (8) of the Girault scheme are actually congruent modulo n.

☆ 8.20. In the Girault scheme, suppose that Trent does *not* verify that Alice knows d_A. Show how Mallory can successfully launch a man-in-the-middle attack that dupes Bob into thinking he is communicating with Alice when, in fact, he is communicating with Mallory.

In Exercises 8.21–8.24, use the Girault scheme to find the shared self-certified key k, where the following values correspond to the variables: $(e, d, p, q, \alpha, I_A, I_B, d_A, d_B, e_A, e_B)$.

8.21. $(5, 9701, 173, 283, 3, 21, 93, 10279, 32773, 151, 37)$.

8.22. $(23, 39239, 317, 409, 21, 517, 944, 13, 2131, 109093, 51547)$.

8.23. $(131, 290315, 499, 593, 7, 156, 1001, 2021, 3011, 14033, 221675)$.

8.24. $(311, 11387, 599, 659, 7, 133, 399, 1123, 1103, 233007, 323563)$.

In Exercises 8.25–8.29, use the Blom key predistribution scheme with $n = 3$ to find the shared keys $k_{A,B}, k_{A,C}, k_{B,C}$ as in Example 8.7, given the values for $(p, k_A, k_B, k_C, r_1, r_2, r_3)$, as follows.

8.25. $(601, 10, 101, 568, 11, 13, 200)$.

8.26. $(701, 12, 113, 686, 121, 311, 203)$.

8.27. $(809, 666, 66, 6, 5, 15, 25)$.

8.28. $(907, 270, 307, 47, 12, 13, 800)$.

8.29. $(1009, 12, 901, 108, 5, 808, 700)$.

☆ 8.30. Prove that the simplified Blom scheme is unconditionally secure against an attack by a user, Mallory. In other words, show that with the knowledge Mallory has, namely, $f_M(x) \equiv r_1 + r_2 k_M + (r_2 + r_3 k_M)x \pmod{p}$ sent by Trent, all values of $k \in \mathbb{F}_p$ are possible for $k_{A,B}$, which he is trying to cryptanalyze.

8.31. Show that if Mallory conspires with another user, Eve, in the simplified Blom scheme, then they can determine r_1, r_2, r_3, thereby allowing them to determine any key, resulting in a total break.

8.3   Public-Key Infrastructure (PKI)

If a man will begin with certainties, he shall end in doubts; but if he will be content to begin with doubts, he shall end in certainties.

Francis Bacon (1561–1626)
English lawyer, courtier, philosopher, and essayist

A *public-key infrastructure*, or PKI, embodies a foundation of protocols and standards, which support and enable the secure and transparent use of public-key cryptography, particularly in applications requiring the use of public-key cryptography. Since the term PKI is quite a recent phenomenon, the reader may not find uniformity of definition throughout the literature, and therefore no "single" PKI (see [2] for a book dedicated to the task of explaining PKI and its various services). The *core* PKI services are authentication, integrity, and confidentiality. We have already devoted the entirety of Chapter 7 to authentication, which includes some integrity establishment techniques such as digital signatures, and the integrity of digital cash (see Section 7.3). We have also addressed confidentiality in Sections 8.1–8.2. A central view of PKI is that it provides protocols for certification of public keys and verification of certificates. The reason is that if the core services are to be provided, then Alice must be assured of being able to associate a public key clearly and verifiably to Bob, with whom she wants to communicate. This is where the role of a certificate comes into play, and this will be our focus. We were introduced to certificates (with Trent as a certification authority) on page 129. We now expand this discussion with a view of PKI as providing key management through the use of a *certification authority* (CA) and a *registration authority* (RA).

The CA is an entity responsible for issuing public-key certificates (PKC) (which are tamperproof data blocks containing, at a minimum, entity identification, CA identifier, and a public key), which are used to bind the individual name to the corresponding public key. The CA accomplishes this by affixing its private key as a digital signature, thereby performing *key registration* via the issuing of a certificate. Think of the certificate as being the analogue of a driver's license.

The RA typically plays the role of assisting the CA by establishing and verifying the identity of entities who wish to register on a network, for instance. These entities are typically called *end-users*. Other functions of the RA may include key predistribution for later online verification; initiation of the certification process with the CA for end-users; and performance of certificate management functions such as *certificate revocation* (meaning the cancellation of a previously issued certificate). Hence, we may come to an agreement as to the probable services of a given PKI: certificate creation, distribution, management, and revocation. There are also PKI-enabled services that are not part of the PKI, but can be built upon the core PKI. These include secure communications; secure timestamping; and non-repudiation. Note that a PKI, in itself, may not involve any cryptographic operations with the keys that it is managing. A common feature of all PKIs is a set of certification and validation protocols, since the fundamental core predicate of PKI is the secure management of public keys.

At the end of Chapter 7, when we were discussing issues surrounding e-commerce, we mentioned SET, the *Secure Electronic Transaction* specification established by Visa and Master Card to support credit card payments over the Internet. Basically SET has adopted the ISO/ITU-T X.509 Version 3 public-key certificate format.[8.6] SET was created to ensure the authenticity, confidentiality, and integrity of e-commerce transactions over the *World Wide Web* (WWW).[8.7] The X.509 Version 3 certificate is sophisticated enough that it can handle e-commerce applications as an international standard.

There exist other network and security protocols that use a profile of X.509 Version 3 certificates such as *Internet Protocol Security* (IPSec). There are some that are proprietary, such as *Pretty Good Privacy* (PGP) introduced by Zimmerman [247] in 1995. PGP does not have the sophistication, nor was it meant to have, as that enjoyed by the SET certificate. It is essentially a method for encrypting and digitally signing e-mail messages and files. There is an improved version, called *Open PGP*, as an IETF[8.8] standards-track specification. Version 6.5 of Open PGP actually supports X.509 certificates, but the trust[8.9] model differences between PGP certificates and X.509 Version 3 PKCs are enormous, so it is unlikely that PGP will ever be used for e-commerce.

In PKIs, the *trust models* are used to describe the relationships of CAs with

---

[8.6]The *International Organization for Standardization* (ISO) is a global federation of national standards bodies, one from each of 140 countries. This non-governmental organization was initiated in 1947 to promote the development of standardization and related activities. Note that ISO is *not* an acronym, but rather it is taken from the Greek *isos* meaning *equal* and this is the prefix *iso-* such as in *isometric*. So *equal* devolved to *standard*, and the ISO name was adopted. Note that it has the feature of not requiring translation in each country, as would an acronym. ISO develops precise criteria for such applications as the formatting of smart cards, for instance. ITU is the *International Telecommunication Union*, which was established on May 17, 1865 (as the *International Telegraph Union*) to manage the first international telegraph networks. The name change came in 1906 to properly reflect the new scope of the Union's mandate. The ITU-T is the ITU *Telecommunication Standardization Section*, one of three sections of ITU. It was established on March 1, 1993. In conjunction ISO and ITU-T form world standards such as the X.509, which is a public-key certificate. Version 3 (as specified in [112]) was developed to correct deficiencies in earlier versions. This version has become the accepted standard so that often the term *certificate* is used to mean this version of X.509. Version 3 contains each of the following fields: version number, certificate serial number, signature algorithm identifier, issuer name, validity period, entity name, entity public key information, issuer unique identifier, entity unique identifier, extensions, and signature. In addition, the extensions field can contain numerous types such as authority key identifier, extended key usage, and private key usage period. There are other widely known joint standards such as ISO/IEC, which are joint between ISO and the *International Electrotechnical Commission* (IEC). For example, [112] is tantamount to ISO/IEC 9594-8, 1997.

[8.7]This is the information network using HTTP (see Footnote 8.5) and HTML on Internet host computers. HTML means *HyperText Markup Language*, which is the text format used for World Wide Web pages.

[8.8]This is the *Internet Engineering Task Force*, which is responsible for making recommendations concerning the progress of so-called "standard-track" specifications from a status called *Proposed Standard* (PS) to a stage called *Draft Standard* (DS) to the final stage called *Standard* (STD). The citation [10] is called a *Request for Comments* (RFC), which are essentially working documents of the Internet R&D community.

[8.9]We take the definition of "trust" to be that used in [2], which is from [112, Section 3.3.23]: *Entity A trusts entity B when A assumes that B will behave exactly as A expects.*

end-users and others. We describe only two of them, the one used by PGP and the one used by X.509 Version 3. The first, used by PGP is called *User-Centric Trust*. In this model, each user makes the decision as to which certificates to accept and which to reject. In the PGP implementation, a user, such as Alice, exchanges certificates which are public keys of those other users with whom she wants to communicate. She protects her certificate from alteration by signing it with her private key. Alice acts as a CA in the following fashion. Upon receipt of, say, Bob's certificate, she will assign it one of three levels: (1) complete trust, meaning that she trusts Bob and anyone whose certificate is signed with Bob's key; (2) partial trust, meaning that Alice does not completely trust Bob, so certificates signed by Bob must also be signed by other users (whom she does trust) before she accepts it; (3) no trust, meaning that Alice does not trust Bob and will not trust any certificate signed by Bob. In some implementations, there is a fourth level of *uncertain*, but this essentially amounts to no trust. In this fashion she builds a *web of trust* with other users. However, this user-centric model is not acceptable for such applications as e-commerce. Nevertheless, it is popular for sending secure e-mail, but for most people it is difficult to manage securely. A more generally secure trust model is described in what follows.

In the PKI trust model called *cross-certification*, first the CAs (in their respective security domains[8.10]) are required to form a *trust path* between themselves. The process called *mutual cross-certification* involves CA_1 signing the certificate of CA_2, and CA_2 signing the certificate of CA_1 (see [1]). If the domains are different, called *interdomain cross-certification*, then *relying parties* (those entities who verify the authenticity of an end-user's certificate) are able to trust end-users in the other domain. This trust model is clearly suited to e-commerce, such as that engaged by two distinct business organizations. If the two CAs are part of the *same* domain, called *intradomain cross-certification*, then this model can be varied to accommodate a hierarchy of CAs where CA_1 can sign the certificate of CA_2 who is at a lower level, without having CA_2 sign CA_1's certificate. This is called *unilateral cross-certification*. An advantage of unilateral cross-certification is that it allows relying parties to trust only the top-level root CA, but have their certificates issued by the authority closest to them.

Clearly, the trust model is a vital part of any PKI. We have described only two of many such models, which is sufficient for our purposes. The reader interested in seeing more of them in greater detail may consult, for instance, [2], [49], or [156].

We have not addressed the PKI issue of certificate storage. After the generation of a certificate, it must be stored for use at a later time. It is not desirable to have the end-users store them on their individual computers, so CAs require what is called a public *certificate directory*, which is a public database or server accessible for read-access by end-users. The CA manages the directory and supplies certificates to it. The directory is a central storage location that provides an individual, public, central location for the administration and distribution of

---

[8.10]A *security domain* is a system governed by a trusted authority.

certificates. As with PKI itself, there is no single standard. Perhaps the most popular is the X.500 series, which is the ISO/ITU-T array of standards with specifications in [111], which is equivalent to ISO/IEC 9594-1 (see Footnote 8.6). In fact, the X.500 series is the underlying structure in which X.509 was originated. Proprietary directories based on X.500 include *Microsoft Exchange*, for instance. An additional benefit of X.500 is that there are standardized protocols for obtaining the data structures, thus allowing any PKI to have access. The reason is that X.500 has a standardization mechanism called a *schema* for the storage of certificates and *certificate revocation list* (CRL)[8.11] data structures in a given entity's directory entry. We will now look at certificate revocation.

Suppose that Alice's private key has been compromised. Thus, the corresponding public key can no longer be used for Alice. The mechanism for alerting the rest of the network of users is *certificate revocation checking*. We can now invoke the earlier analogy in terms of a driver's license. A police officer, upon checking a driver's license, not only verifies the date on the license, but also calls some central police authority to confirm that the license has not be revoked. *Certificate revocation* refers to the act of marking the certificate as revoked by the CA and placing it in a CRL. CAs issue periodic CRLs to ensure relying parties that the most recent CRL is current, so even if there are no changes, a CRL is issued on time according to the schedule. In addition, as we have seen, some certificates are cross-certified between the CAs themselves. To revoke these certificates, we need a separate *authority revocation list* (ARL), which plays the role of CRLs. However, revoking the PKC of a CA is rare and usually occurs when the CAs private key is compromised.

The X.509 Version 2 standard for CRLs, as with the Version 3 certificates, discussed earlier, has extension fields to make the CA's job of revocation easier. They are (1) *reason code*, namely, a specification of the reason for the revocation; (2) *hold instruction code*, which is a mechanism to temporarily suspend a certificate, and contains an *object identifier* (OID), which stipulates the action to be taken if this field is filled; (3) *certificate issuers*, which has the identity of the certificate issuer; (4) *invalidity date*, which contains the date and time of the known or suspected compromise.

There is an alternative online mechanism for certificate revocation, the most popular being the *Online Certificate Status Protocol* (OCSP), documented in [170] with HTTP being the most common practical mechanism (see Footnote 8.5). This is a challenge-response protocol offering a mechanism for online revocation of data from a trusted authority, called an *OCSP responder*. However, as a mere protocol, it does not have the capacity to store revocation data, so the OCSP responder must obtain information from some other source. Thus, latency is involved with its use. Moreover, it is limited to the supplying of information about the revocation status of a given list of certificates, and nothing else. Hence, there is still the need for CRLs.

The next aspect of PKI that we will discuss is the *key backup and recovery*

[8.11]This CRL is a signed data structure embodying a timestamped inventory of revoked certificates.

server. This gives the CA a mechanism for backing up private keys together with a means of recovering them later should end-users lose their private keys. In the literature, the term "key recovery" is often used synonymously with "key escrow". However, the latter is quite different from the former in that key escrow requires that keys are released to a third-party organization, such as a police agency, as evidence in an investigation. On the other hand, key recovery is implemented in an individual PKI by its authorities to provide key recovery for its end-users. The key recovery server is an automated process to relieve the burden on PKI authorities (see [1]). To prevent an adversary from accessing an entity's private key and launching an impersonation attack, a CA may support not one, but two key pairs: one for enciphering and deciphering and the other for signature and verification. For instance, in the DSA, discussed on page 139, the key pair cannot be used for encryption and decryption, whereas the Diffie-Hellman key pair, discussed on page 49, cannot be use for signing and verification. The management of key pairs is paramount in any PKI, and dual key pairs has become a central feature of any in-depth PKI.

First, keys must be generated. This can occur at the end-user's computer, then conveyed securely to the CA or RA. This necessitates the user having software to generate cryptographic keys and securely send them to a central authority. Alternatively, a CA or RA can generate the key pair. The mechanism for so doing may vary. For instance, the ANSI X9.17 standard delineates a method for key generation. See also [1] and [172].

Once multiple key pairs for individual entities have been generated, there is a need for multiple certificates, since X.509, for example, does not support multiple key pairs in a single certificate. A private key used for signing and verification requires secure storage throughout its lifetime. In this case, we should not back up the key pair, since the compromise of the pair necessitates the generation of a new key pair. Moreover, compromise of the key pair makes verification of all signatures associated with that key pair impossible. Such key pairs must always be secured, usually in one module as mandated by [6], since knowledge of the private key needed for non-repudiation will allow the owner of the key to claim the adversary engaged in the non-repudiable act. This defeats the whole point of having the key pair for non-repudiation. Furthermore, once this key pair expires, it should be destroyed within the same module in which it was secured, again as required in [6]. On the other hand, a private key used for decryption must be backed up to enable recovery of enciphered data. Moreover, it should not be destroyed once expired since it may be needed for later decryptions. It should be placed in a *key archive*, which is a long-term storage of keying data including certificates. Typically, archives are appended with timestamp and notarization data in order to resolve any future disputes, as well as for audit purposes.

Another aspect of PKI is the *updating* of keys. Key pairs must be updated at regular intervals, if for no other reason than to thwart compromise threatened by cryptanalytic attacks. Once the key pair expires, the CA can reissue a new certificate based on the new key pair, or a new certificate for the old key pair can be generated. This gives rise what is called a *key history*, consisting principally of old private keys. A key history must be maintained by the PKI for such purposes as later decryptions of old data. Ideally, a key history is stored with a CA who has an automated process available to retrieve the data from the key history as it is needed. This is different from key archiving which meets the need for storing public keys and certificates for digital signature purposes.

To leave the end-user with the task of requesting updates is usually an unacceptable burden, so the process in any comprehensive PKI will be automatic. This can be done by an automatic verification of a certificate each time it is used on the network. Once expiration approaches, the automated system will request a key update from a suitable CA or more likely, an RA. Once the new certificate is created by a CA, it is automatically replaced and the end-user is spared any unnecessary burden of having to interact to make it so.

If private keys are lost by end-users, and they will be, there should also be an optimal automatic process of key recovery in the PKI (see [1]). Note that this means the recovery of private decryption keys only, not private signature keys, for the reasons cited above. An alternative method to the CAs storing public keys and certificates for digital signature purposes is the RSA digital envelope (see page 75). Alice can use a secret symmetric session key to encipher, but also she encrypted it, using an RSA public recovery key, when it was generated. Thus, if Alice loses her key, the CA who owns the private RSA recovery key can open the digital envelope and recover Alice's session key. Key recovery can also be accomplished using secret sharing schemes such as those we discussed in Section 8.1. These key recovery threshold schemes are also very common. The palatable aspect of using threshold schemes is the checks and balances feature. Splitting a private key among shares thwarts attempts by any one entity from capturing private keys in a clandestine fashion. Yet, it allows, as we have seen, a reconstruction of the key shares without one or more of the trusted entities being present to pool the shares.

The future of PKI is open, vigorous, and exciting. It is developing at a fast pace and new standards are emerging. For instance, the reader may consult the *PKI Forum* for further information at: http://www.pkiforum.org. Recall that we have already mentioned the IETF's working group, see: http://www.ietf.cnri.reston.va.us/html.charters/pkix-charter.html. Moreover, there are standard PKIs developed by The Government of Canada: http://www.cse-cst.gc.ca/en/services/pki/pki.html. NIST has a *Federal PKI Technical Working Group* (PKI-TWG) studying PKI infrastructures for use by government agencies: http://csrc.nist.gov/pki/twg/. *The Open Group*, an international vendor and technology-neutral consortium, is developing PKI standards: http://www.opengroup.org/public/tech/security/pki/cki/, and there are numerous others.

In the next chapter we will look at the future of PKI and several revolutionary applications. We have barely opened the door, but what we see is truly a magnificent vista to bring us a better and more secure world.

Chapter 9

Applications and the Future

The distinction between past, present and future is only an illusion, however persistent.

Albert Einstein (1879–1955)
German born theoretical physicist and discoverer of the theory of relativity

9.1   Secrecy and Authentication

Suppose that two adversarial countries enter into an agreement to terminate all underground nuclear weapons testing. Obviously, each wants to verify that the other is complying and not surreptitiously engaging in such testing. In this case, there is no need for secrecy, only guaranteed authentication, called *authentication without secrecy*. What is being sought is authentication without covert channels[9.1] with a goal to verify treaty compliance. This was indeed proposed in the late 1970s and early 1980s by Simmons[9.2] [209]–[211] as a means

---

[9.1]A *covert channel* is not uniformly defined throughout the literature. We will use Lampson's definition [130]: any communication pathway that was neither designed nor intended to transfer information. Hence, only untrustworthy adversaries would find and use them.

[9.2]Gustavus J. Simmons was born on October 27, 1930 in Ansted, West Virginia. He received his B.Sc. in mathematics from New Mexico Highlands University, Las Vegas in 1955, his M.Sc. from the University of Oklahoma at Norman in 1958, and his Ph.D. in 1968 from the University of New Mexico, Albuquerque. He worked for Sandia National Laboratories from which he retired in 1993 as the Director for National Security Studies. His work at Sandia was principally concerned with integrity and authentication issues surrounding national security, especially those involving command and control of nuclear weapons. His wide range of research interests includes a core focus on combinatorics and graph theory, with applications to cryptography. In 1986, he was honoured with the U.S. Government's E.O. Lawrence Award, as well as the Department of Energy Weapons Recognition of Excellence Award for "Contributions to the Command and Control of Nuclear Weapons" in the same year. In 1996 he was made an honourary Lifetime Fellow of the Institute of Combinatorics and Its Applications. Dr. Simmons has not only made such contributions, but also has a warm, collegial, and humanitarian disposition with a humorous foil, which allows him (as he puts it in [221]) "to pillory the National Security Agency, one of my favorite pastimes."

for the U.S.A. and the former U.S.S.R., say, to ban such testing and have a treaty in place to verify compliance using public-key cryptography.

◆ Nuclear Test Ban Treaty Compliance

To detect an underground nuclear test, seismic sensors can be placed in each country in probable sites where such testing would occur, essentially creating a family of seismic nets. However, each country has to have assurance that the other country will not be able to cheat, so there must be confidence in the scheme through authentication, and the assurance of tamper-resistant seismic sensors. The latter can be accomplished by the host (meaning the host country which is allowing the other side to put sensors in their territory) creating their own tamper-proof sensors.[9.3] Suppose the treaty specifies that each host must allow monitors from the other country who can engage in on-site inspections within the their borders, and suppose that the United Nations (UN) is involved in the treaty as an arbitrator.

The authentication equipment, which we will call HAL, in the downhole seismic device, generates primes p and q, and forms $n = pq$, which we assume is in compliance with security issues in Chapter 6. HAL selects an RSA enciphering exponent e, kept secret from all parties, including the monitor, we'll call Monty, and the host, we'll call Hostvania. HAL also computes d from p, q, and e. Only n and d are made public. Also, HAL collects data m, then computes $c \equiv m^e \pmod{n}$. Hostvania first receives (m, c) from HAL's sensors in response to Monty's request (challenge). They calculate $m \equiv c^d \pmod{n}$ and verify that this copy of m matches their copy (bit for bit). Then Hostvania is convinced that there is no hidden information in the cipher, so they send (m, c) to Monty and the UN, who compute $m \equiv c^d \pmod{n}$ independently, and verify that the two copies of m match. (Compare this scheme with the RSA signature scheme on page 135.) Here is what this scheme ensures:

(1) Since no entity knows e, then by the RSA conjecture, they cannot forge messages that would be accepted as authentic (see Exercise 9.1).

(2) Since n and d are public, Monty, the UN, and Hostvania can independently verify the authenticity of messages.

(3) Since e is kept secret from all entities, no unilateral actions are possible by any party that would be capable of lessening the confidence in the authentication of the message.

(4) No part of the message is concealed from Hostvania, the UN, or Monty. Hence, the above is an application of authenticity without secrecy using public-key cryptography, and each of the entities can try to cheat as much as they wish without compromising the system.

As noted by Simmons at the end of [210], the above scheme has a direct analogue for communication between international banks each having branches in the foreign host country (see Section 7.3).

[9.3]This is essentially what the NSA asked Sandia to do in the late 1970s (see [221] and Footnote 9.2).

In 1984, Simmons [213] discovered a problem with the above scheme. Although a built-in feature of the scheme is that it does not allow for a covert channel to be built into the message (since a process is in place for Hostvania to verify this), HAL could still be used to hide (what Simmons calls) a *subliminal channel*. What this means is that a channel can be implanted so that Hostvania could not detect the use of the covert channel and could not read the hidden part. In particular, as noted by Simmons [219] in 1994 (with reference to the Second Strategic Arms Limitation Treaty (SALT II) between the former U.S.S.R. and the U.S.A.) the subliminal channel could be used to reveal to the other country which of those silos in the host country were loaded with missiles and which were empty. The country in possession of this knowledge would be able to successfully launch a first strike! In the early 1990s, Simmons [216]–[217] came up with a proposed solution to the problem (see also [215], [218], [220]). However, in 1996, Desmedt [70] provided a counterexample to this claim, and demonstrated how several other protocols in the literature are susceptible to this problem. This was addressed by Simmons [222] in 1998. The actual details, including the very definition of *subliminal-channel-free protocol* is beyond the scope of this text. For details consult [70]–[71], as well as the aforementioned papers by Simmons.

Now we look at another public-key cryptographic application to authenticity without secrecy. In Chapter 8, we discussed key management issues including PKI itself in Section 8.3. Among the PKI issues that we considered was the use of certificates, with examples such as PGP (see page 174) for the sending of secure e-mail messages. There is another e-mail specification that is sometimes used in conjunction with PGP (see [81]), called *Multipart Internet Mail Extension* (MIME) developed by IETF (see Footnote 8.8) to attach nontextual data, including graphics, in e-mail messages. However, there is no security attached to MIME. In 1995, a group of vendors, including the leader RSA Data Security Inc., enhanced the specifications and in concert with IETF, created S/MIME, or *Secure* MIME. The current version is S/MIME 3 — see [109] and [193]. The S/MIME Version 3 specifications include PKI concepts such as certificate processing and CRLs (see Footnote 8.11). This includes the use of X.509 certificates (see page 174) with extensions specific to S/MIME (see [110]).

The IETF S/MIME working group has developed many such enhancements. The following is a scheme allowing S/MIME to be used for authentication without secrecy via digital signatures only algorithms, such as DSA. We describe it in the following for public-key enciphering algorithms.

◆ S/MIME Authentication Protocol — Without Secrecy

Alice wants to prove to Bob that she is the sender of a message, m, by exchanging e-mail messages. Their e-mail programs both use a hashing algorithm, h and a public-key enciphering algorithm P.

Protocol Steps:

(1) Alice's e-mail program creates a message digest from m using h, to create $h(m)$, then uses her private key d_A from P to encipher and get $d_A(h(m))$.

(2) Her e-mail program sends an *S/MIME e-mail message*,

$$M = (m, d_A(h(m)), C(A))$$

to Bob, where $C(A)$ is Alice's X.509 certificate.

(3) Bob's e-mail program verifies $C(A)$ and obtain's Alice's public key e_A from it, to decipher and obtain $h(m)$.

(4) Bob's e-mail program independently computes $h(m)$ from m and compares the two copies. If they are both the same, Bob is convinced of Alice's identity since the message cannot have been subject to tampering, given that Alice's private key d_A is secure.

The recommended hash functions for the above protocol are either MD5 or SHA-1. The MD5 algorithm developed from an initial version called MD, created by Rivest (see footnote 3.2 on page 61), with the fifth version, developed in 1991, being the strongest. Simply put, it takes an input of arbitrary length and outputs a 128-bit message digest (see [156] for details of the algorithm). Rivest also played a hand in developing the *Secure Hash Algorithm* SHA-1, which produces a digest of 160 bits, but there is a variant that produces 256-bit digests, called SHA-512. This is part of the Draft FIPS 180-2, announced by NIST on May 30, 2001 (see page 139). There are also the proposed variants SHA-256, which produces 256-bit digests and SHA-384, which produces 384-bit digests. Hence, the strength of the SHA-1 algorithm gives it a higher profile in the cryptographic community. Also, see [156] for a detailed description of SHA-1.

Such *signed-only data*, as in the above protocol, may contain any number of certificates and CRLs to help Bob's e-mail program with the job of path construction and certificate verification.

S/MIME is also a good candidate for illustrating authenticity *with* secrecy. In order to accomplish this, we first need the following protocol.

◆ S/MIME Secrecy Protocol — Without Authentication

Alice wants to send a message M to Bob that no other entity can read. The e-mail programs used by Alice and Bob share a common symmetric-key enciphering algorithm E (typically DES or triple DES — see page 22), and a public-key cryptosystem from which Bob has public/private key pair (e_B, d_B).

Protocol Steps:

(1) Alice's e-mail program generates a random key k, called the *session key* (valid only for this e-mail transaction).

(2) Her e-mail program computes $E_k(M)$, then calculates $e_B(k)$, and sends the S/MIME e-mail message

$$C = (E_k(M), e_B(k), C(A))$$

to Bob were $C(A)$ is as in the preceding protocol.

(3) Bob's e-mail program uses d_B to obtain k, which is used to get M.

The above employs a technique known as *enveloping data* (see page 75). It provides secrecy but not authentication. To get both we need the following.

◆ **S/MIME Authentication and Secrecy Protocol**

If Alice and Bob wish to send a message that is both secret and authenticated, they perform what is called a *nesting of protocols*. This means that they first put the message m through steps (1)–(2) in the authentication protocol to output M. Then M is made secret by putting it through steps (1)–(2) of the secrecy protocol, which produces C. When Bob receives C, he first uses d_B to get M, then performs steps (3)–(4) of the authentication protocol (on M) to both verify m and Alice's identity.

The above protocol provides both the desired security and authenticity. The nesting described therein can also be accomplished in the reverse order. In other words, the message can first be made secret, then the secret output can be authenticated. The security risks of either approach are described in [128] and [192].

We briefly mentioned the *Secure Sockets Layer* (SSL) in Footnote 3.10 on page 67. SSL is an Internet protocol that provides authenticity and secrecy for session-based communication. It provides a secure channel on the client/server model using a secret sharing scheme. The security model of SSL is that it encrypts the channel by enciphering the bits that go through that channel. As mentioned in Footnote 3.10, SSL began with Netscape who originated it in 1994. In 1996, Netscape handed over the specifications of SSL to IETF who worked to standardize the SSL version 3 model, which had been released in 1995. The IETF section responsible for working on the SSL project is called the *Transport Layer Security* (TLS) working group. In 1999, they released TLS version 1, which has now become the IETF standards-track variant of the SSL version 3 protocol (see [72]). For details of the mechanisms behind SSL, see [49].

Although well-studied and widely used, the SSL/TLS mechanism is sometimes considered unsuitable for use on the WWW due to the lack of a central PKI, which can contribute to the proliferation of weak authentication schemes. The reason is that each browser must trust the public keys of a large number of CA root keys that are embedded within it. Most of these keys belong to commercial, third-party CAs, such as *Verisign*, which may have different constraints on issuing certificates as required by any particular application or security domain. On the other hand, the cryptographic power of SSL/TLS is that it operates at the transport level so HTTP runs on top of SSL, called HTTPS.

We now look at another Internet authentication and secrecy scheme. Suppose that Alice wants to access a service provider, whom we will call QZZ.com. How does QZZ.com authenticate Alice? One mechanism is to use a *cookie*, which is an HTTP header embodying a textual string of data that is put into a browser's memory. The textual string must have within it, not only the cookie's name and value, but also the domain name,[9.4] path,[9.5] lifetime,[9.6] and a field for a "secure" label.[9.7] Cookies[9.8] are required to keep track of visitors to the given website, such as Alice to QZZ.com. Alice first must be authenticated by some strong authentication scheme. Once she successfully logs in, QZZ.com gives her two cookies L and T. The T cookie is a timestamp together with a MAC (see Footnote 7.3 on page 138). The L and H cookies will be encrypted using a symmetric-key enciphering algorithm in order to keep any sensitive information confidential. The L cookie will be accepted for a low-level security request such as her homepage on QZZ.com. However, if she asks for her e-mail, for instance, then the L cookie will embed her user identification, and some other pertinent data unique to her. If she wants to buy something from QZZ.com using her credit card, then another cookie H is needed for high-level security. The H cookie will only operate in SSL mode, and requires a secure password from Alice. Internally, QZZ.com uses separate systems of computers to authenticate the transactions. Cookies also embed demographics and other information pertinent to marketing that QZZ.com wants to accomplish.

Although the above QZZ.com is contrived, it represents some aspects of existing authentication and security schemes for e-commerce.

Exercises

9.1. Suppose that instead of the nuclear test ban treaty verification scheme we have discussed in this section, we have the following. Monty picks p, q, e, and securely downloads $n = pq$ and e into HAL. Then Hostvania is given n and d, which Monty also has. Show how Monty can generate undetectable forgeries, and how this could compromise the scheme if the UN, say, is an arbitrator.

[9.4]This is the textual equivalent of the numerical IP address (see Footnote 8.3 on page 171) of a given resource on the Internet.

[9.5]This is a specification of the subset of URLs in a domain for which a cookie is valid. A *URL* is a *Uniform Resource Locator*, which is the global address associated with given data. The first part of he URL indicates which protocol to employ, and the second part indicates the domain name. For instance: http://www.math.ucalgary.ca/~ramollin/ indicates that this is a WWW page and the HTTP protocol should be used. The second part is the domain name where my homepage is located.

[9.6]This is a date string specifying the valid lifetime of the cookie. Once expired, the cookie is discarded, and is rejected as invalid if there is an attempt to use it again.

[9.7]If the cookie is marked *secure*, it will be transmitted only if the communication channel within the host is secure. Usually this means over an HTTPS server.

[9.8]The term "cookie" ostensibly comes from the computer science term for "an opaque portion of data held by an intermediary," according to Paul Bonner [42]. Cookies are needed since, without them, HTTP cannot differentiate among the visits to a website by a given entity. The cookie, in a sense, "stamps" the user by embedding unique data in the user's browser. To be able to state the reason for this, we need to define another concept. The *state* of a protocol may be viewed as a description of the protocol at some point in time. Thus, a *stateless* protocol (sometimes called *non-persistent*) is one that is incapable of distinguishing its description from one time to another. HTTP is a stateless protocol (see footnote 8.5 on page 171). It needs a cookie to mark the visits in order to distinguish them one from the other. By contrast, SSL has two states, one for the client side of the protocol and the other for the server side. In the case of SSL, the interaction between the two states is called a *handshake* (see [49]).

9.2 Other Threats to System Security

There are no such things as applied sciences, only applications of science.
Louis Pasteur (1822–1895) French chemist and bacteriologist

In this section, we discuss security of passwords and the general features of logging into a system and ensuring secure access and use. We have mentioned passwords throughout the text, and defined it in our discussion of EKE at the bottom of page 169, along with variants of it. Anyone who has used a computer over a network understands the notion of *logging on* (sometimes called *signing on*), which involves a process of identification (a *username* or *userid*) together with some authentication, a password, for instance. If the process is made secure, then the legitimate user is uniquely identified to the system at hand. Suppose that the user is attempting to login from home to his or her computer at work (*remote login*). Then passwords may have to travel over unsecured channels, which are susceptible to eavesdropping by Eve and interception by Mallory. Also, bad choices of passwords are common so this compromises security. At work, she is probably protected from intrusion by a *firewall* (a mechanism installed at the juncture where network connections enter a network traffic control centre for a given system) that prevents intruders from accessing the system. However, there needs to be some strong authentication mechanism for remote logins. This is where a secure PKI can assist. If, for instance, Alice needs to remotely access several different servers at the corporation where she works, she would normally need several passwords, one for each server. This impedes her ease of use, and if she tries to short-cut the process in order to circumvent the multi-password requirement, this will reduce security. A secure PKI could assist by supporting a strong authentication protocol so that passwords do not travel over the network. Then Alice can login remotely to a central server (securely) from which she may access other servers necessitating only one login.

Another (potential) feature of such a secure PKI (although not all PKIs provide this) is the *black box*, meaning that there is *end-user transparency* in the sense that Alice, for example, does not need to know much of anything about hardware, or software such as how HTTPS works, or IP addresses. This means that virtually all of the security measures in the PKI are invisible to Alice. What she *does* need to understand can be kept on a "need-to-know" basis, such as first time logins.

PKI yields secure communication through various schemes such as secure e-mail via say, S/MIME (see Section 9.1), or secure server access through TLS (see page 183 and [72]).

Now we look at some of the means for ensuring the above security measures. If Alice chooses a password that is easy to guess, the system will have a program called a *password checker*, which tells Alice when she has chosen a "bad" password (in her initial login). Once Alice has chosen a password that satisfies the checker, then at any time in the future, she can have access, but there will be a limitation (usually three times) that she will be able to re-enter her password, if she forgets it — another security device.

So what is a good password? Here is a potential list of parameters for a *good* password:

(1) It must have at least 7 symbols.

(2) It must use at least three of: uppercase letters, lowercase letters, numbers, and symbols such as \$,#,&, and so on.

(3) No symbol should be repeated.

(4) No personal data such as birthdays, or telephone numbers should be used.

(5) Memorize it — don't write it down, and certainly do not type it into your computer to be stored as a file.

So now we have a good password that has been accepted by the system through the checker. How does the system protect your password? We already have seen a method described on page 115, namely, *salting*, by adding some random bitstring as a pad to the password. Then the system can store a hash of the password together with the salt value and Alice's identification. When Alice enters her password, x, the system finds the salt for Alice, applies it to x, hashes it, and if this matches the stored value, she is granted access. This salting of passwords helps to thwart dictionary attacks (see page 171).

◆ **Birthday Attack**

We are given a hash function $h : \mathcal{S} \mapsto \mathcal{T}$, with $|\mathcal{T}| = n$ and $|\mathcal{S}| > n$, so there is at least one collision. If $|\mathcal{S}| \geq 2n$, there are at least n collisions.[9.9] How do we find one? For a given $m \leq n$ of trials, we could randomly choose elements $s_j \in \mathcal{S}$ with $1 \leq j \leq m$ and compute the $h(s_j)$ to see if we get a collision. The analogue of the above is given in Exercise 9.2. In fact upon solving this exercise, the reader will determine that the probability that there do not exist any collisions is $\prod_{j=1}^{m-1}(1 - j/n)$. However, by Exercise 9.4, $1 - x \approx e^{-x}$ for small x values (such as ours). Hence, the probability of no collisions is,

$$\prod_{j=1}^{m-1}\left(1 - \frac{j}{n}\right) \approx \prod_{j=1}^{m-1} e^{-j/n} = e^{-m(m-1)/(2n)}.$$

Therefore, the probability of at least one collision occurring is,

$$p_c \approx 1 - e^{-m(m-1)/(2n)}. \tag{9.1}$$

Moreover, by Exercise 9.5, if $p_c = 1/2$, then $m \approx 1.17\sqrt{n}$. This tells us quite plainly that by hashing over little more than $\sqrt{n}$ random elements of $\mathcal{S}$, we have a greater than 50% chance of finding a collision. This is the birthday attack. It places a lower bound on the number of bits a hash function should possess

---

[9.9]It can be shown that when $\infty > |\mathcal{S}| \geq 2|\mathcal{T}|$, there exists a probabilistic algorithm that finds a collision for h with probability bigger than $1/2$ (see [156, Fact 9.33] and [225, Theorem 7.1, p. 235]).

in order to be secure, since the birthday attack can find a collision in $O(2^{k/2})$ hashings on an k-bit function. Thus, if $k = 64$, then it is not secure against the birthday attack since only 2^{32} hashings are required.

● Alice Cheats Bob Using the Birthday Attack

The hash function has 64 bits. Alice wants Bob to sign a contract that he thinks will benefit him, and later she wants to "prove" that he signed a contract that actually robs him of his life savings.

(1) Alice prepares two contracts, one that is "good" for Bob, C_G, and one, C_B, which will sign away his savings.

(2) Alice makes very minor changes in each of C_G and C_B. Then she hashes 2^{32} modified versions of C_G and 2^{32} modified versions of C_B.

(3) She compares the two sets of hash values until she finds a collision $h(C_G) = h(C_B)$ and recovers the corresponding preimages.

(4) Alice has Bob sign C_G via the hash of its value.

(5) Later Alice substitutes C_B for C_G whose hash value is the same as that signed by Bob, who has now lost all his money.

The above scenario between Alice and Bob was first described by Yuval [246] in 1979. The hash function MD-5, described on page 182, is essentially the foundation for the Secure Hash Algorithm SHA-1, also described therein. Since these are typically employed in S/MIME, we can be relatively certain that they are secure. Moreover, the proposed variants of SHA-1, discussed earlier, provide very secure alternatives for the future.

We have seen already, in Remark 1.30 on page 27, another type of password attack called *keypress snooping* (also called *keypress sniffing*). If your computer is connected to a network, then Eve can *remotely* install a keylogger (via say a virus) and capture any password without detection (and without using any cryptanalytic techniques). Ostensibly, DOS-based PGP is particularly vulnerable to this type of attack.

There is yet another more insidious attack, first described in 1985 by van Eck [78], called *Van Eck Snooping*[9.10] (also called *Van Eck Sniffing*). In this case, there does not need to be any physical contact with the computer (locally or remotely). All an adversary needs to do is read the electromagnetic

---

[9.10]Wim van Eck was a member of the Bio-engineering Group of the Electronics Department at the Twente University of Technology in the Netherlands, from which he had graduated in 1981. In 1982, he became a member of the Propagation and Electromagnetism Compatibility Department of the Dr. Neher Laboratories of the Netherlands. His work revolves around supervising research projects on emission and susceptibility aspects of telecommunications equipment and related areas. In [78], he uses the term *phreaking* to describe the mechanism of reading radiation for reconstruction of information. This is a term that previously applied to methods for using a telephone system without paying.

radiation from the video display screen (as well as SCSI cables,[9.11] printer and network cables, for instance). The reason is that electrical devices generate electromagnetic fields when they operate. In particular, digital computers generate an enormous amount of electromagnetic signals. A trained adversary can intercept, process, and reconstruct information. Essentially, any device with a diode, microchip, or transistor radiates these fields. In this type of surveillance, an adversary can not only get passwords, but also can reconstruct all activity on a given computer.

Perhaps the most notable use of Van Eck Snooping was accomplished by the F.B.I., which used the attack to monitor the activities of C.I.A. intelligence officer, Aldrich Hazen Ames, known as "Rick" Ames, who turned out to be a Soviet-Russian mole. On October 9, 1993, the F.B.I. began monitoring Ames' computer via Van Eck snooping by using an antenna outside Ames' home and gathered enough evidence to arrest him. He was arrested in February 1994, perhaps the most deleterious mole to have tunnelled into the C.I.A. He was charged with conspiracy to commit espionage on behalf of the former Soviet Union and Russia. On April 28, 1994, Ames and his wife, Rosario, pleaded guilty to charges arising from the espionage activities. In an agreed-upon statement of facts, Ames admitted to having received more than $1.8 million U.S. from the KGB and that $900,000 was in an account for his later activities.

The U.S. military thwarts the Van Eck snooping attack via a project called, *Transient ElectroMagnetic Pulse Emanation STandard* (TEMPEST), joint between the U.S. National Security Agency (NSA) and the U.S. Department of Defense (DoD). Although the term "TEMPEST" was coined in the 1960s, the notions behind building security against radiation-type leaks was invented in 1918 when it was discovered that normal unmodified equipment was allowing classified information to be leaked to the enemy though "compromising emanations". This was largely due to Herbert Yardley[9.12] and his group at the *Black Chamber*. TEMPEST equipment essentially provides shielding, grounding, and bonding to avoid compromising emanations. What is known as the *van Eck receiver* was founded on older video monitors using signals with little

---

[9.11]This is pronounced "scuzzy", the acronym for *Small Computer System Interface*. SCSI is an ANSI standard, designed to provide faster data transmission than serial and parallel ports.
[9.12]Herbert Osborne Yardley was born on April 13, 1889 in Worthington, Indiana. Early in life, he recognized his flair for cryptanalysis when he was employed as a "code clerk" in the State Department at the age of twenty-three. After the declaration of war in April 1917 by the U.S.A., Yardley was made head of the newly established cryptology section of the Military Intelligence Division, MI-8. In May of 1919, he submitted a plan for a permanent cryptology organization. This ultimately came to be known as the *American Black Chamber*. The Black Chamber operated very successfully by cryptanalyzing more than 45,000 enciphered telegrams from various countries. By 1929, however, the Black Chamber was shut down by the Secretary of State, Henry L. Stimson, who disapproved of the Chamber saying "Gentlemen do not read each other's mail". In 1931, Yardley published a book entitled *The American Black Chamber* which was an exposé of the U.S.A.'s weak, if not defenceless, status in the arena of cryptology. It caused a furor in many circles. In fact, when he tried to publish a second book *Japanese Diplomatic Secrets*, it was suppressed by the U.S. government. He involved himself in real estate speculations in the late 1930s, and served as enforcement agent in the Office of Price Administration during WWII. He died of a stroke on August 7, 1958 in Silver Spring, Maryland.

or no standard shielding. The term TEMPEST has been updated, and one often hears the modern term EM*ission* SEC*urity* (EMSEC). Nevertheless, one needs to ensure that, certainly in such circles as the military, all equipment including cables are shielded. Otherwise, if a trained adversary can read the radiation, they can read both the encrypted and the plaintext data (both of which radiate) and separate the two, resulting in a total break of any cryptosystem.

A hacker (an entity who attempts to break into a computer system) may use a *password cracking* software package, which might be an automated exhaustive password search mechanism. Their activities are called *hacking*. A *cracker*, on the other hand, might be a hacker, but is often an entity who is hired by an organization to find weaknesses in their system by attempting to *hack* or *break* their system's security.[9.13] A *system administrator* or *sys admin* is some entity responsible for maintaining the systems and network including the management of accounts. Since one aspect of the power of a sys admin is to be able to override password protection in order to deal with problems including users forgetting their passwords, for example, then hackers who can impersonate a sys admin would be their optimal task. Hence, a secure, in-depth, comprehensive PKI that ensures data integrity including password protection is vital. On page 171, we discussed the SRP protocol that eliminates the need to send passwords over a network. These kinds of protocols in conjunction with a well-developed PKI, help to create a secure system. In the next section, we look at another type of modern-day real-world security concern — wireless authentication.

Exercises

☆ 9.2. Suppose there are $n > 1$ balls in a container numbered from 1 to n inclusive and $1 < m \leq n$ of them are drawn one at a time, listed, and replaced each time. Find the probability that one of the balls is drawn at least twice. Use the formulation to prove that in any room of 23 people, the probability that at least two of them have the same birthday is greater than 50%. This is called the *birthday paradox*.

9.3. Let $n \in \mathbb{N}$, x be a variable, and $D_k(x) = \sum_{j=0}^{\lfloor k/2 \rfloor} \frac{k}{k-j} \binom{k-j}{j} x^{k-2j}$, which is a special case of a *Dickson polynomial* of the first kind, the brackets denoting the binomial coefficient. Alice and Bob have RSA key pairs (e_A, d_A), (e_B, d_B), respectively, with an RSA modulus $n = pq$. Alice encrypts her message m as $D_{e_B}(m) \equiv c \pmod{n}$, where $\gcd(e_B, (p^2-1)(q^2-1)) = 1$, and $D_{d_A}(c) \equiv s \pmod{n}$, is her signed message, which she sends to Bob. How can Bob decipher to get m and verify Alice's identity? (*This is a result by Lidl and Mullen — see [141]–[142].*)

9.4. If e is the natural exponential base, and $x \in \mathbb{R}$ is a small real number, show that $1 - x \approx e^{-x}$.

9.5. Prove that if $p_c = 1/2$ in (9.1), then $m \approx 1.17\sqrt{n}$.

[9.13]However, the debate is open. Some would define a hacker as an entity with benign intent, and a cracker as a malicious entity. On can avoid this by using the term "attacker".

9.3   Wireless Security

Along the electric wire the message came: He is not better — he is just the same.

Anonymous

Data communication is being revolutionized by advances in wireless (radio) mobile communications that augment Internet, terrestrial fiber networks, and satellite communications advances. However, wireless communication is notoriously (and inherently) insecure. Wireless networking was deployed long before the means for making it secure were put in place — and they are still not there. Nearly five million *Wireless Local Area Network* (WLAN) nodes were shipped globally in 2000 with more than ten times that predicted for 2006. Thus, wireless privacy and security standards are fundamental goals to be achieved for a successful multifarious global communications network of the future.

WLANs provide numerous advantages over wired networks, one of the most important being the elimination of cabling costs. The obvious advantage to the user is mobility, but attached to that benefit comes the disadvantage of serious security issues. Broadcasting in the clear leads to the problem of determining whether or not the messages are actually coming from the mobile device in question. Moreover, a serious problem occurs if Mallory can interfere with WLANs transmission so that there is a prevention of the normal communication facilities, called a *denial of service attack* (DOS)-attack. Many computationally intensive operations such as those required for some security operations can be problematic on handheld[9.14] devices such as mobile (cell) phones. Moreover, existing security attempts at encryption schemes have proved to be woefully inadequate. For example, the current WLAN standard, IEEE[9.15] 802.11, called *Wired Equivalent Privacy* (WEP) is very easy to break, and has no key management scheme — a major flaw for any system, as we saw in Chapter 8.[9.16] Furthermore, the fact that so many handheld devices, mobile phones, and PDAs[9.17] are lost or stolen each year (conservative estimates put the total at more than a quarter million in 2001 at airports alone!) then to secure these devices physically is a serious priority. However, we will be concerned herein with cryptographic security — a necessity once physical security is established. The unfortunate

---

[9.14] A *handheld* device, usually a phone or computing mechanism, is one that will conveniently fit into a suitable pocket and can be used while you are holding it in hand.

[9.15] This is the *Institute of Electrical and Electronics Engineers Inc.*, with predecessors the AIEE *American Institute of Electrical Engineers* and the IRE *Institute of Radio Engineers*, dating to 1884.

[9.16] However, RSA Security Incorporated has a "fix" for WEP, called *Fast Packet Keying* (see http://www.rsasecurity.com/rsalabs/technotes/wep-fix.html).

[9.17] A PDA is a *Personal Digital Assistant*, a handheld device that combines computing, telephone, and networking functions. Typically a PDA will have a built-in keyboard or an electronically sensitive pad where a stylus can be used for handwriting to be electronically sent. The first of these was the now defunct *Apple Newton*. The PDA is an acronym now used generically to reference any of a range of such products including Hewlett-Packard's (HP's) *Palmtop* and 3Com's *Palm Pilot*.

tendency of vendors is to focus on features before security and quality issues have been addressed.

PDAs have low power capacity, slow processors, and limited storage capacity, so they are not well suited to carry out cryptographic calculations. Since PDAs were meant to support personal applications, this was not seen, initially, as a problem. Yet, most of these devices depend upon password-based access control, which typically are not secure. Once an adversary gets hold of the password, consider what happens when the PDA user has the same password for many other access points. Corporate employees could thereby subject their corporate resources to compromise. One solution is a smart card to be inserted in order to access a PDA (see Section 9.4). *Global System for Mobile Communication* (GSM) phones already have this capacity and have had it or some time. A feature called the *Subscriber Identity Module* (SIM) is a removable, hardware security module, separate from the mobile phone. The SIM can have an associated *Wireless Application Protocol Identity Module* (WIM), which can be used for application level security.

There exist methods for organizations to "fix" the problem with WEP. For instance, *Cisco Systems*™ employs a CA (see Section 8.3) that generates and distributes RSA key pairs at the client level for authentication. The CA also sends out RC4[9.18] keys for encryption of data. There exist user-based authentication servers sold by vendors for wireless communication, which will not allow a client to access a system before validation. This must be seen as a basic security requirement. Yet most WLANs do not have this minimal security. At a minimum, a proprietary WLAN should only be accessed through a *Virtual Private network* (VPN) to a central server with secure authentication mechanisms. Defenders of WEP say that it was only meant to be a deterrent, not a VPN replacement. WEP essentially uses RC4 in a fashion similar to a one-time-pad, and of course the "one-time-pad" is used significantly more than once. In any case, the fix is needed, whatever the initial intention, since most modern corporations, medical facilities, and military installations, for example, require secure WLANs. The IEEE 802.11i group is working on enhanced security for WLANs and is fixing the security in 802.11. They will have two modes, one updates WEP and addresses its insecurities. The other uses AES and provides for modern authentication and key management using PKI or Kerberos, for instance.

The first step in wireless security, to be carried out before encryption can be used, is authentication. When a mobile phone is used in a commercial network, the initial connection is authenticated by a *base station* in order to ensure proper billing. In military networks, mobile devices have to be authenticated by base stations to ensure that the transmitted data is legitimate. Conversely, the mobile device has to ensure the authenticity of the base station to guarantee validity of

---

[9.18]RC4 is a stream cipher developed at RSA Data Security Inc. by Ron Rivest in 1987 (accounting for the "R" in it as is the "R" in RSA). RSA kept RC4 secret, but in 1994, it was somehow leaked to the *cyberpunks mailing list* on the Internet. RC4 is actually part of SSL (see page 183), among other applications. RC4 is very fast, about 700 times faster than RSA using a 1204-bit modulus.

the receiver. A hybrid system is best suited to the task in these communications where some computations can be carried out at the base station, which will have some higher computing power than the mobile units, thereby solving one of the problems (albeit not being considered in IEEE 802.11i).

We now look at two wireless protocols and examine their suitability. The first [21] was introduced in 1993.

◆ **The Beller-Yacobi Wireless Authentication/Key Agreement Protocol**

Alice has a mobile phone and she wants to authenticate with Bob at the base station who also needs verification of Alice's identity. Their second goal to establish a symmetric session key.

Background Assumptions: A symmetric-key algorithm E is used and a public-key algorithm P is employed (keys to be determined below). Also, there is a digital signature scheme and an associated hash function h. Trent acts as the CA.

Setup Stage:

(1) A prime n_S and a primitive root α modulo n_S are chosen as ElGamal parameters (see page 67).

(2) Trent chooses primes p and q and forms the RSA modulus $n_T = pq$ (which will be used for RSA signatures — see page 135).

(3) Trent's public key is $e_T = 3$, and his private key d_T is computed so that

$$e_T d_T \equiv 1 \,(\mathrm{mod}\ (p-1)(q-1)).$$

(4) Alice and Bob are each given a copy of the public exponent $e_T = 3$ by Trent, as well as the public-key modulus n_T and the public (n_S, α).

(5) Trent generates identity strings I_A and I_B for Alice and Bob, respectively.

Mobile Phone Initialization:

(1) Alice chooses a random integer a (as the ElGamal private signature key — see page 136) in the range $1 < a \le n_S - 2$, and calculates the ElGamal signature public key $u_A \equiv \alpha^a \,(\mathrm{mod}\ n_S)$. She keeps a private, but sends u_A to Trent.

(2) Trent creates the certificate $C(A) = (I_A, u_A, \mathrm{sig}_T(I_A, u_A))$, where $\mathrm{sig}_T \equiv (h(I_A, u_A))^{d_T} \,(\mathrm{mod}\ n_T)$ is Trent's RSA signature. He sends $C(A)$ to Alice.

Base Station Initialization:

(1) Bob chooses an RSA modulus n_B, creates a public key $e_B = 3$ and the corresponding private key d_B. Then he sends n_B to Trent.

(2) Trent creates a certificate $C(B) = (I_B, n_B, \text{sig}_T(I_B, n_B))$, where

$$\text{sig}_T(I_B, n_B) \equiv (h(I_B, n_B))^{d_T} \pmod{n_T},$$

is Trent's signature. He sends $C(B)$ to Bob.

Protocol Steps:

(1) **Precomputation Step**: Alice chooses a random integer x such that $1 < x \leq n_S - 2$ and computes,

$$v \equiv \alpha^x \pmod{n_S}; \quad x^{-1} \pmod{n_S - 1}; \quad \text{and} \quad av \pmod{n_S - 1},$$

where $\gcd(x, n_S - 1) = 1$.

(2) Bob sends $C(B)$ to Alice.

(3) Alice verifies the authenticity of n_B by computing

$$h(I_B, n_B) \equiv \text{sig}_T^3(I_B, n_B) \pmod{n_T}.$$

If valid, she selects a random key $k > 1$ such that $k < n_B - 1$ and sends $P_B(k) \equiv k^3 \pmod{n_B}$ to Bob.

(4) Bob computes

$$k \equiv \text{sig}_B(P_B(k)) \equiv P_B(k)^{d_B} \pmod{n_B}.$$

Then he chooses a random challenge number m, pads it with $t \geq 50$ least significant zeros, to form m', and enciphers it as $E_k(m')$, which he sends to Alice.

(5) Alice uses E_k^{-1} to decipher m', verifies the format, to get m and if valid, she accepts that k came from Bob. Alice then forms $M = (m, I_B)$ and computes

$$w \equiv (M - av)x^{-1} \pmod{n_S - 1}.$$

She sends $E_k((v, w), C(A))$ to Bob, where (v, w) is her ElGamal signature on M and $C(A)$.

(6) Bob uses E_k^{-1} to get (v, w) and $C(A)$. He verifies the authenticity of u_A by computing,

$$h(I_A, u_A) \equiv \text{sig}_T^3(I_A, u_A) \pmod{n_T}.$$

If valid, he creates $M = (m, I_B)$. He verifies Alice's signature on m by calculating

$$\alpha^M \equiv u_A^v v^w \pmod{n_S},$$

(see Exercise 9.6). If all is valid, Bob accepts that k came from Alice.

In step (3), Alice uses a cubic RSA exponentiation to encrypt the message, and in step (5), she uses an ElGamal signature scheme to sign the message. In step (4), Bob uses his corresponding private RSA exponent to decrypt it, and in step (6) uses the cubic RSA public exponent to authenticate. The cubic RSA exponent is an inexpensive public-key algorithm, and the ElGamal signature is also inexpensive in that there is one modular exponentiation, the one in step (5), excluding the *precomputation* in step (1). Moreover, mutual authentication requires both Alice and Bob to demonstrate knowledge of their respective private keys (a for Alice from the ElGamal signature scheme, and d_B for Bob from the RSA signature scheme). Hence, this provides a unique mingling of the two public-key schemes and the exploitation of economy involved in their base operations — a nice feature.

The second hybrid algorithm, first developed [12] in 1994, is given as follows. The protocol provides link security only, meaning that the mobile device and the base station are mutually authenticated to one another which allows the formation of a communication link over which the key agreement can be exercised. This provides a seamless embedding into existing WLANs (not considered in IEEE 802.11i). Another feature is the flexibility in the protocol that allows a choice of the symmetric-key algorithm to be chosen, which therefore has an eye to the future since future developments in symmetric-key technology can be implemented.

◆ Aziz-Diffie Wireless Authentication/Key Agreement Protocol

Alice has a mobile phone and she wants to authenticate with Bob at the base station who also needs verification of Alice's identity.

Background Assumptions: We assume that L_S is a list of symmetric-key cryptosystems containing information on key sizes. Also, Alice and Bob have public/private key pairs (e_A, d_A) and (e_B, d_B), respectively, from a public-key cryptosystem. Moreover, certificates are issued by the CA, Trent. Alice's certificate, for instance, is $C(A) = \text{sig}_T(S_A, V_A, I_A, e_A, I_T)$, where S_A is the serial number of the certificate, V_A is the lifetime (or validity period) of it, I_A is Alice's unique identifier string, e_A is Alice's public key, and I_T is Trent's identifier. Also, ver_T is Trent's public verification algorithm, and sig_T is his signature. As well, Alice and Bob have digital signature schemes sig_{d_A} and sig_{d_B}, respectively.

Authentication Protocol Steps:

(1) Alice selects a nonce n_A, which is a random 128-bit string, and sends $(C(A), n_A, L_S)$ to Bob at the base station.

(2) Bob verifies $C(A)$ using ver_T. If valid, he generates a nonce n_B (128-bit) and selects a symmetric-key cryptosystem E from the intersection of L_S with the set of them supported by his base station. The key sizes are determined by the minimum of what Alice's list suggests and what the base station can support. He extracts e_A from Alice's certificate and sends her $(C(B), e_A(n_B), E, \text{sig}_{d_B}(e_A(n_B), E, n_A, L_S))$.

(3) Alice verifies $C(B)$ using ver_T, and if valid, the signature is verified using e_B. She compares n_A and L_S with what she sent in her message. If they match, she is convinced of Bob's authenticity at the base station. She generates another nonce, n'_A, and sends $(e_B(n'_A), \text{sig}_{d_A}(e_B(n'_A), e_A(n_B)))$ to Bob.

(4) Bob uses e_A to verify the signature. He verifies Alice's identity by comparing $e_A(n_B)$ with what he sent Alice in step (2). If they match, then Alice has been identified to Bob.

Now that they have established a communication link, they can proceed to establish a shared symmetric key through an exchange as follows.

Key Agreement Protocol Steps:

(1) Bob computes $d_B(e_B(n'_A)) = n'_A$ and forms a session key, $n_B \oplus n'_A = n_{A,B}$ by addition modulo 2. He sends, $(e_A(n_B), \text{sig}_{d_B}(e_A(n_{A,B})))$, to Alice.

(2) Alice computes $d_A(e_A(n_B)) = n_B$, and forms $n_B \oplus n'_A = n_{A,B}$, the session key. She sends, $(e_B(n'_A), \text{sig}_{d_A}(e_B(n_{A,B})))$, to Bob, and they now have agreement on a session key.

If Mallory were to attempt to modify L_S by intercepting Alice's message from step (1), Bob will detect this in step (2). When Alice sends her message in step (3), she completes her authentication and sends the second half of the key n'_A to Bob as well. This initializes the key agreement protocol should the authentication protocol be successful. The fact that there is an XORing step by the both of them rather than simply choosing n_B, say, for the key is that the addition of the two key parts reduces the damage that would otherwise occur if one of Alice or Bob's private key is compromised. Hence, with the setup as above, Mallory would have to obtain *both* of Alice and Bob's private keys.

The key exchange is provided in case there is a breach of security and Alice and Bob need to change the session key. This might occur even if the initial connection remains unbroken. Moreover, the key agreement protocol may be initiated by Alice, in which case steps (1) and (2) are interchanged. Since n_B and n'_A are from previous key exchanges, the signed values in the key agreement protocol thwart any attempt by Mallory to launch a replay attack (see page 133).

The Aziz-Diffie protocol has less storage requirements than the Beller-Yacobi protocol, and it is a three-pass protocol, whereas Beller-Yacobi is a four-pass protocol. Also, Aziz-Diffe requires no precomputation step, whereas Beller-Yacobi does (each time it sets up a call). Lastly, Beller-Yacobi offers forward secrecy (see page 171) to Alice but not to Bob, whereas in Aziz-Diffie there is forward secrecy for both.

There exist a number of other wireless protocols that have been developed over the years. For instance, the *Future Public Land Mobile Telecommunications Systems* (FPLMTS) wireless protocol [244] was developed in 1995, and updated to ICG IMT-2000 (see http://www.itu.int/itudoc/itu-t/icg/imt2000/). This is a symmetric-key protocol with three passes, so there are key management problems. There is the *Groupe Spécial Mobile* (GSM), developed in Europe in the early 1990s as a standard for mobile phones (see [166] and [168]). It was the first WLAN architecture to provide user authentication, confidentiality, and key agreement. It is currently used in Europe, Australia, and other countries. These are but a few of many. There are more recent developments proposing the use of elliptic curve cryptography. For instance see [11], which involves the implementation of ECDSA (see Footnote 7.5 on page 139). They demonstrate that their method requires less bandwidth than either the Aziz-Diffie protocol or the Beller-Chang-Yacobi protocol [20], which was a precursor to the Beller-Yacobi protocol given above. Moreover it is shown to have low storage requirements, making it suitable for PDAs and smart cards (see section 9.4). Lastly, they show that there are modest computational requirements for the user. In fact, there is a current brand of opinion that advances elliptic curve cryptography as the new wave for wireless and smart cards. For example, Certicom maintains that its ECC software is better suited than RSA's BSAFE[9.19] in terms of less computational overhead, smaller key sizes, and lower bandwidth. Moreover, RSA Security has a multi-prime proposal, which is specially suited for wireless environments. (See ftp://ftp.rsasecurity.com/pub/pkcs/pkcs-1/pkcs-1v2-1.doc for the latest version of PKCS#1, which includes multi-prime RSA). Not to be outdone, NTRU claims its cryptographic software makes "dinosaurs" of the older standards since theirs is faster than both RSA and ECC. Time will, as always, tell who will be proven to be correct; or as is often the case, perhaps some hybrid will win the day.

From what we have seen, certain observational guidelines emerge for wireless security protocols:

(1) There must be mutual authentication.

(2) Session keys should be securely established as needed.

(3) Public-key computations within protocols should be minimized when used.

(4) Message digests should be used when possible (without reducing efficiency).

(5) Nonrepudiation should be guaranteed by the use of Trent, as well as for authentication, but preferably outside the protocol actions between Alice and Bob.

(6) Challenge-response protocols provide authentication and access control services.

(7) Public-key certificates mutually authenticate Alice and Bob.

(8) There should be authentication between the base station and the CA.

[9.19]RSA Security Inc. has *BSAFE* encryption software for e-commerce, available in a wireless version, (see http://www.rsasecurity.com/products/bsafe/) which is used, for instance, by Palm Inc. in its Palm OS. There are other vendors offering competing software including Certicom (see http://www.certicom.com/) and NTRU (see http://www.ntru.com/).

What does the future hold for wireless communications? The ITU has coordinated an effort to support a framework for what is called *third-generation* (3G) wireless communications services, called *Universal Mobile Telecommunications System* (UMTS), which is part of a larger framework called IMT-2000. The goal is to create an edifice for digital content distribution in mobile communications that can be seamlessly incorporated into the existing second-generation (2G) wireless environment. The first UMTS services were launched in 2001, with over one hundred 3G licenses awarded to date. The goals of UMTS include the retention of the robust features of 2G technology; improvement of 2G security systems; and the provision of new systems that do not exist in 2G systems. For instance, the 3G analogue of a SIM is called a USIM, or *Universal Smart Card*. The advance in 3G technology over SIM cards would be the use of dual slot phones for WIMs not located on the SIM card. UMTS is capable of data rates as high as 2 Mbps (see Footnote 7.1 on page 131) in conjunction with global roaming. Some of the problems that we discussed with 2G technology above will be addressed by providing advanced support for security and data encryption; increased key sizes; enhancement of user identity; confidentiality by employing group keys; basic UMTS algorithms will be public; and advanced enhancements provided for integrity and confidentiality. There are a wealth of other features that we cannot begin to describe here. The interested reader may find more at **http://www.umtsworld.com/**.

The bottom line is that the future of wireless technology depends upon improvement of security services across wireless networks on a global scale, from end to end. There should be a robust mix of public and symmetric-key management techniques to ensure optimal security.

Exercises

9.6. Verify that in (Protocol) step (6) of the Beller-Yacobi Wireless Scheme, Bob's computation at the end actually verifies Alice's signature.

9.7. The Beller-Yacobi is a 4-pass protocol and Aziz-Diffie is a 3-pass protocol. The following is a 2-pass protocol. In what follows (e_A, d_A) and (e_B, d_B) are Alice and Bob's respective public/private key pairs. T_A is a timestamp from Alice's mobile phone, and $C(A), C(B)$ are their respective certificates issued by Trent. Also, α is a primitive root modulo a prime n_S as ElGamal parameters, while a and b are Alice and Bob's respective private ElGamal signature keys.

(1) Alice sends $(C(A), d_A(T_A, \alpha^a))$ to Bob.

(2) Bob checks $C(A)$, and verifies using e_A to get T_A and α^a. If valid, he sends $(C(B), e_A(d_B(T_A, \alpha^b)))$ to Alice. Alice decrypts first using d_A, then e_B. If valid, they have mutually authenticated each other and have agreement on the key α^{ab}, which is a Diffie-Hellman exchange.

What is a potential weakness of this scheme, and what protects against this weakness?

9.4   Smart Cards and Biometrics

The firmness makes my circle just, and makes me end, where I begun.
John Donne (1572–1631) English poet

We have encountered smart cards in our cryptographic travels in Section 7.3 when we discussed digital cash, and in the presentation of the timing attack on page 113, as well as the mention in Footnote 7.10 on page 151 when we described them as stored value cards. Now we are going to look at smart cards in greater detail and depth. First of all, we should settle upon a formal definition of the term, namely, a plastic card containing an embedded integrated microprocessor chip. Smart cards have both memory and computational capacity, and there are numerous kinds of smart cards. The international standard for smart cards is ISO 7816 (see Footnote 8.6 on page 174) *Integrated Circuit Cards with Electrical Contact* (which has six parts relating to everything from the card size to its electrical attributes).

In Section 7.1, we looked into the issue of identification. Smart cards can be used to identify. An identification scheme that we did not discuss yet, but one that is ideally suited for smart card applications is the following [103] introduced in 1988.

◆ **Guillou-Quisquater Identification Scheme**

Background Assumptions: Alice is a smart card that wants to access a bank account at Bob the Bank. She has her identification string I_A stored as a hash $h(I_A)$. Trent has a public RSA modulus $n = pq$, where p and q are private, and the public RSA exponent is e_T. Alice's private key is computed via

$$h(I_A)d_A^{e_T} \equiv 1 \,(\mathrm{mod}\ n).$$

She sends $h(I_A)$ to Bob and must now convince him that she knows d_A without revealing it.

Protocol Steps:

(1) Alice picks a random nonnegative integer $k \leq n - 1$ and computes

$$\alpha \equiv k^{e_T} \,(\mathrm{mod}\ n),$$

which she sends to Bob.

(2) Bob chooses a random nonnegative integer $\ell \leq e_T - 1$, which he sends to Alice.

(3) Alice computes, $\beta \equiv kd_A^\ell \,(\mathrm{mod}\ n)$, which she sends to Bob.

(4) Bob calculates, $\gamma \equiv h(I_A)^\ell \beta^{e_T} \,(\mathrm{mod}\ n)$, and if $\gamma \equiv \alpha \,(\mathrm{mod}\ n)$, then by Exercise 9.8, Alice has been identified to Bob.

Since Alice is able to identify herself to Bob without revealing d_A, this is an example of what is called a *zero-knowledge proof of identity*. We encountered another example of this notion in Section 2.1 (see Remark 2.1 on page 35). The above scheme can be modified to provide a digital signature mechanism that is also well suited for smart card applications (see [104]–[105]). On page 138, we talked about how identification schemes can be converted into signature schemes, and from where the idea originated. The following is another application of that idea.

◆ **Guillou-Quisquater Signature Scheme**

The setup remains in effect from the identification scheme above. The goal is for Alice to sign a message m and have Bob verify the signature. We assume that H is a cryptographic hash function.

Protocol Steps:

(1) Alice selects a random integer r with $1 < r < n - 1$, and computes both: $s \equiv r^{e_T} \pmod{n}$, and $t = H(m, s)$.

(2) Alice then computes $c \equiv r d_A^t \pmod{n}$. She sends $(m, t, c, h(I_A))$ to Bob.

(3) Bob computes $s' \equiv c^{e_T} h(I_A)^t \pmod{n}$, then $t' \equiv H(m, s') \pmod{\phi(n)}$. By Exercise 9.9, if $t' \equiv t \pmod{\phi(n)}$, then he accepts Alice's signature.

Identification and signatures are but two of many applications for smart card environments. We have already discussed their use in e-commerce. There is also the use in health care since smart cards can carry vital medical information, which might include data such as drug allergies, current medications, emergency contact numbers, and other facts that could save a person's life when brought unconscious to a hospital after an accident, say. Perhaps a kidney patient would have a smart card with dialysis and treatment data — already a reality in France and Japan. Naturally, such data would have to be made secure on the card. Also, another important application of smart cards in health care is as an identification method to high security systems that contain confidential patient records or information.

Smart cards that interface with information technology are being developed. Imagine the day when every PC comes with a smart card reader.[9.20] It may not be far away. For instance, the PC/SC workgroup is setting standards for smart cards and their readers to be integrated into computing environments (see http://www.pcscworkgroup.com/).

Another aspect of smart card applications that combines identification with e-commerce is the so-called *Loyalty card* used by retailers globally to reward their loyal customers with points to collect for redemption of gifts. Another developing area is the use of smart cards in *mass transit*. A smart card can be used to store transportation value, and ideally do not need contact to open fare gates. These *contactless cards* may use electrical inductive coupling with the reader and card roughly a millimeter apart. This allows for a large volume of

---

[9.20]Smart cards can be plugged into a variety of hardware devices called readers. These are distinguished by the means with which they interface with a PC, such as say *Universal Serial Bus* (USB). However, we will not be concerned with the hardware issues herein.

traffic to be processed quickly. For instance, Hong Kong initiated such a smart card system in September of 1997. Since that time 2.5 million transactions per day have taken place. When they are done they expect 7.5 million cards to be in use. Currently there are more than four million distributed. As evidence of the international recognition for the state of the art of their system, they were awarded the 1998 *Sesames Award* for the best smart card application at *Cartes* '98. An astonishing 1.6 billion transactions occur in the Hong Kong public transit system yearly. Thus, this smart card application is an amazing achievement, especially when one learns that the system can handle as much as 10 million transactions per day.

In Section 9.3, we discussed wireless aspects of telephony. Smart cards can also be used in pay phones instead of coins, and hundreds of millions of such cards are in use worldwide.

Of course, we are all aware of *debit cards*. However, most in use today have magnetic stripes. Smart cards have three major advantages over magnetic-stripe cards. One is that they can carry as much as 100 times more information and do so more robustly. The second is that they can carry out relatively complex tasks when interfaced with a terminal. One of these tasks might be the above identification and digital signature scheme used to convince a bank that Alice is legitimate in her request to withdraw cash. Third, magnetic-stripe cards are far less secure than smart cards. In particular, this was the driving force in France in its switch to smart cards in the mid-1980s since fraud rates were incredibly high with no end in sight to the escalation. The chips on smart cards are more secure against tampering than magnetic stripes. Recall that we mentioned the term "phreaking" in Footnote 9.10 on page 187, having originally meant the using of telephony without paying. In Europe, where telephone rates are significantly higher than in North America, there was incentive to switch to smart cards as did the French. In fact, the French are responsible for the term "smart card", which has been in development since the 1970s when the French invested a large amount of money into this R&D technology. They originally called these cards *Carte a memoire* or *memory card* in the 1970s. The French government's marketing arm, *Intelimatique*, coined the term *smart card* in 1980.

Although we initially defined smart cards in the outset of this section, there are four types of cards, with differing capabilities, that are typically referenced with this term, so we'll now sort this out. There are those with the capacity to store data, but do not possess data processing capabilities — *standard memory cards*. There are those which contain some built-in logic circuit to access the memory of the card, wired logic, called *intelligent memory cards*. Sometimes these cards can be configured to restrict access via a password. Third, there are *stored-value cards* which have security features hard-wired into the chip at the point of manufacture. Examples of such cards are prepaid phone cards, wherein a terminal inside the pay phone will write a declining balance into the card's memory. The card can be discarded or recharged when the balance is zero. Fourth, there are those which contain memory, a processor, and have data processing capabilities — *processor cards*. Typically, the processor is employed for cryptographic calculations and, for instance, public/private key pairs can be stored and used. Moreover, the chip is more sophisticated and can manage data much like your PC can do. It is the latter that is the one most deserving of the term *smart card*. This is an integrated circuit card with ISO 7816 interface. In fact, the first three are often taken under the heading of *memory cards* and the fourth under the heading of *microprocessor multifunction cards*. Clearly, smart cards have greater security and convenience for transactions of any kind. They are vulnerable to numerous attacks, and we discussed some of those in section 6.1. Although price was initially a problem, this is correcting itself with time and more widespread use. Lastly, the PKS#11 and PKS#15 standards are important for storing and accessing information on smart cards. See http://www.rsasecurity.com/rsalabs/pkcs/index.html.

Now we turn to *biometrics* — the science and its attendant technology for measuring and analyzing biological data, such as fingerprints, eye retinas and irises, voice patterns, and facial geometry, especially for authentication. The digital storage of these characteristics is called a *template*. Later, when an individual's characteristic is measured, it can be compared to the template, and if the comparisons match, the entity is authenticated. The device that measures a biometric is called a *sensor*, such as a retinal scanner or a microphone. This requires an analog-to-digital converter, necessitating the taking of multiple samples to account for measurement variance.

Very few financial institutions use biometrics, but they are often used for physical access to various sites. For instance, the authorities responsible for the security of the *Statue of Liberty* in the U.S.A. are testing, at the time of this writing, a process whereby a picture is taken of each individual (before they are allowed access to the statue) and compared to a database of known criminals and terrorists. If the facial geometry matches any of them, the individual is held. Of course, these matches cannot be exact, so there are algorithms for deciding an *acceptable range* of comparison, which is necessarily somewhat subjective. Sensitivity levels for this range can be set in a similar fashion to the scanners at airport facilities.

One drawback in the use of biometrics is that it requires, say, for a bank, that the customer submit several biometric indicators over a period of time for security, whereas a PIN can be established once for account initiation. Other problems can arise. Suppose that the biometric is a voice identifier and the customer has a severe cold or laryngitis. Then the customer cannot be authenticated. Thus, more than one biometric is needed and some mechanism has to be in place to select a significant subset of them for authentication purposes. For example, an individual's fingerprints, facial geometry, and voice pattern can be embedded into a smart card. Then the entity can present both the smart card, the fingerprints and a check of voice patterns say, can be verified.

Most of us know about fingerprint biometrics, which is the most accepted. Moreover, it is relatively inexpensive, and perhaps the most trusted. However, this trust appears to be misplaced. At Eurocrypt 2001, there was a rump session talk in which it was shown that using a small amount of money a replica of a finger could be created from "gummy", the substance used for gummy bears. These replicas were accepted by all fingerprint readers tested, often with a higher acceptance rate than real fingers! Consequently, this throws serious doubt onto the amount of trust that can be placed on fingerprint scanners. See http://www.counterpane.com/crypto-gram-0205.html#5.

We mentioned two types of optical biometrics, retinal and iris. These are the most common. Retinal and iris scanners are more expensive than fingerprint scanners, but also more reliable in that the error rate is typically 1 in $2 \cdot 10^6$ for the former and 1 in 10^5 for the latter. Facial geometry is perhaps the most subjective of the biometrics, but also less expensive than fingerprint or optical biometric scanners. Voice biometrics are attractive since the technology already exists in the computer industry, and the costs are minimal. Yet error rates are high, and we mentioned earlier if the entity has a cold or some other voice-altering condition, the biometric is unreliable. There are several other types of biometrics such as signatures and hand geometry, but the technology is young in general and needs time to work out the bugs. Accuracy is an important factor, so if costs descend in optical scanners and they become more widespread, this may become the medium of choice. Yet there are factors that can render this biometric unreliable such as cataracts. Again, the optimal solution may require several biometrics to ensure proper identification and accuracy beyond merely subjective limits. It would appear at this juncture that smart cards with several biometrics embedded from which an arbitrary subset can be extracted may be the best compromise. Only the future will give us the answer.

Exercises

9.8. Prove that in the Guillou-Quisquater identification scheme, Alice is identified to Bob if $\gamma \equiv \alpha \pmod{n}$.

9.9. In the Guillou-Quisquater signature scheme, prove that n Bob may accept that Alice's signature is valid, provided that $t' \equiv t \pmod{\phi(n)}$.

9.10. The following is a signature scheme.[9.21] Let $\ell, \ell' \in \mathbb{N}$ such that $\ell + 1 < \ell'$, and assume that H is a cryptographic hash function (see Footnote 3.1), generated using say, SHA-1 (see page 182). Choose two ℓ'-bit safe primes p and q (see page 120) and let $n = pq$. Also, choose quadratic residues t_1, t_2 modulo n and a random $(\ell + 1)$-bit prime p'. The public key is (n, t_1, t_2, p') and the private key is (p, q). Suppose that m is the message to be signed. Select a random $(\ell + 1)$-bit prime $p'' \neq p'$ and a random quadratic residue t_3 modulo n. Then solve $u^{p''} \equiv t_2 t_1^{H(u')} \pmod{n}$, for u where u' satisfies, $t_3^{p'} \equiv u' t_1^{H(m)} \pmod{n}$. Thus, the signature on m is (p'', u, t_3). Show how this signature can be verifies once it is checked that p'' is an $(\ell + 1)$-bit integer different from p'.

[9.21]In 2000, Cramer and Shoup proved that, under reasonable assumptions, this signature scheme is provably secure against adaptive chosen-plaintext attacks. They call their values ℓ, ℓ' *security parameters*. Also, in 2000, the authors of [96] presented an undeniable signature scheme that can be used in conjunction with the RSA scheme given on page 135 — typically a hash and sign mechanism. See also [118].

Appendix A: Letter Frequency Analysis

First of all, to ensure that we know what we are discussing by *frequency count* of a symbol in a given text, such as a letter in ciphertext, we mean the number of occurrences of it therein. Then one looks at *frequency distribution* of a given symbol (or group of symbols, such as digrams and trigrams) by which we mean the ratio of the frequency count of the symbol(s), such as in a cryptogram, to the total number of symbols in a large body of text under consideration. For instance, we will concentrate upon English language texts and look at frequency count of given groups of symbols in a cryptogram in ratio with a frequency count of those symbols over all English texts. However, it must be stressed that no table (and there are many of them) can definitively contain conclusive information on such frequency distributions since no table is capable of taking into account every kind of English text. Nevertheless, there are some commonalities which will serve as lampposts to guide us in our cryptographic journey throughout the text. For instance, the following are the most common words in order of frequency distribution.

THE, OF, ARE, I, AND, YOU, A, CAN, TO, HE,

HER, THAT, IN, WAS, IS, HAS, IT, HIM, HIS

The following are the most common letters to *end* a word, in order of frequency distribution, which is an example of *positional frequency*, wherein the frequency count of the position of a given letter is taken in ratio with the total number of letters occurring in that position over all English texts.

E, T, S, D, N, R, Y

However, the frequency distribution of letters at the *beginnings* of words is different. Most English words begin with the letter **S**, whereas the letter **E** is about halfway into the list and **X** is last. The most common *digrams* in the English language, ordered by frequency distribution, are:

TH IN ER RE AN HE AR EN TI

TE AT ON HA OU IT ES ST OR

The most common *trigrams* are given as follows.

THE AND THA HAT ENT ION FOR TIO HAS

EDT TIS ERS RES TER CON ING MEN THO

Consider the following table, where letters are ordered by frequency count in sets of printer's type. The row below the letters gives the number of frequency count of the individual letters.

Table A.1

E	T	A	I	N	O	S	H	R
$12,000$	$9,000$	$8,000$	$8,000$	$8,000$	$8,000$	$8,000$	$6,400$	$6,200$
D	L	U	C	M	F	W	Y	G
$4,400$	$4,000$	$3,400$	$3,000$	$3,000$	$2,500$	$2,000$	$2,000$	$1,700$
P	B	V	K	Q	J	X	Z	
$1,700$	$1,600$	$1,200$	800	500	400	400	200	

Table A.1 was originally given by Samuel Morse.[A.1] He was primarily concerned with knowing the frequency of letters so that he could give the simplest codes to the most frequently used letters. However, it should be noted that Table A.1 gives the frequency of letters in an English text, which is dominated by a relatively small number of common words. In various tables, the order of the letters varies, in terms of their frequency distributions. However, **E** is always the first and **T** is always the second. In general, the letters

E, T, A, I, N, O, S, H, R

the first row in table A.1, make up more than 70% of English text. In the case of Morse's Table A.1 it is greater than 91%! Some tables are better for specialized situations as that encountered by Morse.

With all this being said, for our purposes, the above tables will provide a working template. Moreover, the concrete facts, such as the predominance of **E** and **T**, as well as the predominance of the letters displayed above will serve us well in our trip through the text.

[A.1]Samuel Finley Breese Morse (1791–1872) was born on April 27 in Charlestown, Massachusetts to Reverend Jedidiah Morse and Elizabeth Breese. Jedidiah was also known as the "father of American geography" and was author of the first text on the subject "Geography Made Easy", published in 1784, which saw twenty-five editions in his lifetime. Samuel attended Phillips Academy in Andover, Massachusetts, then entered Yale College in 1805, graduating in 1810. In 1811, he left for England to study painting, and when he returned in 1815, he became a well-known wayfaring portrait painter, settling in New York in 1825. He founded the "National Academy of Design" and served as its first president from 1826 to 1845. Although, Morse had no formal training in electricity, he nevertheless came to realize that electrical current pulses could be used to convey information over wires. In 1832, he first conceived of a telegraph and had a complete working model by 1837. This was, incidentally, independently and almost simultaneously discovered by the two British inventors, Sir William Cook and Sir Charles Wheatstone (see Footnote 1.8 on page 19). They took out a joint patent in 1837 for the first electric telegraph put into practical use by the British railway system. By 1838, Samuel had invented the *Morse Code*, and in 1854, he was granted patent rights by the U.S. Supreme Court. The first telegraph line in America was established between Baltimore and Washington, and the first message was sent May 24, 1844: "What hath God wrought?" By 1861, the U.S. was linked coast-to-coast by telegraph. Morse died April 2, 1872 in New York City.

Appendix B: Elementary Complexity Theory

First, we need to define some basic concepts. An *algorithm* is a well-defined computational procedure which takes a variable input and halts with an output. An algorithm is called *deterministic* if it follows the same sequence of operations each time it is executed with the same input. A *randomized algorithm* is one that makes random decisions at certain points in the execution, so the execution paths may differ each time the algorithm is invoked with the same input.

The amount of time required for the execution of an algorithm on a computer is measured in terms of *bit operations*, which are defined as follows: addition, subtraction, or multiplication of two binary digits; the division of a two-bit integer by a one-bit integer; or the shifting of a binary digit by one position. The number of bit operations to perform an algorithm is called its *computational complexity* or simply its *complexity*. This method of estimating the amount of time taken to execute a calculation does not take into account such things as memory access or time to execute an instruction. However, these executions are very fast compared with a large number of bit operations, so we can safely ignore them. These comments are made more precise by the introduction of the following notation.

Definition B.1 (**Asymptotic or Big O Notation**) *Suppose that f and g are positive real-valued functions. If there exists a positive real number c such that*

$$f(x) < cg(x) \tag{B.2}$$

for all sufficiently large x, then we write[B.1]

$$f(x) = O(g(x)) \text{ or simply } f = O(g). \tag{B.3}$$

(In the literature $f << g$ is also often used to denote $f = O(g)$.)

The *asymptotic notation* Big O is *the order of magnitude of the complexity*, an *upper bound* on the number of bit operations required for execution of an algorithm in the *worst-case scenario*, namely, the largest running time[B.2] of all inputs of a particular length. The greatest merit of this method for estimating execution time is that it is machine-independent. In other words, it does not rely upon the specifics of a given computer, so the order of magnitude complexity remains the same, irrespective of the computer being used.

[B.1]Here sufficiently large means that there exists some bound $B \in \mathbb{R}^+$ such that $f(x) < cg(x)$ for all $x > B$. We just may not know the value of B explicitly. Often f is defined on $\mathbb{N}$ rather than $\mathbb{R}$, and occasionally over a subset of $\mathbb{R}$.

[B.2]The *running time* is defined as the number of bit operations executed for a given input. Hence, *asymptotic running time* is defined as the measure of how the running time of an algorithm increases as the size of the input increases without bound. The term *expected running time* is an upper bound on the running time for each input (with expectation in the probability sense taken over all inputs of the random number generator used by the algorithm), expressed as a function of the input size (see Remark 1.13 on page 8).

Example B.4 *A simple example to illustrate the asymptotic notation is given by $f(n) = 12n^{21} + 21n^{17} + \pi n^5 + 12n - 1$, wherein we ignore all lower terms except n^{21} and even ignore the coefficient 12 since $f = O(n^{21})$, given that the highest order term dominates the other terms on large inputs, and $O(cf) = O(f)$ for any constant c. The above can be generalized as follows. If $f(n) = O(n^m) + O(n^k)$ where $m \geq k$, then $f(n) = O(n^m)$. When dealing with logarithms, this has an application as follows. Since we have the identity $\log_b(n) = \ln(n)/\ln(b)$ for any base b with $\ln(n) = \log_e(n)$ for the natural base e, then $O(\log_b(n)) = O(\ln(n))$ for any such base b since the constants, as we have seen, are irrelevant in the asymptotic notation. Also, the amount of time taken by a computer to perform a task is (essentially) proportional[B.3] to the number of bit operations. In the simplest possible terms, the constant of proportionality, which we define as the number of nanoseconds[B.4] per bit operation, depends upon the computer being used. This accounts for the machine-independence of the asymptotic method of estimating complexity since the constant of proportionality is of no consequence in the determination of Big O.*

A fundamental *time estimate* in executing an algorithm is *polynomial time* (or simply *polynomial*); that is, an algorithm is polynomial when its complexity is $O(n^c)$ for some constant $c \in \mathbb{R}^+$, where n is the bitlength of the input to the algorithm, and c is independent of n. (As we observed in Example B.4, any polynomial of degree c is $O(n^c)$.) In general, these are the desirable algorithms, since they are the fastest. Therefore, roughly speaking, the polynomial-time algorithms are the *good* or *efficient* algorithms. For instance, the algorithm is constant if $c = 0$; if $c = 1$, it is linear; if $c = 2$, it is quadratic, and so on. Examples of polynomial time algorithms are those for the ordinary arithmetic operations of addition, subtraction, multiplication, and division. On the other hand, those algorithms with complexity $O(c^{f(n)})$ where $c > 1$ is constant and f is a polynomial on $n \in \mathbb{N}$ are *exponential time algorithms* or simply *exponential*. A *subexponential* time algorithm is one for which the complexity for input $n \in \mathbb{N}$ is

$$L_n(r, c) = O(\exp((c + o(1))(\ln n)^r (\ln \ln n)^{1-r})) \qquad (B.5)$$

where $r \in \mathbb{R}$ with $0 \leq r < 1$ and c is a positive a constant (see Footnote 5.2 for a definition of $o(1)$). Subexponential time algorithms are faster than exponential time algorithms but slower than polynomial time algorithms. (Note that the case which is the case $r = 0$ in (B.5) is polynomial in $\ln n$.) For instance, it can be shown that the DLP discussed on page 39 is $L_p(1/3, c)$ where p is the prime modulus under consideration. Exponential time algorithms are, again roughly speaking, the *inefficient* algorithms. For instance, the method of trial-division as a test for primality of $n \in \mathbb{N}$ uses $\sqrt{n}$ steps to prove that n is prime, if indeed it is. If we take the maximum bitlength $N = \log_2 n$ as input, then $\sqrt{n} = 2^{(\log_2 n)/2} = 2^{N/2}$, which is exponential. Algorithms with complexity

---

[B.3]To say that a is proportional to b means that $a/b = c$, a constant, called the constant of proportionality. This relationship is often written as $a \propto b$ in the literature.

[B.4]A nanosecond is $1/10^9$ of a second — a billionth of a second.

$O(c^{f(n)})$ where c is constant and $f(n)$ is more than constant but less than linear are called *superpolynomial*. It is generally accepted that modern-day cryptanalytic techniques for breaking known ciphers are of superpolynomial time complexity, but nobody has been able to prove that polynomial time algorithms for cryptanalyzing ciphers do *not* exist.

The following table gives some illustrations of time complexity analysis of algorithms in terms of what we might call a "real-world scenario".

Suppose that the unit of time on the computer at our disposal is a microsecond (a millionth $(1/10^6)$ of a second), and assume an input of $n = 10^6$ bits. Then the following gives us some time complexities for various values.

complexity of algorithm	time to execute	number of bit ops.
$O(n)$	one second	10^6
$O(n^2)$	$11.5741 = 10^{12}/(10^6 \cdot 24 \cdot 3600)$ days	10^{12}
$O(n^3)$	$31,709 = 10^{18}/(10^6 \cdot 24 \cdot 3600 \cdot 365)$ years	10^{18}
$O(c^{f(n)})$	longer than the known life of the universe	

Complexity theory also divides problems into classes based upon the algorithms required to solve them. To understand this type of complexity, we need to define some basic notions. Let's start with the notion of a *problem* which means a general question to be answered, defined by a description of its *parameters* (sometimes called *free variables*) and the properties the answer, called a *solution*, is required to satisfy. If the parameters are specified, this is called an *instance* of the problem. A *decision problem* is one whose solution is "yes" or "no". A problem is called *intractable* if no polynomial time algorithm could possibly solve it, whereas one can be solved by a polynomial time algorithm is called *tractable*. (See Remark 1.9 on page 7 on *unsolvable, computationally easy* and *computationally infeasible* problem. Also, see Footnote B.9.)

Now we need the notion of a *Turing*[B.5] *machine*, which is a finite state machine having an infinite read-write tape, i.e., our theoretical computer has infinite memory and the ability to search for and retrieve any data from memory.

[B.5] Alan Mathison Turing (1912–1954) was born on June 23, 1912 in Paddington, London, England. By 1933 Turing had already read Russell and Whitehead's *Principia Mathematica* and was deeply immersed in the study of logic. By 1938, he had received his Ph.D. under Alonzo Church at Princeton for a thesis entitled *Systems of Logic Based on Ordinals*. He later returned to England and worked in the British Foreign Office, where he got involved in cryptanalyzing the German Enigma cryptosystem. Toward this end, he invented a machine called the BOMBE which was operational on March 18, 1939. In August 1939, the British government seconded the Code and Cypher School to Bletchley Park, a Victorian country mansion in Buckinghamshire, halfway between Oxford and Cambridge. From 1939 to 1942, the researchers at Bletchley Park, including Turing, were cryptanalyzing U-boat Enigma cryptograms, helping to win the battle for the Atlantic. In 1945, he began work at the National Physical Laboratory in London, where he helped to design the *Automatic Computing Engine*, which led the world at the time as a design for a modern computer. In 1948, Turing became the deputy director of the Computing Laboratory at Manchester, where the first running example of a computer using electronically stored programs was being built. His contributions include pioneering efforts in artificial intelligence. In 1950, he wrote a paper on machine intelligence, presenting what has come to be known as the *Turing test*. By 1951, he had been elected as Fellow of the Royal Society. In 1952, he was arrested for violation of the British homosexuality statutes. His death from potassium cyanide poisoning occurred while he was doing experiments in electrolysis. It is uncertain whether this was an accident or deliberate.

More specifically, a (deterministic, one-tape) Turing Machine has an infinitely long magnetic tape (as its unlimited memory) on which instructions can be written and erased. It also has a processor that carries out the instructions: (1) move the tape right, (2) move the tape left, (3) change the state of the register based upon its current value and a value on the tape, and write or erase on the tape. The Turing Machine runs until it reaches a desired state causing it to halt. A famous problem in theoretical computer science is to determine when a Turing Machine will halt for a given set of input and rules. This is called the *Halting Problem*. Turing proved that this problem is *undecidable*, meaning that there does not exist any algorithm whatsoever for solving it. The *Church-Turing Thesis* (which came out of the 1936 papers of Turing and Church)[B.6] essentially says that the Turing Machine as a model of computation is equivalent to any other model for computation. (Here we may think of a "model" naively as a simplified mathematical description of a computer system.) Therefore, Turing Machines are realistic models for simulating the running of algorithms, and they provide a powerful computational model. However, a Turing Machine is not meant to be a practical design for any actual machine, but is a sufficiently simple model to allow us to prove theorems about its computational capabilities while at the same time being sufficiently complex to include any digital computer irrespective of implementation. (See [223] for the Church-Turing thesis, and for a more number-theoretic approach see [245].)

Complexity theory designates a decision problem to be in class **P** if it can be solved in polynomial time, whereas a decision problem is said to be in class **NP** if it can be solved in polynomial time on a *nondeterministic* Turing Machine, which is a variant of the normal Turing Machine in that it *guesses* solutions to a given problem and checks its guess in polynomial time. Another way to look at the class **NP** is to think of these problems as those for which the *correctness of a guess* at an answer to a question can be proven in polynomial time. Another equivalent way to define the class **NP** is the class of those problems for which a "yes" answer can be verified in polynomial time using some extra information, called a *certificate* (see Chapter 7). For instance, the problem of answering whether or not a given $n \in \mathbb{N}$ is composite is a problem in **NP** since, we can verify in polynomial time if it is composite given the certificate of a nontrivial divisor a of n. However, it is not known if this problem is in **P**. See [156].

The class **P** is a subset of the class **NP** since a problem that can be solved in polynomial time on a *deterministic* machine can also be solved, by eliminating the guessing stage, on a nondeterministic Turing Machine. It is an open problem in complexity theory to resolve whether or not **P** = **NP**. However, virtually everyone believes that they are unequal. It is generally held that most modern

---

[B.6] Alonzo Church (1903–1995) was born June 14, 1903 in Washington, D.C. He held a position as a professor of mathematics at Princeton from 1929 to 1967. Then he went to UCLA to join the faculty as a professor of mathematics and philosophy. Among his many achievements is the 1936 result, called *Church's Theorem*, which says that arithmetic is *undecidable*. His interests involved not only mathematical logic, but also recursion theory and theoretical computer science. He is certainly regarded as one of the greatest logicians of the twentieth century. He died Friday, August 11, 1995 in Hudson, Ohio.

ciphers can be cryptanalyzed in nondeterministic polynomial time. However, in practice it is the deterministic polynomial-time algorithm that is the end-goal of modern-day cryptanalysis. Defining what it means to be a "computationally hard" problem is a *hard problem*. One may say that problems in **P** are *easy*, and those not in **P** are considered to be *hard* (for instance, see [198, pp. 195–196].) However, there are problems that are regarded as computationally easy, yet are not known to be in **P**. For instance, the Miller-Rabin-Selfridge Test, which we is studied in Section 4.4, is such a problem. It is in the class **RP**, called *randomized polynomial time* or *probabilistic polynomial time*. Here, $\mathbf{P} \subseteq \mathbf{RP} \subseteq \mathbf{NP}$. A practical (but mathematically less satisfying) way to define "hard" problems is to view them as those which have continued to resist solutions after a concerted attack by competent investigators for a long time up to the present.

Another classification in complexity theory is the **NP**-complete problem, which is a problem in the class **NP** that can be proved to be as difficult as any problem in the class. Should an **NP**-complete problem be discovered to have a deterministic polynomial time algorithm for its solution, this would prove that $\mathbf{NP} \subseteq \mathbf{P}$, so $\mathbf{P} = \mathbf{NP}$. Hence, we are in the position that there is no proof that there are *any* hard problems in this sense, namely, those in **NP** but not in **P**. Nevertheless, this has not prevented the flourishing of research in complexity theory. The classical **NP**-complete problem is the *Travelling Salesman Problem*: A travelling salesman wants to visit $n \in \mathbb{N}$ cities. Given the distances between them, and a bound B, does there exist a tour of all the cities having total length B or less? The next in the hierarchy of complexity classification is **EXPTIME**, problems that can be solved in exponential time.

Thus far, we have been concerned with the *time* it takes for an algorithm to execute, measured (asymptotically — the worst-case scenario) in terms of the number of bit operations required. Another component of complexity is the amount of computer memory (storage required) for the computation of a given algorithm, called the *space requirement*. Time calculation on a Turing Machine is measured in terms of the number of steps taken before it enters a halt state, as we have discussed above. The *space* used is defined as the number of tape squares visited by the read-write head (where we think of the tape as having infinitely many squares read, written or erased by a "read-write head"). Thus, the notion of *polynomial space* takes on meaning, and since the number of steps in a computation is at least as large as the number of tape squares visited, then any problem solvable in polynomial time is solvable in polynomial space. Thus, we define **PSPACE** as those problems that can be solved in polynomial space, but not necessarily in polynomial time. Hence, **PSPACE**-complete problems are those such that if any one of them is in **NP**, then **PSPACE=NP**, and if any one of them is in **P**, then **PSPACE=P**. At the top of the hierarchy of the classification of problems in terms of complexity is **EXPSPACE**, those problems solvable in exponential space, but not necessarily in exponential time. It is known that $\mathbf{P} \neq \mathbf{EXPTIME}$ and $\mathbf{NP} \subseteq \mathbf{PSPACE} \neq \mathbf{EXPSPACE}$. (There are also the nondeterministic versions, **NPSPACE** and **NEXPSPACE** that we will not discuss here (see [94]).) The following provides an illustration of the above discussion on the hierarchy of problems in complexity theory.

Figure 9.5: **Hierarchy of Problems in Complexity Theory**

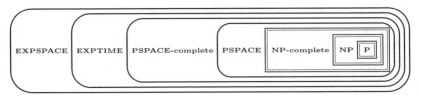

There are other types of complexity such as *circuit complexity*, which looks at the connection between Boolean circuits and Turing Machines as a computational model for studying **P** vis-à-vis **NP** and affiliated problems. We will not discuss these more advanced themes here (see [223] for deeper considerations).

Roughly speaking, complexity theory can be subdivided into two categories: (a) structural complexity theory, and (b) the design and analysis of algorithms. Essentially, category (a) is concerned with lower bounds, and category (b) deals with upper bounds. Basically, the primary goal of structural complexity theory is to classify problems into classes determined by their intrinsic computational difficulty. In other words, how much computing time (and resources) does it take to solve a given problem? As we have seen, the fundamental question in structural complexity theory remains unanswered, namely, does **P** = **NP**? We have been primarily concerned with the analysis of algorithms, which is of the most practical importance to cryptography.

The foundations of complexity theory were laid by the work done starting in the 1930s by Turing and Church, among others (see Footnote B.5). As we have seen in this appendix, the first goal was to formalize the notion of a computer (or realistic model thereof such as the Turing Machine). Then the goal was whether such devices could solve various mathematical problems. One of the outcomes of this research, again as we have seen, is that there are problems that cannot be solved by a computer. This was contrary to the program set out by David Hilbert in 1900, which sought to show that all mathematical problems could be answered in a deterministic fashion. For example, *Hilbert's tenth problem* was to find an algorithm for testing whether a polynomial has an integral root. The Church-Turing thesis, discussed above, provides the precise definition of an algorithm necessary to resolve Hilbert's tenth problem. In 1970, Matiyasevich[B.7]

[B.7]Yuri Matiyasevich was born March 2, 1947 in Leningrad, the USSR, and is currently a Russian citizen. He obtained his Doctor of Sciences in Physics and Mathematics at the Steklov Institute of Mathematics, Moscow in 1972 (approved by the Higher Attestation Committee in 1973). He became a professor at St. Petersburg State University in 1995, was awarded a *Docteur honoris Causa* by the University of Auvergne, France in 1996, and became a Correspondent Member of the Russian Academy of Sciences in 1997. Among his other honours are the A.A. Markov Prize in Mathematics from the Academy of Sciences of the USSR in 1980; a Humboldt Research Award in 1997; and the Pacific Institute of Mathematical Sciences Distinguished Chair held in 1999–2000. He is currently head of the Laboratory of Mathematical Logic at the St. Petersburg Department of the Steklov Institute of Mathematics (see:

proved that there does not exist an algorithm for testing whether a polynomial has integral roots. Hence, Hilbert's tacit assumption, that an algorithm existed and merely had to be found, was flawed. In fact, the Church-Turing thesis may be considered to be the cement between informal and formal definitions of an algorithm.

It may be argued that one of the pinnacles of complexity theory is the above classification of problems according to computational difficulty. In cryptography, the primary goal is to establish cryptosystems that are computationally easy (to use), whereas all algorithms for cryptanalysis are computationally hard.

The establishment of complexity theory may be credited to the pioneering work of Stephen Cook,[B.8] Richard Karp,[B.9] Donald Knuth,[B.10] and Michael Rabin.[B.11] Each of these individuals has since been awarded the highest honour in computer science research — the Turing Award.

---

http://www.pdmi.ras.ru/). He has more than sixty publications and one book *Hilbert's Tenth Problem* first published in 1993.

[B.8]Stephen Cook (1939–) was born in Buffalo, New York. In 1966, he received his Ph.D. from Harvard after which he spent four years as an Assistant Professor at the University of California, Berkeley (UC Berkeley). In 1970, he was appointed Associate Professor to the Department of Computer Science at the University of Toronto. By 1975, he had attained full professorship and by 1985 the rank of University Professor. In 1982, he won the Turing award for his work in complexity theory. Among numerous other awards, he has also been recognized as a Fellow of the Royal Society of Canada.

[B.9]Richard M. Karp (1935–) received his Ph.D. in Applied Mathematics from Harvard in 1959, after which he joined the IBM Thomas J. Watson Research Center until 1968. From then until 1994, he held the position of Professor of Computer Science, Mathematics, and Operations Research at UC Berkeley. From 1994 until June of 1999 (when he returned to UC Berkeley), he held a position as Professor of Computer Science and Adjunct Professor of Molecular Biotechnolgy at the University of Washington. His fundamental research activities are glued together by his interest in combinatorial algorithms. Among his numerous achievements is the Turing award, received in 1985, for his extensions of Stephen Cook's ideas on **NP**-completeness. The *Cook-Karp thesis* says that if a problem is in **P**, then it is said to be *computationally tractable (feasible)*, whereas a problem which is not in **P** is said to be *computationally intractable (infeasible)*, albeit no proof of this thesis exists.

[B.10]Donald Knuth (1938–) first saw a computer (an IBM 650) in his freshman year at Case Institute of Technology (CIT), now called Case Western Reserve, in 1956. He was so fascinated that he stayed up all that night reading the manual on the 650, teaching himself some fundamentals for programming it, and was able to improve upon the programming techniques illustrated therein. By 1960, he was graduated from CIT, and in 1963, he achieved his Ph.D. from the California Institute of Technology (Cal Tech), after which he joined the faculty there as a professor of mathematics. In 1968, he went to Stanford University and in 1977 was endowed with its first chair in computer science. By 1993, he was Professor Emeritus of (the art of) Computer Programming. By his own admission, his life's work is his magnum opus: *The Art of Computer Programming*, which he actually started writing in 1966 and which today has three volumes (of a projected seven volumes, the fourth of which is slated to be out by 2007). Among his numerous recognitions is the Turing award given to him in 1974. He now lives on the Stanford University campus with his wife and two children.

[B.11]Michael Rabin (1931–) was born in Breslau, Germany (now Wroclaw, Poland) in 1931. In 1956, he obtained his Ph.D. from Princeton University where he later taught. In 1958, he moved to the Hebrew University in Jerusalem. He is known for his seminal work in establishing a rigorous mathematical foundation for finite automata theory. For such achievements, he was co-recipient of the 1976 Turing award, along with Dana S. Scott, both students of Church (see Footnote B.6). He now divides his time between positions at Harvard and the Hebrew University in Jerusalem.

Appendix C: Fundamental Facts

Throughout the text, we will be using the notational devices for multiplication and addition.

◆ **The Sum Notation**

In general, if we have numbers $a_m, a_{m+1}, \cdots, a_n$ $(m \leq n)$, we may write their sum as

$$\sum_{i=m}^{n} a_i = a_m + a_{m+1} + \cdots a_n.$$

The letter i is the *index of summation* (and any letter may be used here), n is *the upper limit of summation*, m is *the lower limit of summation*, and a_i is a *summand*.

The close cousin of the summation notation is the following symbol.

◆ **The Product Symbol**

The multiplicative analogue for the summation notation is the *product symbol* denoted Π, upper case Greek *pi*. Given $a_m, a_{m+1}, \ldots, a_n \in \mathbb{R}$ $(m \leq n)$, their product is denoted:

$$\prod_{i=m}^{n} a_i = a_m a_{m+1} \cdots a_n.$$

The letter i is the *product index*, m is the *lower product limit*, n is the *upper product limit*, and a_i is a *multiplicand*.

The notation $m \mid n$ for integers m, n means that m *divides* n, namely, there exists an integer ℓ such that $n = \ell m$. Behind the basic notion of divisibility is a special kind of divisor that we will use throughout.

Definition C.1 (The Greatest Common Divisor)

If $a, b \in \mathbb{Z}$ are not both zero, then the greatest common divisor *or* gcd *of a and b is the natural number g such that $g \mid a$, $g \mid b$, and g is divisible by any common divisor of a and b, denoted by $g = \gcd(a, b)$.*

We have a special term for the case where the gcd is 1.

Definition C.2 (Relative Primality)

If $a, b \in \mathbb{Z}$, and $\gcd(a, b) = 1$, then a and b are said to be relatively prime *or* coprime. *Sometimes the phrase a is prime to b is also used. If $\{a_1, a_2, \ldots, a_n\}$ has the property that $\gcd(a_i, a_j) = 1$, for each $i \neq j$, we say that the elements are* pairwise relatively prime.

Of particular importance for divisibility is the following algorithm.

Theorem C.3 (The Division Algorithm)

If $a \in \mathbb{N}$ and $b \in \mathbb{Z}$, then there exist unique integers $q, r \in \mathbb{Z}$ with $0 \leq r < a$, and $b = aq + r$.

Proof. See [163, Theorem 1.3.1, p. 33]. □

Euclid's celebrated algorithm tells us how to find the gcd based upon the division algorithm.

Theorem C.4 (The Euclidean Algorithm)

Let $a, b \in \mathbb{Z}$ $(a \geq b > 0)$, and set $a = r_{-1}, b = r_0$. By repeatedly applying the Division Algorithm, we get $r_{j-1} = r_j q_{j+1} + r_{j+1}$ with $0 < r_{j+1} < r_j$ for all $0 \leq j < n$, where n is the least nonnegative number such that $r_{n+1} = 0$, in which case $\gcd(a, b) = r_n$.

Proof. See [163, Theorem 1.3.3, p. 37]. □

We will have occasion to use the following generalization of Theorem C.4 throughout the text.

Theorem C.5 (The Extended Euclidean Algorithm)

Let $a, b \in \mathbb{N}$, and let q_i for $i = 1, 2, \ldots, n + 1$ be the quotients obtained from the application of the Euclidean Algorithm to find $g = \gcd(a, b)$, where n is the least nonnegative integer such that $r_{n+1} = 0$. If $s_{-1} = 1$, $s_0 = 0$, and

$$s_i = s_{i-2} - q_{n-i+2}s_{i-1},$$

for $i = 1, 2, \ldots, n + 1$, then

$$g = s_{n+1}a + s_n b.$$

Proof. See [163, Theorem 1.3.4, p. 38]. □

The elementary fundamental fact behind factorization algorithms studied in Chapter 5 is the following.

Theorem C.6 (The Fundamental Theorem of Arithmetic)

Let $n \in \mathbb{N}, n > 1$. If $n = \prod_{i=1}^{r} p_i = \prod_{i=1}^{s} q_i$, where the p_i and q_i are primes, then $r = s$, and the factors are the same if their order is ignored.

Proof. See [163, Theorem 1.4.2, pp. 44–45]. □

Also in Chapter 5, we will need the following basic concept related to the gcd.

Definition C.7 (The Least Common Multiple)

If $a, b \in \mathbb{Z}$, then the smallest natural number which is a multiple of both a and b is the least common multiple of a and b, denoted by $\mathrm{lcm}(a, b)$.

Throughout the text, we will be working within the following structure.

Definition C.8 (The Ring $\mathbb{Z}/n\mathbb{Z}$)
  For $n \in \mathbb{N}$, the set

$$\mathbb{Z}/n\mathbb{Z} = \{\overline{0}, \overline{1}, \overline{2}, \ldots, \overline{n-1}\}$$

is called the Ring of Integers Modulo n, *where $\overline{m}$ denotes the congruence class of m modulo n.*[C.1]

A special kind of division is given a special notation as follows. If $n \in \mathbb{N}$ and $a, b \in \mathbb{Z}$, we say that a is *congruent to b* modulo n if $n \mid (a - b)$ and we denote this by $a \equiv b \,(\mathrm{mod}\ n)$. If $n \nmid (a - b)$, then we write $a \not\equiv b \,(\mathrm{mod}\ n)$. With the basic notion of congruential arithmetic comes the following notion that is essential in many cryptographic algorithms such as RSA.

Definition C.9 (Modular Multiplicative Inverses)
  Suppose that $a \in \mathbb{Z}$, and $n \in \mathbb{N}$. A multiplicative inverse of the integer a modulo n *is an integer x such that $ax \equiv 1 \,(\mathrm{mod}\ n)$. If x is the least positive such inverse, then we call it the* least multiplicative inverse of the integer a modulo n, *denoted $x = a^{-1}$.*

Since cryptosystems related to RSA are essentially predicated upon the following celebrated result, we include it here for convenience.

Theorem C.10 (Fermat's Little Theorem)
  If $a \in \mathbb{Z}$, and p is a prime such that $\gcd(a, p) = 1$, then $a^{p-1} \equiv 1 \,(\mathrm{mod}\ p)$.

Proof. See [163, Theorem 2.1.4, p. 80].   □

Knowing when the integers modulo n form a field is essential as is the structure of the units modulo n in what follows.

Theorem C.11 (The Field $\mathbb{Z}/p\mathbb{Z}$)
  If $n \in \mathbb{N}$, then $\mathbb{Z}/n\mathbb{Z}$ is a field if and only if n is prime.

Proof. See [165, Theorem 2.13, p. 62].   □

We employ the notation F^* to denote the multiplicative group of nonzero elements of a given field F. In particular, when we have a finite field $\mathbb{Z}/p\mathbb{Z} = \mathbb{F}_p$ of p elements for a given prime p, then $(\mathbb{Z}/p\mathbb{Z})^*$ denotes the multiplicative group of nonzero elements of $\mathbb{F}_p$. This is tantamount to saying that $(\mathbb{Z}/p\mathbb{Z})^*$ is the group of units in $\mathbb{F}_p$, and $(\mathbb{Z}/p\mathbb{Z})^*$ is cyclic. Thus, this notation and notion may

---

[C.1]When the context is clear and no confusion can arise when talking about elements of $\mathbb{Z}/n\mathbb{Z}$, we will eliminate the *overline bars*.

be generalized as follows. Let $n \in \mathbb{N}$ and let the group of units of $\mathbb{Z}/n\mathbb{Z}$ be denoted by $(\mathbb{Z}/n\mathbb{Z})^*$. Then

$$(\mathbb{Z}/n\mathbb{Z})^* = \{\bar{a} \in \mathbb{Z}/n\mathbb{Z} : 0 \leq a < n \text{ and } \gcd(a, n) = 1\}. \tag{C.12}$$

Numerous times we will need to solve systems of congruences for which the following result from antiquity is most useful.

Theorem C.13 (Chinese Remainder Theorem)
Let $n_i \in \mathbb{N}$ for natural numbers $i \leq k \in \mathbb{N}$ be pairwise relatively prime, set $n = \prod_{j=1}^{k} n_j$ and let $r_i \in \mathbb{Z}$ for $i \leq k$. Then the system of k simultaneous linear congruences given by:

$$x \equiv r_1 \pmod{n_1},$$

$$x \equiv r_2 \pmod{n_2},$$

$$\vdots$$

$$x \equiv r_k \pmod{n_k},$$

has a unique solution modulo n.

Proof. See [165, Theorem 2.29, p. 69]. □

The natural generalization of Fermat's idea is the following, which provides the modulus for the RSA enciphering and deciphering exponents, for instance.

Definition C.14 (Euler's ϕ-Function)
For any $n \in \mathbb{N}$ the Euler ϕ-function, also known as Euler's Totient $\phi(n)$ is defined to be the number of $m \in \mathbb{N}$ such that $m < n$ and $\gcd(m, n) = 1$.

Theorem C.15 (The Arithmetic of the Totient)
If $n = \prod_{j=1}^{k} p_j^{a_j}$ where the p_j are distinct primes, then

$$\phi(n) = \prod_{j=1}^{k} \phi(p_j^{a_j}) = \prod_{j=1}^{k} (p_j^{a_j} - p_j^{a_j-1}) = \prod_{j=1}^{k} (p_j - 1)p_j^{a_j-1}.$$

Proof. See [165, Theorem 2.22, p. 65]. □

Theorem C.16 (Euler's Generalization of Fermat's Little Theorem)
If $n \in \mathbb{N}$ and $m \in \mathbb{Z}$ such that $\gcd(m, n) = 1$, then $m^{\phi(n)} \equiv 1 \pmod{n}$.

Proof. See [163, Theorem 2.3.1, p. 90]. □

Example C.17 *Let $n \in \mathbb{N}$. Then the cardinality of $(\mathbb{Z}/n\mathbb{Z})^*$ is $\phi(n)$. Hence, if G is a subgroup of $(\mathbb{Z}/n\mathbb{Z})^*$, $|G| \mid \phi(n)$.*

The caclulus of integer orders and related primitive roots is an underlying fundamental feature of cryptographic problems such as the discrete log problem.

Definition C.18 (Modular Order of an Integer)

Let $m \in \mathbb{Z}$, $n \in \mathbb{N}$ and $\gcd(m, n) = 1$. The order of m modulo n is the smallest $e \in \mathbb{N}$ such that $m^e \equiv 1 \pmod{n}$, denoted by $e = \operatorname{ord}_n(m)$, and we say that m belongs to the exponent e modulo n.

Note that the modular order of an integer given in Definition C.18 is the same as the element order in the group $(\mathbb{Z}/n\mathbb{Z})^*$.

Proposition C.19 (Divisibility by the Order of an Integer)

If $m \in \mathbb{Z}$, $d, n \in \mathbb{N}$ such that $\gcd(m, n) = 1$, then $m^d \equiv 1 \pmod{n}$ if and only if $\operatorname{ord}_n(m) \mid d$. In particular, $\operatorname{ord}_n(m) \mid \phi(n)$.

Proof. See [165, Proposition 4.3, p. 161].   □

Definition C.20 (Primitive Roots)

If $m \in \mathbb{Z}$, $n \in \mathbb{N}$ and

$$\operatorname{ord}_n(m) = \phi(n),$$

then m is called a primitive root modulo n. In other words, m is a primitive root if it belongs to the exponent $\phi(n)$ modulo n.

Theorem C.21 (Primitive Root Theorem)

An integer $n > 1$ has a primitive root if and only if n is of the form $2^a p^b$ where p is an odd prime, $0 \leq a \leq 1$, and $b \geq 0$ or $n = 4$. Also, if m has a primitive root, then it has $\phi(\phi(n))$ of them.

Proof. See [165, Theorem 4.10, p. 165].   □

The following will be most useful in Chapter 4. The result can be easily proved using Definition C.20 and Theorem C.10.

Proposition C.22 (Primitive Roots and Primality)

(1) If m is a primitive root modulo an odd prime p, then for any prime q dividing $(p-1)$, $m^{(p-1)/q} \not\equiv 1 \pmod{p}$.

(2) If $m \in \mathbb{N}$, p is an odd prime, and $m^{(p-1)/q} \not\equiv 1 \pmod{p}$ for all primes $q \mid (p-1)$, then m is a primitive root modulo p.

The following generalization of the Legendre symbol will be required for understanding, for example, the primality test on page 84.

Definition C.23 (The Jacobi Symbol)

Let $n > 1$ be an odd natural number with $n = \prod_{j=1}^{k} p_j^{e_j}$ where $e_j \in \mathbb{N}$ and the p_j are distinct primes. Then the Jacobi Symbol of a with respect to n is given by

$$\left(\frac{a}{n}\right) = \prod_{j=1}^{k} \left(\frac{a}{p_j}\right)^{e_j},$$

for any $a \in \mathbb{Z}$, where the symbols on the right are Legendre Symbols.

Theorem C.24 (Properties of the Jacobi Symbol)

Let $m, n \in \mathbb{N}$, both odd, and $a, b \in \mathbb{Z}$. Then

(1) $\left(\dfrac{ab}{n}\right) = \left(\dfrac{a}{n}\right)\left(\dfrac{b}{n}\right)$.

(2) $\left(\dfrac{a}{n}\right) = \left(\dfrac{b}{n}\right)$ if $a \equiv b \pmod{n}$.

(3) $\left(\dfrac{a}{mn}\right) = \left(\dfrac{a}{m}\right)\left(\dfrac{a}{n}\right)$.

(4) $\left(\dfrac{-1}{n}\right) = (-1)^{(n-1)/2}$.

(5) $\left(\dfrac{2}{n}\right) = (-1)^{(n^2-1)/8}$.

(6) If $\gcd(a, n) = 1$ where $a \in \mathbb{N}$ is odd, then

$$\left(\frac{a}{n}\right)\left(\frac{n}{a}\right) = (-1)^{\frac{a-1}{2} \cdot \frac{n-1}{2}},$$

which is the Quadratic Reciprocity Law for the Jacobi Symbol.

Proof. See [165, Theorem 4.40, p. 175]. □

Theorem C.25 (Euler's Criterion for Power Residue Congruences)

Let $e, c \in \mathbb{N}$ with $e \geq 2$, $b \in \mathbb{Z}$, $p > 2$ is prime with $p \nmid b$, and $g = \gcd(e, \phi(p^c)) \mid b$. Then the congruence

$$x^e \equiv b \pmod{p^c} \tag{C.26}$$

is solvable if and only if

$$b^{\phi(p^c)/g} \equiv 1 \pmod{p^c}.$$

Proof. See [165, Theorem 4.17, p. 168]. □

When $b = 1$ and we are concenrned with quadratic residues this becomes the following.

Corollary C.27 (Euler's Criterion for Quadratic Residues)
  Let p be an odd prime. Then

$$c^{(p-1)/2} \equiv \left(\frac{c}{p}\right) \pmod{p}.$$

Theorem C.28 (Solutions of Power Residue Congruences)
  *Let n be an odd natural number with $n = \prod_{j=1}^{r} p_j^{m_j}$ for distinct primes p_j
$(1 \le j \le r)$. Then*

$$x^t \equiv 1 \pmod{n}$$

has exactly $g = \prod_{j=1}^{r} \gcd(t, \phi(p_j^{m_j}))$ solutions. Also, if

$$x^t \equiv -1 \pmod{n}$$

has a solution, then it has g solutions.

Proof. See [163, Theorem 3.3.2, p. 155].   □

  Of course, the gem of number theory will come into play in our cryptographic
journey, so we state it here for finger-tip reference.

Theorem C.29 (Prime Number Theorem)
  Let the number of primes less than x be denoted by $\pi(x)$. Then

$$\lim_{x \to \infty} \frac{\pi(x)}{x/\log(x)} = 1,$$

denoted by $\pi(x) \sim x/\log(x)$.[C.2]

Proof. See [229].   □

  We now provide a brief reminder of vector space basics and elementary
matrix theory that we will need to describe other concepts to be used in the
text.

◆ **Vector Spaces**
  A *vector space* consists of an additive abelian group V and a field F together
with an operation called *scalar multiplication* of each element of V by each
element of F on the left, such that for each $r, s \in F$ and each $\alpha, \beta \in V$ the
following conditions are satisfied:

C.30. $r\alpha \in V$.

C.31. $r(s\alpha) = (rs)\alpha$.

[C.2]In general, if f and g are functions of a real variable x, then $f(x) \sim g(x)$ means
$\lim_{x \to \infty} f(x)/g(x) = 1$. Such functions are said to be *asymptotic*.

C.32. $(r + s)\alpha = (r\alpha) + (s\alpha)$.

C.33. $r(\alpha + \beta) = (r\alpha) + (r\beta)$.

C.34. $1_F\alpha = \alpha$.

The set of elements of V are called *vectors* and the elements of F are called *scalars*. The generally accepted abuse of language is to say that V is a *vector space over F*. If V_1 is a subset of a vector space V that is a vector space in its own right, then V_1 is called a *subspace of V*.

Definition C.35 (Bases, Dependence, and Finite Generation)

If S is a subset of a vector space V, then the intersection of all subspaces of V containing S is called the subspace generated *by S, or* spanned *by S. If there is a finite set S, and S generates V, then V is said to be* finitely generated. *If $S = \varnothing$, then S generates the zero vector space. If $S = \{m\}$, a singleton set, then the subspace generated by S is said to be the* cyclic subspace generated *by m.*

A subset S of a vector space V is said to be linearly independent *provided that for distinct $s_1, s_2, \ldots, s_n \in S$, and $r_j \in F$ for $j = 1, 2, \ldots, n$,*

$$\sum_{j=1}^{n} r_j s_j = 0 \text{ implies that } r_j = 0 \text{ for } j = 1, 2, \ldots, n.$$

If S is not linearly independent, then it is called linearly dependent. *A linearly independent subset of a vector space that spans V is called a* basis *for V. The number of elements in a basis is called the* dimension *of V. A* hyperplane *H is an $(n - 1)$-dimensional subspace of an n-dimensional vector space V.*

Example C.36 For a given prime p, $m, n \in \mathbb{N}$, the finite field $\mathbb{F}_{p^n}$ is an n-dimensional vector space over $\mathbb{F}_{p^m}$ with p^{mn} elements.

◆ **Basic Matrix Theory**

If $m, n \in \mathbb{N}$, then an $m \times n$ *matrix* (read "m by n matrix") is a rectangular array of entries with m rows and n columns. For simplicity, we will assume that the entries come from a field F. (For instance, in Section 8.1, we will be using $\mathbb{F}_p$ to describe a secret sharing scheme base on vectors and matrices.) If A is such a matrix, and $a_{i,j}$ denotes the entry in the ith row and j^{th} column, then

$$A = (a_{i,j}) = \begin{pmatrix} a_{1,1} & a_{1,2} & \cdots & a_{1,n} \\ a_{2,1} & a_{2,2} & \cdots & a_{2,n} \\ \vdots & \vdots & & \vdots \\ a_{m,1} & a_{m,2} & \cdots & a_{m,n} \end{pmatrix}.$$

Two $m \times n$ matrices $A = (a_{i,j})$, and $B = (b_{i,j})$ are equal if and only if $a_{i,j} = b_{i,j}$ for all i and j. The matrix $(a_{j,i})$ is called the *transpose* of A, denoted by

$$A^t = (a_{j,i}).$$

Addition of two $m \times n$ matrices A and B is done in the natural way.

$$A + B = (a_{i,j}) + (b_{i,j}) = (a_{i,j} + b_{i,j}),$$

and if $r \in F$, then $rA = r(a_{i,j}) = (ra_{i,j})$, called *scalar multiplication*.

Matrix products are defined by the following.

If $A = (a_{i,j})$ is an $m \times n$ matrix and $B = (b_{j,k})$ is an $n \times r$ matrix, then the *product* of A and B is defined as the $m \times r$ matrix:

$$AB = (a_{i,j})(b_{j,k}) = (c_{i,k}),$$

where

$$c_{i,k} = \sum_{\ell=1}^{n} a_{i,\ell} b_{\ell,k}.$$

Multiplication, if defined, is associative, and distributive over addition. If $m = n$, then the identity $n \times n$ matrix:

$$I_n = \begin{pmatrix} 1_F & 0 & \cdots & 0 \\ 0 & 1_F & \cdots & 0 \\ \vdots & \vdots & \vdots & \vdots \\ 0 & 0 & \cdots & 1_F \end{pmatrix},$$

is called *the $n \times n$ identity matrix*, where 1_F is the identity of F.

Another important aspect of matrices that we will need throughout the text is motivated by the following. Consider the 2×2 matrix with entries from F:

$$A = \begin{pmatrix} a & b \\ c & d \end{pmatrix},$$

then $ad - bc$ is called the *determinant* of A, denoted by $\det(A)$. More generally, we may define the determinant of any $n \times n$ matrix with entries from F for any $n \in \mathbb{N}$. The determinant of any $r \in F$ is just $\det(r) = r$. Thus, we have the definitions for $n = 1, 2$, and we may now give the general definition inductively. The definition of the determinant of a 3×3 matrix

$$A = \begin{pmatrix} a_{1,1} & a_{1,2} & a_{1,3} \\ a_{2,1} & a_{2,2} & a_{2,3} \\ a_{3,1} & a_{3,2} & a_{3,3} \end{pmatrix}$$

is defined in terms of the above definition of the determinant of a 2×2 matrix, namely, $\det(A)$ is given by

$$a_{1,1} \det \begin{pmatrix} a_{2,2} & a_{2,3} \\ a_{3,2} & a_{3,3} \end{pmatrix} - a_{1,2} \det \begin{pmatrix} a_{2,1} & a_{2,3} \\ a_{3,1} & a_{3,3} \end{pmatrix} + a_{1,3} \det \begin{pmatrix} a_{2,1} & a_{2,2} \\ a_{3,1} & a_{3,2} \end{pmatrix}.$$

Therefore, we may inductively define the determinant of any $n \times n$ matrix in this fashion. Assume that we have defined the determinant of an $n \times n$ matrix. Then we define the determinant of an $(n + 1) \times (n + 1)$ matrix $A = (a_{i,j})$ as follows. First, we let $A_{i,j}$ denote the $n \times n$ matrix obtained from A by deleting the ith row and j^{th} column. Then we define the *minor* of $A_{i,j}$ at position (i, j) to be $\det(A_{i,j})$. The *cofactor* of $A_{i,j}$ is defined to be

$$\text{cof}(A_{i,j}) = (-1)^{i+j} \det(A_{i,j}).$$

We may now define the determinant of A by

$$\det(A) = a_{i,1}\text{cof}(A_{i,1}) + a_{i,2}\text{cof}(A_{i,2}) + \cdots + a_{i,n+1}\text{cof}(A_{i,n+1}). \qquad \text{(C.37)}$$

This is called the *expansion of a determinant by cofactors* along the ith row of A. Similarly, we may expand along a column of A.

$$\det(A) = a_{1,j}\text{cof}(A_{1,j}) + a_{2,j}\text{cof}(A_{2,j}) + \cdots + a_{n+1,j}\text{cof}(A_{n+1,j}),$$

called the *cofactor expansion along the jth column of A*. Both expansions can be shown to be identical. Hence, a determinant may be viewed as a function that assigns a real number to an $n \times n$ matrix, and the above gives a method for finding that number.

If A is an $n \times n$ matrix with entries from F, then A is said to be *invertible*, or *nonsingular* if there is a unique matrix denoted by A^{-1} such that

$$AA^{-1} = I_n = A^{-1}A.$$

Another important fact, that we will need for instance in Chapter 8, is a result which follows from cofactor expansions.

Theorem C.38 (Cramer's Rule)
Let $A = (a_{i,j})$ be the coefficient matrix *of the following system of n linear equations in n unknowns:*

$$a_{1,1}x_1 + a_{1,2}x_2 + \cdots + a_{1,n}x_n = b_1$$

$$a_{2,1}x_1 + a_{2,2}x_2 + \cdots + a_{2,n}x_n = b_2$$

$$\vdots \qquad \vdots \qquad \vdots \quad \vdots \qquad \vdots$$

$$a_{n,1}x_1 + a_{n,2}x_2 + \cdots + a_{n,n}x_n = b_n,$$

over a field F. If $\det(A) \neq 0$, then the system has a solution given by:

$$x_j = \frac{1}{\det(A)} \left(\sum_{i=1}^{n} (-1)^{i+j} b_i \det(A_{i,j}) \right), \qquad (1 \leq j \leq n).$$

Of particular importance is a special matrix called the *Vandermonde matrix*, of order $t > 1$, which is given as follows.

$$A = \begin{pmatrix} 1 & x_1 & \cdots & x_1^{t-1} \\ 1 & x_2 & \cdots & x_2^{t-1} \\ \vdots & \vdots & \vdots & \vdots \\ 1 & x_t & \cdots & x_t^{t-1} \end{pmatrix},$$

where

$$\det(A) = \prod_{1 \le i < k \le t} (x_k - x_i)$$

(see Exercise 8.17 on page 159).

A notion that uses vector spaces and matrix theory is the following, which we will use in Chapters 5 and 8, for instance.

◆ Gaussian Elimination

The term *Gaussian elimination* refers to an efficient algorithm for finding linear dependency relations among vectors in a vector space over a suitable field. Suppose that we have vectors $v_j = (v_{1,j}, v_{2,j}, \ldots, v_{n,j})$ for $j = 1, 2, \ldots, m$ over a field F. What we seek field elements $c_1, c_2, \ldots, c_m \in F$ such that

$$\sum_{i=j}^{m} c_j v_j = \overrightarrow{0}, \tag{C.39}$$

where $c_j v_j = (c_j v_{1,j}, c_j v_{2,j}, \ldots, c_j v_{n,j})$, and $\overrightarrow{0} = (0, 0, \ldots, 0)$ is the *zero vector* of length n. Since $\sum_{i=j}^{m} c_j v_j$, the relation given in (C.39) where *not all coefficients* are 0, is a *linear dependency relation* (see Definition C.35 on page 219). Gaussian elimination uses the basic notions of linear algebra to define matrices with the vectors v_j as columns, then performs elementary row operations to put them into a form to determine the dependency relations therefrom, if there are any. The basic point from elementary linear algebra is that *if the number of vectors is greater than the dimension of the vector space over the field, then there must be a dependency relation*. For instance, if $m > n$ in (C.39), then at least one of the $c_j \ne 0$. This idea will be used in the text, especially for factoring techniques in Chapter 5.

The following will be needed in our discussions on secret sharing in Section 8.1, for instance.

Theorem C.40 (The Lagrange Interpolation Formula) *Let F be a field, and let a_j for $j = 0, 1, 2, \ldots, n$ be distinct elements of F. If c_j for $j = 0, 1, 2, \ldots, n$ are any elements of F, then*

$$f(x) = \sum_{j=0}^{n} \frac{(x - a_0) \cdots (x - a_{j-1})(x - a_{j+1}) \cdots (x - a_n)}{(a_j - a_0) \cdots (a_j - a_{j-1})(a_j - a_{j+1}) \cdots (a_j - a_n)} c_j$$

is the unique polynomial in $F[x]$ such that $f(a_j) = c_j$ for all $j = 0, 1, \ldots, n$.

We will need to address the notion of convergents of continued fractions, for example in Section 6.2, so we present some basics here.

◆ Continued Fractions

Definition C.41 (Continued Fractions)

If $q_j \in \mathbb{R}$ where $j \in \mathbb{Z}$ is nonnegative and $q_j \in \mathbb{R}^+$ for $j > 0$, then an expression of the form

$$\alpha = q_0 + \cfrac{1}{q_1 + \cfrac{1}{q_2 +}}$$

$$\ddots$$

$$+ \cfrac{1}{q_k + \cfrac{1}{q_{k+1}}}$$

$$\ddots$$

is called a continued fraction. *If $q_k \in \mathbb{Z}$ for all $k \geq 0$, then it is called a simple continued fraction,*[C.3] *denoted by $\langle q_0; q_1, \ldots, q_k, q_{k+1}, \ldots \rangle$. If there exists a nonnegative integer n such that $q_k = 0$ for all $k \geq n$, then the continued fraction is called* finite. *If no such n exists, then it is called* infinite.

Definition C.42 (Convergents) *Let $n \in \mathbb{N}$ and let α have continued fraction expansion $\langle q_0; q_1, \ldots, q_n, \ldots \rangle$ for $q_j \in \mathbb{R}^+$ when $j > 0$. Then*

$$C_k = \langle q_0; q_1, \ldots, q_k \rangle$$

is the kth convergent *of α for any nonnegative integer k.*

Theorem C.43 (Finite Simple Continued Fractions Are Rational)

Let $\alpha \in \mathbb{R}$. Then $\alpha \in \mathbb{Q}$ if and only if α can be written as a finite simple continued fraction.

[C.3]Note that the classical definition of a *simple* continued fraction is a continued fraction that arises from the reciprocals as in the Euclidean Algorithm, so the "numerators" are all 1 and the denominators all integers. This is to distinguish from more general continued fractions in which the numerators and denominators can be functions of a complex variable, for instance. Simple continued fractions are also called *regular* continued fractions in the literature.

Solutions to Odd-Numbered Exercises

Section 1.1 (*In the solutions, we will violate the convention of having plaintext in lower case in order to increase readability due to the smaller font size we use herein.*)

1.1   **I THINK THEREFORE I AM**

This is the phrase coined by the seventeenth century philosopher-mathematician René Descartes, and may be said to be the signature of the basis of his reasoning. It was originally given in Latin as: Cogito ergo sum.

1.3   **BEHOLD THE SIGN**

1.5   **NON SEQUITUR**

This is the Latin phrase for a conclusion that does not follow logically from the premises.

1.7   **TRUTH CONQUERS ALL THINGS**

1.9   **WAR IS IMMINENT**

1.11   *NEVER SAY ANYTHING*

1.13   **VANITY**

1.15 Since

$$m = (0111001101001001001101000011000010001110000000011) =$$
$$(01110)(01101)(00100)(10011)(01000)$$
$$(01100)(00100)(01111)(00000)(00011),$$

then the decimal equivalents are

$$14, 13, 4, 19, 8, 12, 4, 15, 0, 3,$$

to which correspond the letters $O, N, E, T, I, M, E, P, A, D$, to give us the English plaintext:

ONE TIME PAD

1.17 Since $k + c =$

$$(11010111101111010101111011101011010) +$$

$$(10010100111110111011010011101101001) =$$
$$(01000011010001101110101000000110011) =$$
$$(01000)(01101)(00011)(01110)$$
$$(10100)(00001)(10011),$$

then the decimal equivalents are 8, 13, 3, 14, 20, 1, and 19. Hence, via Table 1.2, we get the English equivalents: I, N, D, O, U, B, T, to give us the English plaintext:

IN DOUBT

1.19 Since $k + c =$

$$(11010111101111010101111011101011010) +$$

$$(11010100111110111110101011111001000) =$$

$$(00000011010001101011010000010010010) =$$

$$(00000)(01101)(00011)(01011)$$

$$(01000)(00100)(10010),$$

then the decimal equivalents are 0, 13, 3, 11, 8, 4, and 18. Hence, via Table 1.2, we get the English equivalents: A, N, D, L, I, E, S, to give us the English plaintext: **AND LIES**

1.21 Since

$$k + c = (11010111101111010101111011101011010) +$$

$$(11111100000111110001010001111001011) =$$

$$(00101011101000100100101010010010001) =$$

$$(00101)(01110)(10001)(00100)$$

$$(10101)(00100)(10001),$$

then the decimal equivalents are 5, 14, 17, 4, 21, 4, and 17. Hence, via Table 1.2, we get the English equivalents: F, O, R, E, V, E, R, to give us the English plaintext:

FOREVER

Section 1.2

1.23	**SEARCH THE CAVES**
1.25	**BOMB THE CAMPS**
1.27	**SURROUND THE CITY**
1.29	**SHE CREATED A STATE**
1.31	**HE DESTROYED TOWNS**
1.33	**BOTH STRUGGLES END**
1.35	**FIND THE SECRET NOW**
1.37	**HAVE ANOTHER ONE**
1.39	**TRY TO REMEMBER**
1.41	**WHERE IS THE GOLD**

Note that we discard the Z at the end since it was a filler to make the last triplet.

1.43   **SUMMERTIME**

Note that we discard two copies of Z at the end since they were used as filler to make the last triplet.

1.45   **LET'S ROLL**

Note that the apostrophe is tacitly understood.

1.47   **ALL EVIL DOERS FAIL**

1.49   **TRUST HIM**

1.51   **SPLENDID**

1.53   **FORGED SIGNATURES**

1.55   **GOOD DEEDS PREVAIL**

1.57   **FILLED WITH REGRET**

1.59 We use induction on n. If $n = 2$, then $b_2 > 2b_1 > b_1$, so $S = \{b_1, b_2\}$ is a superincreasing sequence. Assume that

$$S = \{b_1, b_2, \ldots, b_{n-1}\},$$

which satisfies

$$b_{j+1} > 2b_j \text{ for } 1 \le j \le n - 2$$

is a superincreasing sequence. Suppose that $b_n > 2b_{n-1}$. Since S is a superincreasing sequence, then

$$b_n > b_{n-1} + b_{n-1} > b_{n-1} + \sum_{j=1}^{n-2} b_j = \sum_{j=1}^{n-1} b_j.$$

Hence, by induction all such sets are superincreasing sequences.

Section 1.3

1.61 The key is **MATH**, and the plaintext is:

NEVER SAY NEVER AGAIN UNDER ANY

CIRCUMSTANCES WHATSOEVER

1.63 The key is **FIX**, and the plaintext is:

CRYPTOGRAPHERS MAKE VERY HIGH SALARIES

Note that we throw away two copies of Z tacked on the end as filler to make up the last trigram.

1.65 The key is **FAIR**, and the plaintext is:

JOHN NASH WAS A PIONEER OF GAME THEORY

Note that we throw away a Z tacked on the end as filler to make up the last trigram.[S1]

[S1] John Forbes Nash (1928–) was born June 13, 1928 in Bluefield, West Virginia. He first became interested in mathematics at the age of fourteen. He was inspired by Bell's book *Men of Mathematics* [18]. However, the mathematics that he learned at school bored him, and coupled with his lack of social skills, caused his teachers to brand him as obtuse. In 1941, Nash began the study of mathematics at Bluefield College where his mathematical talent began to shine through to his educators. John won a Westinghouse scholarship and entered Carnegie Tech (now Carnegie-Mellon University) in June 1945, where he intended to study chemical engineering. Yet, his absorption into mathematics deepened, so he took courses in tensor calculus and relativity, the latter being taught by the new head of department, John Synge, who along with other professors in mathematics, saw Nash's considerable talent and

1.67 The key is **GOLD**, and the plaintext is:

ALL THAT GLITTERS IS NOT GOLD BUT

PLATINUM IS ALWAYS PRECIOUS

Note that we throw away a Z tacked on the end as filler to make up the last trigram.

Section 2.1

2.1 There are at most two solutions modulo a given prime. Thus, by the Chinese Remainder Theorem there will be at most four modulo pq. We now exhibit those four. We need only show that

$$\pm(xpa^{(q+1)/4} + yqa^{(p+1)/4})$$

are square roots of a modulo pq since the other case has the same argument. We have

$$(xpa^{(q+1)/4} + yqa^{(p+1)/4})^2 \equiv x^2p^2a^{(q+1)/2} + y^2q^2a^{(p+1)/2} \equiv$$

$$x^2p^2z^{q+1} + y^2q^2z^{p+1} \equiv z^2(x^2p^2z^{q-1} + y^2q^2z^{p-1}) \,(\mathrm{mod}\ n),$$

and since $z^{p-1} \equiv 1 \,(\mathrm{mod}\ p)$ and $z^{q-1} \equiv 1 \,(\mathrm{mod}\ q)$, then this is congruent to:

$$z^2(x^2p^2 + y^2q^2) \equiv z^2(xp + yq)^2 \equiv z^2 \equiv a \,(\mathrm{mod}\ n).$$

2.3 Alice selects $m = 21$ and sends $w \equiv m^2 \equiv 441 \,(\mathrm{mod}\ n)$ to Bob. Bob selects $c = 0$ and sends it to Alice who computes $r = 21$, which gets sent back to Bob, who computes $r^2 \equiv 441 \equiv w \cdot t_A^c \,(\mathrm{mod}\ n)$. Thus we set $a = 1$ and execute another round.

---

countenanced him to continue his mathematical journey. In 1948, he received a B.A. and an M.A. in mathematics. In September 1948, he accepted a prestigious fellowship from Lefshetz, head of the Mathematics Department at Princeton, to take up doctoral studies. During his studies in 1949, he wrote a paper that would, forty-five years later, earn him the Nobel prize in economics. In 1950, he was granted his Ph.D. for a thesis called *Non-Cooperative Games*. In the summer of that year, he was working for RAND Corporation, where he worked from time to time over the later years. In 1952, he was teaching at MIT where he met Eleanor Stier with whom he had a son on June 19, 1953. However, despite her persuasions, he did not marry her. He did marry one of his students, Alicia Larde, in February 1957. By the end of 1958, with Alicia pregnant, he began a decline into schizophrenia. (The term *Schizophrenia* was coined by Eugene Bleuler in 1908 to mean a "specific type of alteration of thinking, feeling and relation to the external world".) An example of his descent into mental illness is given by the following anecdote. One winter morning at MIT in 1959 he entered the lounge carrying the *New York Times* and commented that the story in the upper left-hand corner of the front page contained an enciphered message from extra-terrestrials, a message that only he could decrypt. He saw hidden messages in everyday life that were symptoms of his delusions. For decades, Nash was in and out of hospitals for treatment. Amazingly, his mathematical output and academic success continued. By the early 1990s he had made a recovery from his battle with the disease. In 1994, he won (jointly with Harsanyi and Selten) the Nobel Prize in Economics for his research in game theory. By 1999, he was also honoured with the I. P. Steele Prize by the American Mathematical Society for a "seminal contribution to research".

For more on his life, the book: *A Beautiful Mind* [171] is recommended as is the movie of the same name starring Russell Crowe who won a Golden Globe award on January 20, 2002, for his penetrating portrayal of Nash's life.

Alice selects $m = 12$ and sends $w = 144$ to Bob. Bob selects $c = 1$, which Alice subsequently uses to compute $r \equiv 12 \cdot 111 \equiv 1332 \pmod{n}$ and sends r to Bob, who then B verifies that

$$r^2 \equiv 757855 \equiv 144 \cdot 12321 \equiv w \cdot t_A^c \pmod{n}.$$

Since a is now set to zero, Bob accepts Alice's proof.

Section 2.2

2.5 First, note that we have for each $i = 0, 1, \ldots, a_j - 1$,

$$x_i = \sum_{k=i}^{a_j - 1} b_k^{(j)} p_j^k \tag{S1}$$

and

$$\beta_i = \alpha^{x_i},$$

so

$$\beta_i^{(p-1)/p_j^{i+1}} \equiv \alpha^{(p-1)x_i/p_j^{i+1}} \pmod{p}.$$

Thus, it suffices to prove:

$$\alpha^{(p-1)x_i/p_j^{i+1}} \equiv \alpha^{(p-1)b_i^{(j)}/p_j} \pmod{p},$$

which holds precisely when

$$\frac{(p-1)x_i}{p_j^{i+1}} \equiv \frac{(p-1)b_i^{(j)}}{p_j} \pmod{p-1}$$

by Fermat's Little Theorem. From (S1),

$$\frac{(p-1)x_i}{p_j^{i+1}} - \frac{(p-1)b_i^{(j)}}{p_j} = \frac{(p-1)(x_i - p_j^i b_i^{(j)})}{p_j^{i+1}} =$$

$$\frac{p-1}{p_j^{i+1}} \left(\sum_{k=i}^{a_j - 1} b_k^{(j)} p_j^k - b_i^{(j)} p_j^i \right) = (p-1) \left(\sum_{k=i+1}^{a_j - 1} b_k^{(j)} p_j^{k-i-1} \right),$$

which is congruent to 0 modulo $p - 1$.

2.7 $\log_2(19) \equiv 35 \pmod{37}$.

2.9 $\log_3(31) \equiv 1176 \pmod{1579}$.

2.11 $\log_3(7) \equiv 1227 \pmod{1721}$.

2.13 $\log_{10}(3) \equiv 813 \pmod{1783}$.

2.15 $\log_{14}(5) \equiv 718 \pmod{1871}$.

2.17 $\log_{23}(3) \equiv 490 \pmod{2161}$.

2.19 $\log_{13}(2) \equiv 1300 \pmod{2351}$.

2.21 $\log_{19}(3) \equiv 1402 \pmod{2689}$.

2.23 $\log_{22}(3) \equiv 2314 \pmod{3361}$.

Section 2.3

2.25 69.

2.27 91.

2.29 7.

2.31 First, we determine d, the deciphering key, by using the Euclidean algorithm to solve $69d + 166x = 1$, and $d = 77$, $x = -32$ yields the least positive value of d, so we decipher via $85^{77} \equiv 12 \,(\text{mod } 167)$, $50^{77} \equiv 4 \,(\text{mod } 167)$, $96^{77} \equiv 18 \,(\text{mod } 167)$, $0^{77} \equiv 0 \,(\text{mod } 167)$, and $27^{77} \equiv 6 \,(\text{mod } 167)$. Thus, via Table 1.2, we get the letter equivalents of $12, 4, 18, 18, 0, 6, 4$ to be

MESSAGE.

2.33 As in Exercise 2.31, we determine that $d = 1053$ and decipher to get plaintext numerical equivalents $8, 13, 19, 0, 2, 19$, which decrypt to

INTACT.

2.35 $k = 2245$.

2.37 $k = 2902$.

2.39 $k = 871$.

2.41 $k = 1876$.

2.43 $k = 571$.

2.45 $k = 637$.

2.47 $k = 2425$.

Section 3.1

3.1 By solving $7d + 16600y = 1$, where $(p-1)(q-1) = 16600$, we get that $d = 4743$ with $y = -2$. Thus, $x = 8081$.

3.3 As in Exercise 3.1 above, $7d + 33820y = 1$, yields $d = 9663$, $y = -2$, and $x = 723$.

3.5 As above, $7d + 1082400y = 1$, yields $d = 773143$, $y = -5$, and $x = 315043$.

3.7 As above, $7d + 3706560y = 1$, yields $d = 2647543$, $y = -5$, and $x = 168536$.

3.9 As above, $5d + 4726896y = 1$, yields $d = 3781517$, $y = -4$, and $x = 4598308$.

3.11 Since $ed \equiv 1 \,(\text{mod } (p-1)(q-1))$, there exists a $g \in \mathbb{Z}$ such that

$$ed = 1 + g(p-1)(q-1).$$

If $p \nmid x$, then by Fermat's Little Theorem, $x^{p-1} \equiv 1 \,(\text{mod } p)$. Hence,

$$x^{ed} \equiv x^{1+g(p-1)(q-1)} \equiv x(x^{g(q-1)})^{p-1} \equiv x \,(\text{mod } p). \tag{S2}$$

If $p \,\big|\, x$, then (S2) holds again since $x \equiv 0 \,(\text{mod } p)$. Hence,

$$x^{ed} \equiv x \,(\text{mod } p)$$

for any x. Similarly, $x^{ed} \equiv x \,(\text{mod } q)$. Since $p \neq q$, $x^{ed} \equiv x \,(\text{mod } n)$. Thus,

$$(x^e)^d \equiv x^{ed} \equiv x \,(\text{mod } n).$$

3.13 Since $x' \equiv c'^d \equiv c^d(y^e)^d \equiv xy \pmod{n}$, then Mallory computes

$$x \equiv x'y^{-1} \pmod{n}.$$

Section 3.2

3.15 Let $\ell \in \mathbb{N}$ be the least value such that $n < N^{\ell+1}$. Such an ℓ must exist since $n \in \mathbb{N}$ and $N > 1$. Moreover, ℓ is unique. If $n < N^\ell$, then this contradicts the minimality of $\ell + 1$ in this regard. Thus, $N^\ell \le n < N^{\ell+1}$ since $n > N$.

3.17 Let $k > \ell$ be chosen, so that $N^k \ge N^{\ell+1} > n$, and suppose that we have a plaintext message blocks m, m_1 such that $m > n$. Then if

$$m^e \equiv m_1^e \pmod{n},$$

we get, upon decryption $m \equiv m^{ed} \equiv m_1^{ed} \equiv m_1 \pmod{n}$. Thus, there is a nonnegative integer r such that $m = m_1 + nr$. However, $m > n$, so $r > 0$. Hence, the same ciphertext block $c \equiv m^e \equiv m_1^e \pmod{n}$ will yield (at least) two different plaintext messages, m and $m - nr$, only one of which will be correct. As an illustration, if $k = 3 = \ell + 1$ in Example 3.10, then **POW** has 3-digit base 26 numerical equivalent $15 \cdot 26^2 + 14 \cdot 26 + 23 = 10526$, which enciphers as $10526^{701} \equiv 1420 \pmod{1943}$. However, deciphering yields

$$1420^{29} \equiv 811 \pmod{1943},$$

and of the values $811 + 1943j$ for $j = 0, 1, 2, 3, 4, 5$, only $j = 5$ yields 10526.

3.19 Using the solution of Exercise 3.14, we know that $p = 3371$ and $q = 3449$.

3.21 As in Exercise 3.19, we have $p = 4651$ and $q = 5003$.

3.23 As in Exercise 3.19, we have $p = 5657$ and $q = 6397$.

3.25 As in Exercise 3.19, we have $p = 9203$ and $q = 9533$.

3.27 We write the cryptogram as 5-digit base 26 integers since $\ell = 4$:

$$\mathbf{EGSIO} = 4 \cdot 26^4 + 6 \cdot 26^3 + 18 \cdot 26^2 + 8 \cdot 26 + 14 = 1945750,$$

$$\mathbf{XEWXG} = 23 \cdot 26^4 + 4 \cdot 26^3 + 22 \cdot 26^2 + 23 \cdot 26 + 6 = 10596228,$$

and

$$\mathbf{DPXMA} = 3 \cdot 26^4 + 15 \cdot 26^3 + 23 \cdot 26^2 + 12 \cdot 26 + 0 = 1650428.$$

Then determine the deciphering key via $ed + \phi(n)x = 11d + 10758720 = 1$, which gives $d = 1956131$ (for $x = -2$). Thus, deciphering:

$$1945750^d \equiv 111414 \pmod{n}, \quad 10596228^d \equiv 213617 \pmod{n},$$

$$\text{and } 1650428^d \equiv 301506 \pmod{n}.$$

Also, since

$$111414 = 6 \cdot 26^3 + 8 \cdot 26^2 + 21 \cdot 26 + 4 = \mathbf{GIVE}$$

$$213617 = 12 \cdot 26^3 + 4 \cdot 26^2 + 0 \cdot 26 + 1 = \mathbf{MEAB},$$

$$301506 = 17 \cdot 26^3 + 4 \cdot 26^2 + 0 \cdot 26 + 10 = \mathbf{REAK},$$

we have the plaintext: **GIVE ME A BREAK**.

3.29 As in Exercise 3.27, we write:

$$\mathbf{FENFL} = 5 \cdot 26^4 + 4 \cdot 26^3 + 13 \cdot 26^2 + 5 \cdot 26 + 11 = 2364113,$$

$$\mathbf{PLNMZ} = 15 \cdot 26^4 + 11 \cdot 26^3 + 13 \cdot 26^2 + 12 \cdot 26 + 25 = 7057101,$$

and

$$\mathbf{XLMPS} = 23 \cdot 26^4 + 11 \cdot 26^3 + 12 \cdot 26^2 + 15 \cdot 26 + 18 = 10712304.$$

Thus, deciphering:

$$2364113^d \equiv 2122640 \,(\mathrm{mod}\ n), \quad 7057101^d \equiv 199958 \,(\mathrm{mod}\ n),$$

$$\text{and } 10712304^d \equiv 339408 \,(\mathrm{mod}\ n),$$

and since,

$$2122640 = 4 \cdot 26^3 + 16 \cdot 26^2 + 20 \cdot 26 + 0 = \mathbf{EQUA}$$

$$199958 = 11 \cdot 26^3 + 9 \cdot 26^2 + 20 \cdot 26 + 18 = \mathbf{LJUS},$$

$$339408 = 19 \cdot 26^3 + 8 \cdot 26^2 + 2 \cdot 26 + 4 = \mathbf{TICE},$$

we have the plaintext: **EQUAL JUSTICE**.

3.31 As above:

$$\mathbf{BOTDT} = 1 \cdot 26^4 + 14 \cdot 26^3 + 19 \cdot 26^2 + 3 \cdot 26 + 19 = 715981,$$

and

$$\mathbf{ICBYJ} = 8 \cdot 26^4 + 2 \cdot 26^3 + 1 \cdot 26^2 + 24 \cdot 26 + 9 = 3692269,$$

and deciphering:

$$715981^d \equiv 93634 \,(\mathrm{mod}\ n), \text{ and } 3692269^d \equiv 321207 \,(\mathrm{mod}\ n).$$

Then,

$$93634 = 5 \cdot 26^3 + 8 \cdot 26^2 + 13 \cdot 26 + 8 = \mathbf{FINI}$$

$$321207 = 18 \cdot 26^3 + 7 \cdot 26^2 + 4 \cdot 26 + 3 = \mathbf{SHED},$$

so the plaintext is **FINISHED**.

Section 3.3

3.33 $(\alpha^b)^{p-1-a} = 32^{409-1-6} \equiv 379 \,(\mathrm{mod}\ 409)$, and

$$(\alpha^b)^{-a} m \alpha^{ab} \equiv 379 \cdot 12 \equiv 49 = m \,(\mathrm{mod}\ 409).$$

3.35 $(\alpha^b)^{p-1-a} = 512^{941-1-14} \equiv 864 \,(\mathrm{mod}\ 941)$, and

$$(\alpha^b)^{-a} m \alpha^{ab} \equiv 864 \cdot 303 \equiv 194 = m \,(\mathrm{mod}\ 941).$$

3.37 $(\alpha^b)^{-a} = (3,3,2)^{-44} = (1,4,1)$, so

$$\alpha^{-ab} m \alpha^{ab} = (1,4,1)(0,2,1) = (4,4,4) = m.$$

3.39 $(\alpha^b)^{-a} = (0,3,1)^{-24} = (1,0,0)$, so

$$\alpha^{-ab} m \alpha^{ab} = (1,0,0)(4,0,4) = (0,1,0) = m.$$

3.41 Mallory intercepts m^{e_A} and encrypts with his own enciphering key $m^{e_A e_M}$. Then he sends it back to Alice, impersonating Bob. Alice, thinking it is Bob, sends back $m^{e_A e_M d_A} = m^{e_M}$, which Mallory intercepts. He then easily decrypts via his deciphering key to get $m^{e_M d_M} = m$. This demonstrates how easily the system is compromised by such an attack and why it requires more security, in terms of authentication of communicating entities, if it is to be used.

3.43 Bob has public key $(p, \alpha, \alpha^a) = (15485863, 6, 7776)$, which Alice obtains. She converts the English plaintext via Table 1.2 on page 3 to the numerical equivalents: $19, 14, 3, 0, 24$. Since $26^5 < p < 26^6$, she can represent the plaintext message as a single 5-digit base 26 integer:

$$19 \cdot 26^4 + 14 \cdot 26^3 + 3 \cdot 26^2 + 0 \cdot 26 + 24 = 8930660.$$

She first computes $\alpha^b = 6^{69} \equiv 13733130 \pmod{p}$, then

$$m\alpha^{ab} \equiv 8930660 \cdot 7776^{69} \equiv 4578170 \pmod{p}.$$

She sends $c = (13733130, 4578170)$ to Bob. He uses his private key to compute

$$(\alpha^b)^{p-1-a} \equiv 13733130^{1548585863-1-5} \equiv 2620662 \pmod{p}$$

and

$$(\alpha^b)^{-a} m\alpha^{ab} \equiv 2620662 \cdot 4578170 \equiv 8930660 \equiv m \pmod{p},$$

and using Table 1.2, he converts back to the English plaintext.

Section 3.4

3.45 Since each entity needs both the enciphering key and deciphering key to be kept secret, this is "n choose 2", namely, the binomial coefficient:

$$\binom{n}{2} = n!/((n-2)!2!) = n(n-1)/2.$$

3.47 Since
$$(k')^d \equiv 4019872^{802607} \equiv 234561 \pmod{n},$$
then $k = (2, 3, 4, 5, 6, 1)$, from which we deduce that $k^{-1} = (6, 1, 2, 3, 4, 5)$, so $k^{-1}(c) = (18, 8, 11, 21, 4, 17) = $ **SILVER**.

3.49 $(k')^d \equiv 1525853^{802607} \equiv 421653 \pmod{n}$, $k = (4, 2, 1, 6, 5, 3)$, and $k^{-1} = (3, 2, 6, 1, 5, 4)$. Thus, $k^{-1}(c) = (1, 0, 3, 9, 14, 1) = $ **BAD JOB**.

3.51 $(k')^d \equiv 7155548^{802607} \equiv 624315 \pmod{n}$, $k = (6, 2, 4, 3, 1, 5)$, and $k^{-1} = (5, 2, 4, 3, 6, 1)$, so $k^{-1}(c) = (5, 17, 8, 4, 13, 3) = $ **FRIEND**.

3.53 $(k')^d \equiv 371155^{802607} \equiv 214365 \pmod{n}$, $k = (2, 1, 4, 3, 6, 5)$, and $k^{-1} = (2, 1, 4, 3, 6, 5)$, so $k^{-1}(c) = (12, 14, 13, 8, 4, 18) = $ **MONIES**.

3.55 $(k')^d \equiv 8182887^{802607} \equiv 462135 \pmod{n}$, $k = (4, 6, 2, 1, 3, 5)$, and $k^{-1} = (4, 3, 5, 1, 6, 2)$, so $k^{-1}(c) = (0, 17, 0, 1, 8, 2) = $ **ARABIC**.

3.57 $(k')^d \equiv 4125753^{802607} \equiv 246351 \pmod{n}$, $k = (2, 4, 6, 3, 5, 1)$, and $k^{-1} = (6, 1, 4, 2, 5, 3)$, so $k^{-1}(c) = (7, 8, 19, 12, 0, 13) = $ **HITMAN**.

3.59 $(k')^d \equiv 1968543^{802607} \equiv 613245 \pmod{n}$, $k = (6, 1, 3, 2, 4, 5)$, and $k^{-1} = (2, 4, 3, 5, 6, 1)$, so $k^{-1}(c) = (3, 0, 17, 10, 4, 17) = $ **DARKER**.

3.61 $(k')^d \equiv 7066510^{802607} \equiv 415263 \,(\text{mod } n)$, $k = (4,1,5,2,6,3)$, and $k^{-1} = (2,4,6,1,3,5)$, so $k^{-1}(c) = (25,4,1,17,0,18) = $ **ZEBRAS**.

Section 4.2

4.1 Let $p|n$ be prime and set $c = m^{(n-1)/q^e}$ where q is a prime, $e \in \mathbb{N}$ and $q^e||b$. Therefore, since $\gcd(m^{(n-1)/q} - 1, n) = 1$, $c^{q^e} \equiv 1 \,(\text{mod } p)$, but $c^{q^{e-1}} \not\equiv 1 \,(\text{mod } p)$. Thus, $\text{ord}_p(c) = q^e$, so $q^e \mid (p-1)$ by Proposition C.19. Since q was arbitrarily chosen, $p \equiv 1 \,(\text{mod } b)$. For the last assertion of the exercise, assume $b > \sqrt{n} - 1$, but n is composite. Let p be the smallest prime dividing n. Then $p \leq \sqrt{n}$, so $\sqrt{n} \geq p > b \geq \sqrt{n}$, a contradiction. Hence, n is prime.

4.3 If n is prime, then by Exercise 4.1 we have the result with $m = m_q$ for all $q \mid (n-1)$. Conversely, if such m_q exist, then by the Chinese Remainder Theorem, we may find a solution $x = m$ to the system of congruences

$$x \equiv m_q \,(\text{mod } q^e)$$

for all primes q such that $q^e||(n-1)$. Thus, the result now follows from Exercise 4.1.

4.5 If $\mathfrak{F}_n$ is prime, the result follows from Exercise 4.4. Conversely, if p be a prime divisor of $\mathfrak{F}_n$, then $3^{\frac{\mathfrak{F}_n-1}{2}} \equiv -1 \,(\text{mod } p)$, so $3^{\frac{\mathfrak{F}_n-1}{2}} \equiv -1 \,(\text{mod } p)$. If $b = \text{ord}_p(3)$, then $b \mid (\mathfrak{F}_n - 1)$ by Proposition C.19. Thus $b = 2^m$ for some integer m with $1 \leq m \leq 2^n$. Suppose that $m \neq 2^n$. Then

$$-1 \equiv 3^{2^{2^n-1}} = (3^{2^m})^{2^{2^n-m-1}} \equiv 1 \,(\text{mod } p),$$

so $p = 2$, a contradiction. Hence,

$$\text{ord}_p(3) = 2^{2^n} = \mathfrak{F}_n - 1.$$

By Fermat's Little Theorem and Proposition C.19, $\text{ord}_p(3)|p-1$. Hence, $p = \mathfrak{F}_n$.

4.7 Assume first that $(p_j - 1) \mid (n - 1)$ for all $j = 1, 2, \ldots, r$. If $\gcd(a, n) = 1$, then $\gcd(a, p_j) = 1$ for all $j = 1, 2, \ldots, r$. Thus, $a^{p_j - 1} \equiv 1 \,(\text{mod } p_j)$, by Fermat's Little Theorem. Since $n - 1 = m_j(p_j - 1)$ for some $m_j \in \mathbb{N}$,

$$a^{n-1} \equiv a^{m_j(p_j-1)} \equiv (a^{p_j-1})^{m_j} \equiv 1 \,(\text{mod } p_j).$$

Hence, $a^{n-1} \equiv 1 \,(\text{mod } \prod_{j=1}^{r} p_j)$, namely, $a^{n-1} \equiv 1 \,(\text{mod } n)$.

Conversely, suppose that $a^{n-1} \equiv 1 \,(\text{mod } n)$ for each a with $\gcd(a, n) = 1$. In particular, if a is a primitive root modulo p, for any prime divisor p of n, then $(p - 1) \mid (n - 1)$, by Proposition C.19 on page 216.

Section 4.3

4.9 If $b^2 \equiv 1 \,(\text{mod } p^a)$, then $p^a \mid (b + 1)(b - 1)$. Notice that p cannot divide both $(b-1)$ and $b+1$ since that would mean $b-1 \equiv b+1 \,(\text{mod } p)$. so $-1 \equiv 1 \,(\text{mod } p)$, forcing $p = 2$, a contradiction. Thus, either $p^a \mid (b - 1)$ or $p^a \mid (b + 1)$. In other words, $b \equiv \pm 1 \,(\text{mod } p^a)$. Conversely, if $b \equiv \pm 1 \,(\text{mod } p^a)$, then $b^2 \equiv 1 \,(\text{mod } p^a)$.

4.11 They are both Carmichael numbers for the following reasons, using Exercise 4.7. For $n = 8911 = 7 \cdot 19 \cdot 67$, each of 6, 18 and 66 divide $n - 1$. For $n = 10585 = 5 \cdot 29 \cdot 73$, all of 4, 28, and 72 divide $n - 1$. Also, 10585 is an Euler pseudoprime to base 2 since

$$\left(\frac{2}{10585}\right) \equiv 2^{5292} \pmod{10585}.$$

4.13 By Exercise 4.8, $E(n)$ is a subgroup of $(\mathbb{Z}/n\mathbb{Z})^*$. Since the cardinality of $(\mathbb{Z}/n\mathbb{Z})^*$ is $\phi(n)$ by Example C.17, and since n is composite, then $E(n) \neq (\mathbb{Z}/n\mathbb{Z})^*$ by Exercise 4.12, so $E(n)$ is a proper subgroup of $(\mathbb{Z}/n\mathbb{Z})^*$. Thus, by Example C.17, $|E(n)| \mid \phi(n)$. Hence, $|E(n)| \leq \phi(n)/2$ since $|E(n)| \neq \phi(n)$ and $\phi(n)$ is even when $n > 2$.

4.15 Since, by repeated squaring, we get $2^{14670} \equiv -1 \pmod{29341}$ and $(\frac{2}{29341}) = -1$, then 29341 is an Euler pseudoprime to base 2 since $29341 = 13 \cdot 37 \cdot 61$.

4.17 As above, we get, $2^{31372} \equiv 1 \equiv (\frac{2}{62745}) \pmod{62745}$, and since $62745 = 3 \cdot 5 \cdot 47 \cdot 89$, then 62745 is an Euler pseudoprime to base 2.

4.19 Let $n = 2821$. Then by the repeated squaring method, we determine that $3^{1410} \equiv 1 \equiv (\frac{3}{2821}) \pmod{2821}$. Since $2821 = 7 \cdot 13 \cdot 31$, then 2821 is an Euler pseudoprime to base 3. Moreover, since 6, 12, and 30 all divide $n - 1 = 2820$, then by Exercise 4.7, 2821 is a Carmichael number.

Section 4.4

4.21 If $n \in \mathrm{spsp}(a)$, then $a^d \equiv 1 \pmod{n}$ for some divisor d of $n - 1$. Thus, $(a^d)^{(n-1)/d} \equiv 1^{(n-1)/d} \equiv 1 \pmod{n}$, whence $n \in \mathrm{psp}(a)$.

4.23 Let $g = \gcd((x - y), n)$. We need only show that $g \neq 1, n$. If $g = 1$, then since $n \mid (x - y)(x + y)$, we must have $n \mid (x + y)$, contradicting the hypothesis that $x \not\equiv -y \pmod{n}$. If $g = n$, then $x \equiv y \pmod{n}$, a contradiction to the hypothesis that $x \not\equiv y \pmod{n}$. Hence, since $g \mid n$, it is a nontrivial factor of it.

4.25 Each is a strong pseudoprime to base 2 for the following reasons. For $n = 15841$, $n - 1 = 2^5 \cdot 495$ and $2^{495} \equiv 1 \pmod{n}$. For $n = 29341$, $n - 1 = 2^2 \cdot 7335$ and $2^{2 \cdot 7335} \equiv -1 \pmod{n}$, while $2^{7335} \not\equiv \pm 1 \pmod{n}$. For $n = 52633$, $n - 1 = 2^3 \cdot 6579$ and $2^{6579} \equiv 1 \pmod{n}$. For $n = 252601$, $n - 1 = 2^3 \cdot 31575$, and we have both $2^{2 \cdot 31575} \equiv -1 \pmod{n}$ and $2^{31575} \not\equiv \pm 1 \pmod{n}$.

They are all Carmichael numbers for the following reasons, using Exercise 4.7. For $n = 15841 = 7 \cdot 31 \cdot 73$, all of 6, 30, and 72 divide $n - 1$. For $n = 29341 = 13 \cdot 67 \cdot 61$, all of 12, 66, and 60 divide $n - 1$. For $n = 52633 = 7 \cdot 73 \cdot 103$, all of 6, 72, and 102 divide $n - 1$. For $n = 252601 = 41 \cdot 61 \cdot 101$, all of 40, 60, and 100 divide $n - 1$.

4.27 Let n be a Carmichael number that is a strong pseudoprime to every base a prime to n. Furthermore, suppose that $n - 1 = 2^t m$ where $t \in \mathbb{N}$ and m is odd. Let a be a primitive root modulo any prime p dividing n. If $a^m \equiv 1 \pmod{n}$, then $a^m \equiv 1 \pmod{p}$. However, since a is a primitive root modulo p, then by Proposition C.19, $(p - 1) \mid m$, a contradiction. Suppose that there exists a value of j such that $0 \leq j \leq t - 1$, and $a^{2^j m} \equiv -1 \pmod{n}$. Then $a^{2^{j+1} m} \equiv 1 \pmod{p}$, so $(p - 1) \mid (2^{j+1} m)$ as above. Thus,

$$-1 \equiv a^{2^j m} \equiv (a^{(p-1)/2})^{(2^{j+1} m)/(p-1)} \equiv (-1)^{2^{j+1} m} \equiv 1 \pmod{p},$$

a contradiction since p is odd. Hence, no such Carmichael number can exist.

4.29 For $a = 2$ we have $n - 1 = 2^4 \cdot 35$; $2^{35} \equiv 263 \,(\text{mod } 561)$; $2^{2 \cdot 35} \equiv 166 \,(\text{mod } 561)$; $2^{4 \cdot 35} \equiv 67 \,(\text{mod } 561)$; and $2^{8 \cdot 35} \equiv 1 \,(\text{mod } 561)$. Hence, 561 is composite by the strong pseudoprimality test.

4.31 Since $120 = 2^3 \cdot 15$ and $3^{15} \equiv 1 \,(\text{mod } 121)$, then 121 is a strong pseudoprime to base 3.

4.33 Since $24 = 2^3 \cdot 3$; $7^3 \equiv 18 \,(\text{mod } 25)$; and $18^2 \equiv -1 \,(\text{mod } 25)$, then 25 is a strong pseudoprime to base 7.

4.35 Since $a^{(n-1)/2} \equiv \left(\frac{a}{n}\right) \,(\text{mod } n)$, then $a^{n-1} \equiv 1 \,(\text{mod } n)$.

4.37 If $n = p^b \in \text{psp}(a)$, and $n - 1 = 2^t m$ where m is odd, then

$$(a^{(n-1)/2})^2 \equiv 1 \,(\text{mod } n),$$

so by Exercise 4.9, $a^{(n-1)/2} \equiv \pm 1 \,(\text{mod } n)$. If $a^{(n-1)/2} \equiv 1 \,(\text{mod } n)$, and $n - 1$ is even we repeat the above argument to get $a^{(n-1)/4} \equiv \pm 1 \,(\text{mod } n)$ and keep repeating it which ultimately achieves that $n \in \text{spsp}(a)$. The converse follows from Exercise 4.21.

Section 5.1

5.1 We use induction on a to show that $x^{2^a} \equiv 1 \,(\text{mod } 2^{a+2})$ for all odd $x \in \mathbb{Z}$. If $a = 1$, then it is easy to see that $x^2 \equiv 1 \,(\text{mod } 8)$, so we may assume that

$$x^{2^{a-1}} \equiv 1 \,(\text{mod } 2^{a+1}).$$

Therefore, $x^{2^a} = (1 + 2^{a+1}t)^2$ for some $t \in \mathbb{Z}$. In other words,

$$x^{2^a} \equiv 1 \,(\text{mod } 2^{a+2}).$$

5.3 Let $n = \prod_{j=1}^{r} p_j^{a_j}$ be a prime factorization of n, with $p_1 = 2$. By Exercise 5.2, $\lambda(n)$ is a universal exponent for n. We must prove that it is minimal. Let g_j be a primitive root modulo $p_j^{a_j}$ for each $j = 2, 3, \ldots, r$, which exist by the Primitive Root Theorem C.21 on page 216. Thus, by the Chinese Remainder Theorem C.13, the system of congruences

$$x \equiv 3 \,(\text{mod } 2^{a_1}), \text{ and } x \equiv g_j \,(\text{mod } p_j^{a_j}), \quad (2 \leq j \leq r),$$

has a solution $x = a$ which is unique modulo n. If $a^m \equiv 1 \,(\text{mod } n)$ for some $m \in \mathbb{N}$, then $a^m \equiv 1 \,(\text{mod } p_j^{a_j})$ for each j, so $\text{ord}_{p_j^{a_j}}(a) \,\big|\, m$, by Proposition C.19. Thus, since a satisfies the r congruences above, then $\lambda(p_j^{a_j}) = \text{ord}_{p_j^{a_j}}(a)$. Hence, $\lambda(p_j^{a_j}) \,\big|\, m$ for all j. Therefore, $\lambda(n) \,\big|\, m$. We have shown that $\lambda(n) = \text{ord}_n(a)$.

5.5 We have $e = 2^5 \cdot 405$. Choose $a = 2$ and compute $2^{405} \equiv 1 \,(\text{mod } n)$, so we go to step (1) and choose another base, $a = 3$. We compute

$$x_0 \equiv 3^{405} \equiv 2820 \,(\text{mod } n),$$

then $x_1 \equiv 2820^2 \equiv 218 \,(\text{mod } n)$, and $x_2 \equiv 218^2 \equiv 1 \,(\text{mod } n)$. Since we know that $x_1 \not\equiv \pm 1 \,(\text{mod } n)$, then $\gcd(217, 15841) = 217$ is a factor of n. Indeed $n = 217 \cdot 73$.

5.7 Since $e = 2^2 \cdot 26643$, we may try $a = 2$ and compute,

$$x_0 \equiv 2^{26643} \equiv 25719 \not\equiv \pm 1 \,(\text{mod } n),$$

then $x_1 \equiv 25719^2 \equiv 1 \,(\text{mod } n)$. Hence, $\gcd(25718, 107381) = 167$ is a factor of n. In fact, $n = 167 \cdot 643$.

5.9 Since $e = 2^3 \cdot 1831595$, then we choose a base $a = 3$ and compute as follows where all congruences are assumed modulo n.

$$x_0 \equiv 3^{1831595} \equiv 10750120; \qquad x_1 \equiv 10750120^2 \equiv 13251402;$$

and $x_2 \equiv 13251402^2 \equiv 1$. Since $x_1 \not\equiv \pm 1$, then $\gcd(n, 13251401) = 3371$, and $n = 3371 \cdot 4349$.

5.11 Since $e = 2 \cdot 223713$, then we choose $a = 2$ and compute the following where all congruences ae modulo n. Since $2^{223713} \equiv 1$, we need a new base. We select $a = 3$ and compute

$$x_0 \equiv 3^{223713} \equiv 23944214; \qquad x_1 \equiv 23944214^2 \equiv 1,$$

and since $x_0 \not\equiv \pm 1$, then $\gcd(x_0 - 1, n) = 7103 \,\Big|\, n$. Indeed $n = 6679 \cdot 7103$.

5.13 We have that $e = 2^3 \cdot 1600875$ and we select $a = 2$ to compute the following where all congruences are modulo n. Since $x_0 \equiv 2^{1600875} \equiv 76859538 \equiv -1$, we must choose a new base $a = 3$. Then

$$3^{1600875} \equiv 44940756; \quad x_1 \equiv x_0^2 \equiv 9649071; \quad x_2 \equiv x_1^2 \equiv 1;$$

and since $x_1 \not\equiv \pm 1$, then $\gcd(x_1 - 1, n) = 8539 \,\Big|\, n$, and $n = 8539 \cdot 9001$.

5.15 Since 3 is a primitive root of $\mathfrak{F}_6 = 18446744073709551617$ if n is prime, then we know that $3^{e/2} \equiv -1 \,(\text{mod } n)$ if indeed it is prime. By repeated squaring, we calculate that $3^{e/2} \equiv 3653528722731049759 \not\equiv \pm 1 \,(\text{mod } n)$, so $\gcd(3653528722731049758, n) = 274177 \,\Big|\, n$. Indeed we now have factored the sixth Fermat number since

$$\mathfrak{F}_6 = 274177 \cdot 67280421310721.$$

Section 5.2

5.17 We check that $\gcd(a_j - 1, n) = 1$ for all $j = 1, 2, \ldots, 6$, then

$$a_7 \equiv 8697^7 \equiv 4747 \,(\text{mod } n),$$

$\gcd(4746, n) = 113$, and $n = 113 \cdot 107$. Note that $112 = 2^4 \cdot 7$ is B-smooth.

5.19 We compute that $\gcd(a_j - 1, n) = 1$ for $j = 1, 2, \ldots, 7$. Then we check that

$$a_8 \equiv 32494^8 \equiv 12320 \,(\text{mod } n),$$

$\gcd(12319, 37151) = 97$, and $n = 97 \cdot 383$. Here $96 = 2^5 \cdot 3$ is B-smooth.

5.21 Randomly select $a_1 = 614$, then

$$b_1 \equiv 614^2 \equiv 840 \equiv 2^3 \cdot 3 \cdot 5 \cdot 7 \,(\mathrm{mod}\ n),$$

so b_1 is 7-smooth, and we keep it. Randomly select $a_2 = 51$ and compute $51^2 \equiv 2601 \,(\mathrm{mod}\ n)$, which is not 7-smooth since it is divisible by only 3 from $\mathcal{S}_4$ and is not a perfect power of 3, so we discard it. Randomly select $a_2 = 1009$ and compute

$$b_2 \equiv 1009^2 \equiv 2 \cdot 3 \cdot 5^3 \,(\mathrm{mod}\ n),$$

which is then saved. Randomly select $a_3 = 45$ and compute

$$b_3 \equiv 45^2 \equiv 3^4 \cdot 5^2 \,(\mathrm{mod}\ n),$$

which is saved. Randomly select $a_4 = 56$ and compute

$$b_4 \equiv 56^2 \equiv 2^6 \cdot 7^2 \,(\mathrm{mod}\ n),$$

and it is kept. Randomly select $a_5 = 100$ and compute $100^2 \equiv 1451 \,(\mathrm{mod}\ n)$, which is not divisible by any of the primes in $\mathcal{S}_4$ so it is discarded. Randomly select $a_5 = 983$ and compute

$$b_5 \equiv 983^2 \equiv 2^2 \cdot 3^2 \cdot 7 \,(\mathrm{mod}\ n),$$

and save it. We have reached $r + 1 = 5$ such b_i, so we now search for the subset $\mathcal{J}$. Notice that b_3 and b_4 are squares themselves. However, for b_3, $x = 45$, $y = 3^2 \cdot 5$ and $x - y = 0$. Similarly for b_4, the same thing happens, so we need another subset. We see that if we choose $\mathcal{J} = \{1, 2, 5\}$, then we get

$$\prod_{i \in \mathcal{J}} b_i \equiv 614^2 \cdot 1009^2 \cdot 983^2 \equiv 2^6 \cdot 3^4 \cdot 5^4 \cdot 7^2 \,(\mathrm{mod}\ n).$$

Thus,

$$x \equiv \prod_{i \in \mathcal{J}} a_i \equiv 614 \cdot 1009 \cdot 983 \equiv 6043 \,(\mathrm{mod}\ n),$$

and

$$y \equiv 2^3 \cdot 3^2 \cdot 5^2 \cdot 7 \equiv 4051 \,(\mathrm{mod}\ n).$$

Thus $x^2 \equiv y^2 \,(\mathrm{mod}\ n)$ and

$$\gcd(x - y, n) = \gcd(6043 - 4051, 8549) = \gcd(1992, 8549) = 83.$$

In fact, $8549 = 83 \cdot 103$. Of course, sometimes luck plays a role. Suppose that the first random choice that we made was $a_1 = 744$. Then since

$$744^2 \equiv 2^8 \cdot 5^2 \,(\mathrm{mod}\ n)$$

and $\gcd(744 - 2^4 \cdot 5, n) = \gcd(664, 8549) = 83$, then we have a quick and simple factorization. However, one cannot rely on luck alone, and this is unlikely to happen for the large numbers that are used in practice.

Section 5.3

5.23 $6P = (2238, 2448)$

5.25 Since each point $jP = (x_j, y_j)$ on $E(\mathbb{Z}/n\mathbb{Z})$ has at most n^2 possibilities for any j, then there exists a value k such that $jP = kP$.

5.27 Let $\ell = qk + s$ where $0 \le s < k$. Since $kP = \mathfrak{o}$, then $qkP = \mathfrak{o}$. Therefore, by Exercise 5.26, $(\ell - qk)P = sP = \mathfrak{o}$. However, since $s < k$ and k is the least positive such value, then $s = 0$, and we have proved that $k \mid \ell$.

5.29 $2P \equiv (9^{-1}, -82 \cdot 27^{-1}) \equiv (442, 2501) \pmod{n}$. However, to compute $6P$ we must compute $4P \equiv (-26567/242064, 352876013/119095488) \equiv \mathfrak{o} \pmod{n}$ since $\gcd(242064, 3977) = 41$, and indeed $3977 = 41 \cdot 97$.

5.31 $2P \equiv (4055, 10810) \pmod{n}$ and $4P \equiv (363, 16880) \pmod{n}$, but

$$6P \equiv \left(\frac{18984764783665}{25724631321}, \frac{82719639550389910598}{4125947892943869} \right) \equiv \mathfrak{o} \pmod{n}$$

since $\gcd(25724631321, 18247) = 71$, and $18247 = 71 \cdot 257$.

5.33 We compute that $2P \equiv (4268, 11378) \pmod{n}$ and $4P \equiv (9877, 27743) \pmod{n}$, but $6P$ is as in Exercise 5.31 and $\gcd((25724631321, 38411) = 71$. We have factored $n = 38411 = 71 \cdot 541$.

Section 5.4

5.35 Since (a) holds, then

$$a^2 \approx \prod_{i=1}^{k} \left(\frac{\sqrt{2n}}{M} \right)^{2(1/2k)} = \left(\frac{\sqrt{2n}}{M} \right)^{\sum_{i=1}^{k}(1/k)} = \left(\frac{\sqrt{2n}}{M} \right),$$

so the first condition is satisfied in (5.11). Since $b^2 \equiv n \pmod{a^2}$ given the solution to the system of congruences via the Chinese Remainder Theorem, then $a^2 \mid (b^2 - n)$, so $(b^2 - n)/a^2 = c \in \mathbb{Z}$, which is the second condition. If $b \ge a^2/2$, then replace b by $b - a^2$, and we have $|b| < a^2/2$, which is the last condition.

For the choices given, we set $a = 11 \cdot 17 = 187$, and compute $b_1 = 23$ since $23^2 \equiv n \pmod{11^2}$ and $b_2 = 79$ since $79^2 \equiv n \pmod{17^2}$. Then we use the Chinese Remainder Theorem to solve $b \equiv 23 \pmod{11^2}$ and $b \equiv 79 \pmod{17^2}$ for $b = 11639$. Since $b^2 \equiv 31384 \equiv n \pmod{a^2}$, then $c = (b^2 - n)/a^2 = 1802$.

Section 6.1

6.1 Let $m_1 = \sum_{i=1}^{r} s_i/r$ and $m_2 = \sum_{j=1}^{r} t_j/r$. First we observe that,

$$\mathrm{var}(\{s_i\}_{i=1}^{r}) = \sum_{i=1}^{r}(s_i - m_1)^2/r = \frac{1}{r}\left(\sum_{i=1}^{r} s_i^2 - 2m_1 \sum_{i=1}^{r} s_i \right) + m_1^2 =$$

$$\frac{1}{r}\left(\sum_{i=1}^{r} s_i^2 - 2m_1 \sum_{i=1}^{r} s_i + m_1 \sum_{i=1}^{r} s_i \right) = \frac{1}{r}\left(\sum_{i=1}^{r} s_i^2 - m_1 \sum_{i=1}^{r} s_i \right) = \frac{1}{r} \sum_{i=1}^{r} s_i^2 - m_1^2,$$

and a similar result holds for $\mathrm{var}(\{t_j\}_{j=1}^{r})$ and for $\mathrm{var}(\{s_i\}_{i=1}^{r} + \{t_j\}_{j=1}^{r})$. Therefore, if we let

$$m = \frac{1}{r^2} \sum_{i=1}^{r} \sum_{j=1}^{r} (s_i + t_j) = \frac{1}{r^2}\left(r \sum_{i=1}^{r} s_i + r \sum_{j=1}^{r} t_j \right) = m_1 + m_2,$$

then

$$\text{var}(\{s_i\}_{i=1}^r + \{t_j\}_{j=1}^r) = \frac{1}{r^2}\sum_{i=1}^r\sum_{j=1}^r (s_i + t_j)^2 - (m_1 + m_2)^2 =$$

$$\frac{1}{r^2}\sum_{i=1}^r \left(s_i^2 r + 2s_i \sum_{j=1}^r t_j + \sum_{j=1}^r t_j^2 \right) - (m_1 + m_2)^2 =$$

$$\frac{1}{r}\sum_{i=1}^r s_i^2 + \frac{1}{r}\sum_{j=1}^r t_j^2 + \frac{2}{r^2}\sum_{i=1}^r s_i \sum_{j=1}^r t_j - m_1^2 - m_2^2 - 2m_1 m_2 =$$

$$\frac{1}{r}\sum_{i=1}^r s_i^2 - m_1^2 + \frac{1}{r}\sum_{j=1}^r t_j^2 - m_2^2 = \text{var}(\{s_i\}_{i=1}^r) + \text{var}(\{t_j\}_{j=1}^r).$$

Also, $\text{var}(\{s_i\}_{i=1}^r) = 0 = \sqrt{\text{var}(\{s_i\}_{i=1}^r)}$ for $\{s_i\}_{i=1}^r = \{4, 4, 4, 4, \}$, since $m = 4$, and for $\{t_j\}_{j=1}^r = \{5, 10, 50, 100\}$, $\text{var}(\{t_j\}_{j=1}^r) = 23275/16$ and $\sqrt{\text{var}(\{t_j\}_{j=1}^r)} = 38.14$.

Section 6.2

6.3 Using the Chinese remainder theorem, there is a solution to $x \equiv c_i \equiv m^3 \pmod{n_i}$ for each $i = 1, 2, 3$. Since $m^3 < n_1 n_2 n_3$, then $x = m^3$. By computing the cube root of the integer x, we retrieve m.

6.5 We solve $1 = ed + \phi(n)x = 5d + 5903364x$ and get $d = 1180673$ for $x = -1$.

6.7 As above, we solve $1 = ed + \phi(n)x = 3d + 20734288x$ and get $d = 13822859$ for $x = -2$.

6.9 As above, we solve $1 = ed + \phi(n)x = 3d + 56579188x$ and get $d = 37719459$ for $x = -2$.

6.11 Since $d < \phi(n)$, and $0 < m \le e$, then Eve can multiply $ed - m(n-p-q+1) = 1$ by p and reduce modulo $2^{n/4}$ to get

$$(ed)p - mp(n - p + 1) + mn \equiv p \pmod{2^{n/4}},$$

which may be rewrite as

$$mnp^2 - (mn + m + 1 + ed)p + mn \equiv 0 \pmod{2^{n/4}}.$$

Since Eve knows $n/4$ of the least significant bits of d, she knows ed modulo $2^{n/4}$. Now Eve can try each of the possible values of p and use Theorem 6.8 to test each one. Hence, after at most $e \log_2 e$ trials, she has factored n.

Section 6.3

6.13 By Exercise 5.2 on page 95, $a^{\lambda(n)+1} \equiv a \pmod{n}$ for all a prime to n. If $p \mid n$, then we need only show that p divides $a^{\lambda(n)+1} - a$ for each prime dividing n since n is squarefree. However, $a^{\lambda(n)+1} \equiv a \equiv 0 \pmod{p}$ for each such a. This completes the proof. However, if n is not squarefree, then this does not hold in general. For instance, if $n = 12$, then $\lambda(12) = 2$ and

$$10^{\lambda(12)+1} \equiv 10^3 \equiv 4 \not\equiv 10 \pmod{12}.$$

6.15 Since modular exponentiation is computationally easy using, for instance, the repeated squaring method, but finding $f^{-1}(y) \equiv y^d \pmod{n}$ is computationally infeasible without knowledge of d, then this is a one-way, trapdoor function with d as the trapdoor. See the discussion in Section 5.1, especially the discussion at the bottom of page 94.

6.17 Use the extended Euclidean algorithm C.5 on page 213 on e and $2n'$ to find integers d and m' such that $ed + 2n'm' = 1$. Then destroy all records of p, q, n', and m', and keep d as the private key (trapdoor). Thus, $m^e \equiv c \pmod{n}$ is the enciphering function and $c^d \equiv m \pmod{n}$ is the deciphering function where $ed \equiv 1 \pmod{\phi(n)}$.

Section 6.4

6.19 We have that $n = 2AB + 1 = 307$ and $m^{n-1} \equiv 2^{306} \equiv 1 \pmod{n}$ and $\gcd(m^{(n-1)/p}, n) = \gcd(2^{18} - 1, 307) = 1$, so 307 is a provable prime and since $307 = (100110011)_2$, it is a 9-bit prime.

6.21 Since $n = 2AB + 1 = 2311$, and $2^{2310} \equiv 1 \pmod{n}$ with $\gcd(2^{462} - 1, n) = 1$, then 2311 is a provable prime and $2311 = (100100000111)_2$, which is 12 bits so we are done.

6.23 (a) Since $f : (\mathbb{Z}/n\mathbb{Z})^* \mapsto (\mathbb{Z}/n\mathbb{Z})^*$, then there are only finitely many possibilities for s_{j-1}^a, so eventually $s_{\ell+1}^a \equiv s_j$ for some $j \leq \ell$.

(b) $f(2) = (8, 512, 161, 2)$, so $\ell = 4$ which is the answer to (c).

Section 7.1

7.1 Alice is identified since,

$$\delta \equiv \alpha^y v^r \equiv \alpha^{k+er} v^r \equiv \alpha^{k+er} \alpha^{-er} \equiv \alpha^k \equiv \gamma \pmod{p}.$$

7.3 Alice computes $y \equiv k + er \equiv 1001 + 151 \cdot 512 \pmod{1559}$, and Bob computes

$$\delta \equiv \alpha^y v^r \equiv 49^{363} \cdot 460^{512} \equiv 502 \not\equiv \gamma \equiv 501 \pmod{3119},$$

so Bob rejects Alice's proof of identity.

7.5 As above, $y \equiv 4 + 8 \cdot 3 \equiv 11 \pmod{17}$, and

$$\delta \equiv 3668^{11} \cdot 4508^3 \equiv 4104 \equiv \gamma \pmod{7481},$$

so Bob accepts Alice's proof of identity.

7.7 In step (9), Bob must verify Trent's signature. However, since $v' \neq v$ then $s' \neq s$, so $\text{ver}_{T(k)}((I_A, v', s')) = 0$, and Bob rejects the proof of identity.

7.9 Alice computes

$$y_1 \equiv k_1 + e_1 r \equiv 10 + 25 \cdot 8 \equiv 1 \pmod{q}$$

$$y_2 \equiv k_2 + e_2 r \equiv 9 + 27 \cdot 8 \equiv 5 \pmod{q},$$

and sends them to Bob who computes

$$\delta \equiv \alpha_1^{y_1} \cdot \alpha_2^{y_2} \cdot v^r \equiv 443^1 \cdot 2541^5 \cdot 1768^8 \equiv 2490 \equiv \gamma \pmod{p},$$

so he accepts Alice's proof of identity.

7.11 As above,

$$y_1 \equiv 1007 + 998 \cdot 256 \equiv 1144 \,(\text{mod } q)$$
$$y_2 \equiv 506 + 5 \cdot 256 \equiv 1786 \,(\text{mod } q).$$

Then Bob computes,

$$\delta \equiv 25^{1144} \cdot 1133^{1786} \cdot 771^{256} \equiv 3009 \not\equiv \gamma \,(\text{mod } p),$$

so Bob rejects Alice's proof of identity.

Section 7.2

7.13 Since $\text{sig}_k(m)^e \equiv 15971^{133} \equiv 911 \equiv m \,(\text{mod } 20497)$, then Bob accepts Alice's signature.

7.15 Since $\text{sig}_k(m)^e \equiv 317905^{611} \equiv 116106 \not\equiv 1111 \equiv m \,(\text{mod } 58320)$, then Bob rejects Alice's signature.

7.17 We have, $s^d z^{-1} \equiv z^{ed} m^d z^{-1} \equiv z m^d z^{-1} \equiv m^d \,(\text{mod } n)$, since $ed \equiv 1 \,(\text{mod } \phi(n))$. Alice does not know if she is signing away her life savings or confessing to a murder she did not commit. Hence, for her to sign blindly requires safeguards.

7.19 $\delta \equiv y^\beta \beta^\gamma \equiv 11^{227} 227^{207} \equiv 25 \,(\text{mod } p)$ and $\sigma \equiv \alpha^m \equiv 5^2 \equiv 25 \,(\text{mod } p)$, so it is accepted.

7.21 As above, $\delta \equiv 711^{330} 330^{37} \equiv 495 \,(\text{mod } p)$, and $\sigma \equiv 11^{191} \equiv 495$, so it is accepted.

7.23 We have,

$$y^{\beta_1} \beta_1^{\gamma_1} \equiv \alpha^{a\beta_1} \alpha^{(r_1 + ar_2)\gamma_1} \equiv \alpha^{a\beta_1} \alpha^{(r_1 + ar_2)(-\beta_1 r_2^{-1})} \equiv$$
$$\alpha^{a\beta_1} \alpha^{-\beta_1 r_1 r_2^{-1} - \beta_1 a} \equiv \alpha^{-\beta_1 r_1 r_2^{-1}} \equiv \alpha^{\gamma_1 r_1} \equiv \alpha^{m_1} \,(\text{mod } p).$$

7.25 Since $\beta, \gamma,$ and m are known, then knowledge of r means that he may compute $a \equiv (m - r\gamma)\beta^{-1} \,(\text{mod } p - 1)$.

7.27 When Alice sends her valid signature to Bob, we have:

$$\delta \equiv \alpha^\gamma y^h \equiv \alpha^{r + eh} \alpha^{-eh} \equiv \alpha^r \equiv \beta \,(\text{mod } p).$$

7.29 Bob computes $\delta \equiv \alpha^\gamma \cdot y^h \equiv 1220^6 \cdot 1456^{101} \equiv 913 \equiv \beta \,(\text{mod } p)$, so he accepts Alice's signature.

7.31 As above, Bob computes $\delta \equiv 1079^{25} \cdot 427^{21} \equiv 1217 \equiv \beta \,(\text{mod } p)$, so he accepts.

Section 7.3

7.33 Since we have that A is unique to Alice and

$$fA^k \equiv g_1^{f_1} g_2^{f_2} (g_1^{e_1} g_2^{e_2})^k \equiv g_1^{f_1 + ke_1} g_2^{f_2 + ke_2} \equiv g_1^{\ell_1} g_2^{\ell_2} \,(\text{mod } p),$$

then Alice's identity is indeed verified.

7.35 Since only the bank knows x, then only the bank can send a response satisfying both

$$\alpha^r \equiv \alpha^{xc + w} \equiv (\alpha^x)^c \alpha^w \equiv h^c y_2 \,(\text{mod } p)$$

and

$$A^r \equiv A^{xc + w} \equiv (A^x)^c A^w \equiv m^c y_3 \,(\text{mod } p).$$

7.37 By Exercise 7.34, $XY \equiv y_1' \pmod{p}$, and $y_1' \equiv A^x \pmod{p}$ with $A \not\equiv 1 \pmod{p}$ by step (1) of the protocol for opening Alice's account. Also, $x \in (\mathbb{Z}/q\mathbb{Z})^*$, by step (3) of the setup stage, so $x \not\equiv 0 \pmod{q}$, which completes the proof.

Section 8.1

8.1 First we observe that

$$K_k(x_i) = \prod_{\substack{1 \leq \ell \leq t \\ \ell \neq k}} \frac{x_i - x_\ell}{x_k - x_\ell} \equiv 0 \pmod{p}$$

if $i \neq k$ since $K_k(x_i)$ has a factor $(x_i - x_i)/(x_k - x_i)$. Also,

$$K_k(x_k) \equiv 1 \pmod{p}$$

since all factors are of the form $(x_k - x_\ell)/(x_k - x_\ell) = 1$. We note that

$$1/(x_k - x_\ell) \equiv (x_k - x_\ell)^{-1} \pmod{p},$$

so as long as $k \neq \ell$, such inverses exist. Therefore,

$$f(x_i) \equiv \sum_{k=1}^{t} m_k K_k(x_i) \equiv m_i \pmod{p}$$

for $i = 1, 2, \ldots, t$.

8.3 Although $g(x)$ produces the same values, it does so at only one of the three values of x_i that $f(x)$ produces them. We have

$$f(1) \equiv g(1) \equiv 7 \pmod{p},$$

but

$$g(2) \equiv f(3) \equiv 1407 \pmod{p}, \quad \text{and} \quad g(3) \equiv f(5) \equiv 334 \pmod{p}.$$

Note that

$$g(x) \equiv 7\frac{(x-2)(x-3)}{(1-2)(1-3)} + 1407\frac{(x-1)(x-3)}{(2-1)(2-3)} + 334\frac{(x-1)(x-2)}{(3-1)(3-2)} \pmod{p},$$

whereas,

$$f(x) \equiv 7\frac{(x-3)(x-5)}{(1-3)(1-5)} + 1407\frac{(x-1)(x-5)}{(3-1)(3-5)} + 334\frac{(x-1)(x-3)}{(5-1)(5-3)} \pmod{p},$$

and, of course, $f(0) \equiv 3301 \pmod{p}$, while $g(0) \equiv 2856$, which is not the original message.

8.5 We calculate $p(x)$ for $x = 1, 3, 5$ and get $(1, 488)$, $(3, 2400)$, and $(5, 1881)$. Then we compute the Lagrange interpolation formula to get,

$$f(x) = -\frac{2431}{8}x^2 + \frac{4343}{2}x - \frac{11037}{8}.$$

Since $8^{-1} \equiv 529 \pmod{p}$, $2^{-1} \equiv 2116 \pmod{p}$, and

$$-2431 \cdot 529 \equiv 225; \quad 4343 \cdot 2116 \equiv 56; \quad -11037 \cdot 529 \equiv 207,$$

all modulo p, then we recover the message at $f(0) \equiv p(0) \equiv 207$.

8.7 As above, we calculate $(1, 1087)$, $(3, 1677)$, $(5, 2819)$, and plug this into the Lagrange interpolation formula to get immediately that $f(x) = 69x^2 + 19x + 999$.

8.9 We have, $m + rp = 50909 < 248605 = a_1 \cdot a_2 \cdot a_3$. Thus we compute $s_1 \equiv 4 \pmod{5}$, $s_2 \equiv 5 \pmod{7}$, and $s_3 \equiv 1188 \pmod{7109}$. When the participants poll their shares and use the Chinese Remainder Theorem, they recover

$$50909 \equiv 519 \equiv m \pmod{5039}.$$

8.11 As above, $rp + m = 713371 < a_1 \cdot a_2 \cdot a_3 = 770110$;

$$s_1 \equiv 1 \pmod{10}, s_2 \equiv 10 \pmod{11}, \text{ and } s_3 \equiv 6270 \pmod{7001}.$$

The Chinese Remainder Theorem recovers $713371 \equiv 3071 \equiv m \pmod{p}$.

8.13 We compute,
$$c_1 \equiv 90 - 15 \cdot 59 - 52 \cdot 409 \equiv 69 \pmod{503},$$
$$c_2 \equiv 90 - 11 \cdot 59 - 123 \cdot 409 \equiv 440 \pmod{503},$$
$$c_3 \equiv 90 - 308 \cdot 59 - 400 \cdot 409 \equiv 404 \pmod{503},$$

so the distributed hyperplanes are,

$$\ell_1 \equiv 69 + 15x_1 + 52x_2; \quad \ell_2 \equiv 440 + 11x_1 + 123x_2; \quad \ell_3 \equiv 404 + 308x_1 + 400x_2,$$

all congruences modulo 503. Plugging these values into (8.4), we get,

$$AX = \begin{pmatrix} 15 & 52 & -1 \\ 11 & 123 & -1 \\ 308 & 400 & -1 \end{pmatrix} \begin{pmatrix} x_1 \\ x_2 \\ x_3 \end{pmatrix} = \begin{pmatrix} -69 \\ -440 \\ -404 \end{pmatrix} = C.$$

then solving for $\det(A) = 22195$, and using Cramer's rule, we get

$$(x_1, x_2, x_3) = (105323/22195, -110043/22195, -2610936/22195).$$

However, $22195^{-1} \equiv 8 \pmod{503}$ and

$$105323 \cdot 8 \equiv 59; \quad -110043 \cdot 8 \equiv 409; \quad -2610936 \cdot 8 \equiv 90,$$

all congruences modulo 503. Thus, $(m_1, m_2, m_3) = (59, 409, 90)$, and the secret message $m_1 = 59$ is retrieved.

8.15 As above,
$$c_1 \equiv 718 - 297 \cdot 107 - 306 \cdot 1 \equiv 269 \pmod{719},$$
$$c_2 \equiv 718 - 419 \cdot 107 - 537 \cdot 1 \equiv 645 \pmod{719},$$
$$c_3 \equiv 718 - 698 \cdot 107 - 709 \cdot 1 \equiv 99 \pmod{719},$$

so the distributed hyperplanes are,

$$\ell_1 \equiv 269 + 297x_1 + 306x_2; \quad \ell_2 \equiv 645 + 419x_1 + 537x_2; \quad \ell_3 \equiv 99 + 698x_1 + 709x_2,$$

all congruences modulo 719. Thus,

$$AX = \begin{pmatrix} 297 & 306 & -1 \\ 419 & 537 & -1 \\ 698 & 709 & -1 \end{pmatrix} \begin{pmatrix} x_1 \\ x_2 \\ x_3 \end{pmatrix} = \begin{pmatrix} -269 \\ -645 \\ -99 \end{pmatrix} = C.$$

then solving for $\det(A) = 43465$, we get

$$(x_1, x_2, x_3) = (190798/43465, -171516/43465, 3175039/8693).$$

However, $43465^{-1} \equiv 323 \,(\mathrm{mod}\ 719)$, $8693^{-1} \equiv 177 \,(\mathrm{mod}\ 719)$, and

$$190798 \cdot 323 \equiv 107; \quad -171516 \cdot 323 \equiv 1; \quad 3175039 \cdot 177 \equiv 718,$$

all congruences modulo 719. Thus, $(m_1, m_2, m_3) = (107, 1, 718)$, and the secret message $m_1 = 107$ is retrieved.

8.17 In matrix-theoretic terms, the equation $m_k \equiv m + \sum_{j=1}^{t} c_j x_k^j \,(\mathrm{mod}\ p)$ is written as

$$AC = \begin{pmatrix} 1 & x_1 & \cdots & x_1^{t-1} \\ 1 & x_2 & \cdots & x_2^{t-1} \\ \vdots & \vdots & \vdots & \vdots \\ 1 & x_t & \cdots & x_t^{t-1} \end{pmatrix} \begin{pmatrix} c_0 \\ c_1 \\ \vdots \\ c_t \end{pmatrix} \equiv \begin{pmatrix} m_1 \\ m_2 \\ \vdots \\ m_t \end{pmatrix} \,(\mathrm{mod}\ p).$$

We now show that

$$\det(A) = \prod_{1 \le i < k \le t} (x_k - x_i) \tag{S3}$$

We use induction on t. If $t = 2$, then

$$\det(x_k^j) = \begin{vmatrix} 1 & x_1 \\ 1 & x_2 \end{vmatrix} = x_2 - x_1.$$

This is the induction step. Now assume that the result holds for all such $n \times n$ matrices with $n < t$. If $\mathrm{cof}(A_{j,k})$ denotes the *cofactor* of the matrix $A = (x_k^j)$, then by (C.37),

$$\det(x_k^j) = \sum_{i=1}^{t} x_k^j \mathrm{cof}(A_{j,k}).$$

By induction hypothesis, the result holds for each $A_{j,k}$, so the entire result holds. Since we have shown that (S3) holds, it is clear that $\det(A) \equiv 0 \,(\mathrm{mod}\ p)$ if an only if $x_k \equiv x_i \,(\mathrm{mod}\ p)$ for some $i \ne k$.

Section 8.2

8.19 We have,

$$s_B^{d_A}(p_B^e + I_B)^{e_A} \equiv \alpha^{e_B d_A}((c_B - I_B)^{de} + I_B)^{e_A} \equiv \alpha^{e_B d_A}(c_B - I_B + I_B)^{e_A} \equiv$$

$$\alpha^{e_B d_A}(c_B)^{e_A} \equiv \alpha^{e_B d_A} \alpha^{d_B e_A} \equiv \alpha^{e_B d_A + e_A d_B} \equiv k \,(\mathrm{mod}\ n),$$

and

$$s_A^{d_B}(p_A^e + I_A)^{e_B} \equiv \alpha^{e_A d_B}((c_A - I_A)^{de} + I_A)^{e_B} \equiv \alpha^{e_A d_B}(c_A - I_A + I_A)^{e_B} \equiv$$

$$\alpha^{e_A d_B}(c_A)^{e_B} \equiv \alpha^{e_A d_B} \alpha^{d_A e_B} \equiv \alpha^{e_A d_B + d_A e_B} \equiv k \,(\mathrm{mod}\ n).$$

8.21 All congruences are modulo $n = 48959 = pq$. We perform the calculations within the scheme. Alice and Bob, respectively, compute, $c_A \equiv \alpha^{d_A} \equiv 3^{10279} \equiv 38953$ and $c_B \equiv \alpha^{d_B} \equiv 3^{32773} \equiv 11653$, and Trent computes

$$p_A \equiv (c_A - I_A)^d \equiv (38953 - 21)^{9701} \equiv 35736,$$

and

$$p_B \equiv (c_B - I_B)^d \equiv (11653 - 93)^{9701} \equiv 30197.$$

Then Alice and Bob compute, respectively,

$$s_A \equiv \alpha^{e_A} \equiv 3^{151} \equiv 3211,$$

and

$$s_B \equiv \alpha^{e_B} \equiv 3^{37} \equiv 26897.$$

Lastly, Alice and Bob, respectively, compute,

$$k \equiv s_B^{d_A}(p_B^e + I_B)^{e_A} \equiv 26897^{10279}(30197^5 + 93)^{151} \equiv 28578,$$

and

$$k \equiv s_A^{d_B}(p_A^e + I_A)^{e_B} \equiv 3211^{32773}(35736^5 + 21)^{37} \equiv 28578,$$

the self-certified shared key.

8.23 As above, we do the calculations. All congruences are modulo $n = 295907 = pq$.

$$c_A \equiv 7^{2021} \equiv 204856, \text{ and } c_B \equiv 7^{3011} \equiv 59114,$$

$$p_A \equiv (204856 - 156)^{290315} \equiv 26211,$$

$$p_B \equiv (59114 - 1001)^{290315} \equiv 217630,$$

$$s_A \equiv 7^{14033} \equiv 90187, \text{ and } s_B \equiv 7^{221675} \equiv 15244,$$

$$k \equiv 15244^{2021}(217630^{131} + 1001)^{14033} \equiv 124394,$$

and lastly Bob's calculation,

$$k \equiv 90187^{3011}(26211^{131} + 156)^{221675} \equiv 124394,$$

to establish the shared key.

8.25 Trent computes, with all congruences modulo p,

$$p(x, y) \equiv 11 + 13(x + y) + 200xy,$$

$$f_A(x) \equiv 141 + 210x; \quad f_B(x) \equiv 122 + 380x; \quad f_C(x) \equiv 183 + 24x;$$

from which are computed the session keys,

$$k_{A,B} \equiv 316; \quad k_{A,C} \equiv 423; \quad k_{B,C} \equiv 203.$$

8.27 As above,

$$p(x, y) \equiv 5 + 15(x + y) + 25xy,$$

$$f_A(x) \equiv 287 + 485x; \quad f_B(x) \equiv 186 + 47x; \quad f_C(x) \equiv 95 + 165x;$$

from which we get,

$$k_{A,B} \equiv 746; \quad k_{A,C} \equiv 770; \quad k_{B,C} \equiv 468.$$

8.29 As above,
$$p(x, y) \equiv 5 + 808(x + y) + 700xy,$$
$$f_A(x) \equiv 620 + 127x; \quad f_B(x) \equiv 524 + 883x; \quad f_C(x) \equiv 495 + 733x;$$
from which we get,
$$k_{A,B} \equiv 21; \quad k_{A,C} \equiv 210; \quad k_{B,C} \equiv 33.$$

8.31 Since Mallory has
$$f_M(x) \equiv r_1 + r_2 k_M + (r_2 + r_3 k_M)x \pmod{p},$$
and Eve has
$$f_E(x) \equiv r_1 + r_2 k_E + (r_2 + r_3 k_E)x \pmod{p},$$
then they have the four modular equations,
$$a_M \equiv r_1 + r_2 k_M \pmod{p},$$
$$b_M \equiv r_2 + r_3 k_M \pmod{p},$$
$$a_E \equiv r_1 + r_2 k_E \pmod{p},$$
and
$$b_E \equiv r_2 + r_3 k_E \pmod{p}.$$
Hence, they have four equations in three unknowns from which elementary algebra will yield a unique solution for r_1, r_2, r_3.

Section 9.1

9.1 Since Monty knows e, he can easily forge messages at will, and they will go undetected due to the RSA conjecture. If Hostvania violates the treaty by engaging in an underground nuclear test, it can point its finger at Monty as the perpetrator, and Monty could not disavow this claim since he knows e, and is capable of forgeries.

Section 9.2

9.3 Bob verifies Alice's signature by using her public key to get,
$$D_{e_A}(D_{d_A}(c)) \equiv c \pmod{n}.$$
Then he uses his private key to obtain,
$$D_{d_B}(c) \equiv m \pmod{n},$$
where
$$e_B d_B \equiv 1 \pmod{(p^2 - 1)(q^2 - 1)}.$$

9.5 Since
$$e^{-m(m-1)/(2n)} \approx 1 - p_c,$$

then
$$-m(m-1)/(2n) \approx \ln(1 - p_c).$$

Hence,
$$m^2 - m \approx -2n \ln(1 - p_c),$$

and so
$$m^2 \approx -2n \ln(1 - p_c) \approx 2n \ln(1/(1 - p_c)),$$

since we can safely ignore the smaller factor of $-m$ in an approximation. Thus,
$$m \approx \sqrt{2n \ln(1/(1 - p_c))}.$$

If $p_c = 1/2$, then
$$m \approx 1.17\sqrt{n}.$$

Section 9.3

9.7 Mallory could intercept, decrypt to get T_A and α^a, and potentially get a. However, the second part of the key α^b is encrypted, so this provides additional security against this weakness.

Section 9.4

9.9 If $t \equiv t' \pmod{\phi(n)}$, then
$$s' \equiv c^{eT} h(I_A)^t \equiv (rd_A^t)^{eT} h(I_A)^t \equiv r^{eT} d_A^{t'eT} h(I_A)^{t'} \equiv$$
$$r^{eT} (d_A^{eT} h(I_A))^{t'} \equiv r^{eT} \equiv s \pmod{n}.$$

Since only Alice knows d_A, then Bob may conclude that her signature is valid.

Bibliography

[1] C. Adams and S. Farrell, *Internet X.509 public key infrastructure: Certificate management protocols*, Internet Request for Comments 2510 (March 1999).

[2] C. Adams and S. Lloyd, **Understanding Public-Key Infrastructure**, New Riders Publishing, Indianapolis, Indiana (1999).

[3] W. R. Alford, A. Granville, and C. Pomerance, *There are infinitely many Carmichael numbers*, Ann. Math. **140** (1994), 703–722.

[4] N.C. Ankeny, *The least quadratic nonresidue*, Ann. of Math. **55** (1952), 65–72.

[5] ANSI X9.17, *American National Standard — Financial institution key management (wholesale)*, ASCX9 Secretariat — American Bankers Association (1985).

[6] ANSI X9.57, *Public key cryptography for the financial service industry — Certificate management* (1997).

[7] C. Asmuth and J. Bloom, *A modular approach to key safeguarding*, IEEE Trans. Inf. Theor. **IT-30** (1983), 208–210.

[8] A.O.L. Atkin and R.G. Larson, *On a primality test of Solovay and Strassen*, Siam J. Comp. **11** (1982), 789–791.

[9] D. Atkins, M. Graff, A.K. Lenstra, and P.C. Leyland, *The magic words are SQUEAMISH OSSIFRAGE* in **Advances in Cryptology**, ASIACRYPT '94, Springer-Verlag, Berlin, LNCS **917**, (1995), 263–277.

[10] D. Atkins, W. Stallings, and P. Zimmerman, *PGP Message Exchange Formats*, Internet Request for Comments 1991 (August 1996).

[11] M. Aydos, T. Yanik, and Ç.K. Koç, *High-Speed implementation of an ECC-based wireless authentication protocol on an ARM microprocessor*, IEEE Proc. Commun.**148** (2001), 273–279.

[12] A. Aziz and W. Diffie, *Privacy and authentication for wireless local area networks*, IEEE Personal Commun. **1** (1994), 25–31.

249

[13] E. Bach, *Analytic methods in the analysis and design of number-theoretic algorithms* (A distinguished ACM dissertation) MIT Press, Cambridge, Massachusetts (1985).

[14] E. Bach, *Explicit bounds for primality testing and related problems*, Math. Comp. **55** (1990), 355–380.

[15] E. Bach and J. Shallit, **Algorithmic Number Theory**, **1**, MIT Press, Cambridge, Massachusetts (1997).

[16] F.L. Bauer, **Decrypted Secrets**, Springer, Berlin (1997).

[17] A. Beimel and B. Chor, *Secret sharing with public reconstruction*, in **Advances in Cryptology**, in CRYPTO '95, Springer-Verlag, Berlin, LNCS **963** (1995), 353–366.

[18] E.T. Bell, **Men of Mathematics**, Simon and Schuster, New York (1937).

[19] M. Bellare and P. Rogaway, *Optimal asymmetric encryption* in *Advances in Cryptology*, EUROCRYPT '94, Springer-Verlag, Berlin, LNCS **950** (1994), 92–111.

[20] M.J. Beller, L.F. Chang, and J. Yacobi, *Privacy and authentication on a portable communications system*, IEEE J. Selected Areas Commun. **11** (1993), 821–829.

[21] M.J. Beller and Y. Yacobi, *Fully-fledged two-way public-key authentication and key agreement for low cost terminals*, Electron. Lett. **29** (1993), 999–1001.

[22] S.M. Bellovin and M. Merritt, *Encrypted key exchange: Password-based protocols secure against dictionary attacks*, in Proceedings of the 1992 IEEE Computer Society Conference on Research in Security and Privacy (1992), 72–84.

[23] S.M. Bellovin and M. Merritt, *Cryptographic protocol for secure communications*, U.S. Patent 5241599, August 31 (1993).

[24] D.J. Bernstein, *Faster square roots in annoying finite fields*, preprint.

[25] D.J. Bernstein, *The multiple-lattice number field sieve*, preprint.

[26] E. Biham and A. Shamir, **Differential Cryptanalysis of the Data Encryption Standard**, Springer-Verlag, New York (1993).

[27] G.R. Blakely, *Safeguarding cryptographic keys*, Proc. National Computer Conf., American Federation of Information Processing Societies **48** (1979), 242–268.

[28] G.R. Blakely and G.A. Kabatianski, *On general perfect secret sharing schemes*, in **Advances in Cryptology**, CRYPTO '95, Springer-Verlag, Berlin, LNCS **963** (1995), 367–371.

[29] R. Blom, *An optimal class of symmetric key generation systems*, EURO-CRYPT '84, Springer-Verlag, Berlin, LNCS **209** (1985), 335–338.

[30] M. Blum, *Coin flipping by telephone: A protocol for solving impossible problems*, in Proceedings of the 24th IEEE Computer Conference, IEEE Press (1982), 133–137.

[31] C. Blundo, A. De Santis, A. Herzberg, S. Kutten, U. Vaccaro, and M. Yung, *Perfectly-secure key distribution for dynamic conferences*, in **Advances in Cryptology**, CRYPTO '92, Springer-Verlag, Berlin, LNCS **740** (1993), 471–486.

[32] H. Boender and H.J.J. te Riele, *Factoring integers with large prime variations of the quadratic sieve*, Exp. Math. **5** (1996), 257–273.

[33] J.-P. Boly, A. Bossealaers, R. Cramer, R. Michelsen, S. Mjølsnes, F. Muller, T. Pedersen, B. Pfitzmann, P. de Rooij, B. Schoenmakers, M. Schunter, L. Vallée, and M. Waidner, *The ESPRIT project CAFE — High security digital payment systems*, in **Computer Security**, ESORICS '94 Springer-Verlag, Berlin, LNCS **875** (1994), 217–230.

[34] D. Boneh, *Twenty years of attacks on the RSA cryptosystem*, Notices American Mathematical Society **46** (1999), 203–213.

[35] D. Boneh, *Simplified OAEP for the RSA and Rabin functions*, in **Advances in Cryptology**, CRYPTO 2001, Springer-Verlag, Berlin, LNCS **2139** (2001), 275–291.

[36] D. Boneh and G. Durfee, *Cryptanalysis of RSA with private key d less than $N^{0.292}$*, IEEE Trans. Inf. Theor. **46** (2000), 1339–1349.

[37] D. Boneh, G. Durfee, and Y. Frankel, *An attack on RSA given a fraction of the private key bits*, in **Advances in Cryptology**, ASIACRYPT 1998, Springer-Verlag, Berlin, LNCS **1403** (1998), 25–34.

[38] D. Boneh, A. Joux, and P.Q. Nguyen, *Why textbook ElGamal and RSA encryption are insecure*, in **Advances in Cryptology**, ASIACRYPT 2000 (Kyoto), Springer-Verlag, Berlin, LNCS **1976** (2000), 30–43.

[39] D. Boneh and R. Lipton, *Algorithms for black-box fields and their applications to cryptography* in **Advances in Cryptology**, CRYPTO '96, Springer-Verlag, Berlin, LNCS **1109** (1996), 283-297.

[40] D. Boneh and H. Shacham, *Fast variants of RSA*, CryptoBytes **5** (2002), 1–9.

[41] D. Boneh and R. Venkatesan, *Breaking RSA may be easier than factoring*, preprint.

[42] P. Bonner, *Adding cookies to your site*, from CNET: Web Building:
http://builder.cnet.com/webbuilding/pages/Programming/Cookies
/?st.bl.pr.10.feat.1140

[43] S. Brands, *An efficient off-line electronic cash system based on the representation problem*, Report **CS-R9323**, Centrum voor Wiskunde en Informatica (1993).

[44] S. Brands, *Untraceable off-line cash in wallets with observers*, in **Advances in Cryptology**, CRYPTO '93, Springer-Verlag, Berlin, LNCS **773** (1994), 302–318.

[45] S. Brands, *Off-line cash transfer by smart cards*, in **Proceedings First Smart Card Research and Advanced Applications Conference**, V. Cordonnier and J.-J. Quisquater, eds. (1994), 101–117.

[46] J. Brillhart, D.H. Lehmer, J.L. Selfridge, B. Tuckerman, and S.S. Wagstaff, Jr., **Factorizations of $b^n \pm 1, b = 2, 3, 5, 6, 7, 10, 11, 12$ up to High Powers**, Contemp. Math. **22**, American Mathematical Society, Providence, R.I. (1983).

[47] J. Brillhart and J. Selfridge, *Some factorizations of $2^n \pm 1$ and related results*, Math. Comp. **21** (1967), 87–96.

[48] J.P. Buhler, H.W. Lenstra, and C. Pomerance, *Factoring integers with the number field sieve*, in **The Development of the Number Field Sieve**, A.K. Lenstra and H.W. Lenstra, Jr., eds., LNM, Springer-Verlag, Berlin **1554** (1993), 50–94.

[49] S. Burnett and S. Paine, **RSA Security's Official Guide to Cryptography**, RSA Press, Osborne/McGraw-Hill, Berkeley, California (2001).

[50] R, Canetti, R. Gennaro, S. Jarecki, and H. Krawczyk, *Adaptive security for threshold cryptosystems*, in **Advances in Cryptology**, CRYPTO '99, Springer-Verlag, Berlin, LNCS **1666** (1999), 98–115.

[51] D. Chaum, *Blind signatures for untraceable payments*, in **Advances in Cryptology**, CRYPTO '82, Plenum Press, New York (1983), 199–203.

[52] D. Chaum, *Security without identification: Transaction systems to make big brother obsolete*, Commun. ACM **28** (1985), 1030–1044.

[53] D. Chaum, *Blinding for unanticipated signatures*, in **Advances in Cryptology**, EUROCRYPT '87, Springer-Verlag, Berlin, LNCS **304** (1988), 227–233.

[54] D. Chaum, *Online cash checks*, in **Advances in Cryptology**, EUROCRYPT '89, Springer-Verlag, Berlin, LNCS **434** (1990), 288–293.

[55] C.Y. Chen, C.C. Chang, and W.P. Yang, *Cryptanalysis of the secret exponent of the RSA scheme*, J. Inf. Sci. Eng. **12** (1996), 277–290.

[56] B. Chor and R.L. Rivest, *A knapsack-type public-key cryptosystem based on arithmetic in finite fields*, in **Advances in Cryptology**, CRYPTO '84, Springer-Verlag, Berlin, LNCS **196** (1985), 54–65.

[57] B. Chor and R.L. Rivest, *A knapsack-type public-key cryptosystem based on arithmetic in finite fields*, IEEE Trans. Inf. Theor. **IT-34** (1988), 901–909.

[58] C.C. Cocks, *A note on "non-secret" encryption*, GCHQ/CESG publication, November 20 (1973), 1 page.

[59] H. Cohen, **A Course in Computational Algebraic Number Theory**, Springer-Verlag, Berlin (1993).

[60] D. Coppersmith, *Small solutions to polynomial equations, and low exponent RSA vulnerabilities*, J. Cryptol. **8** (1995), 101–114.

[61] D. Coppersmith, *Small solutions to polynomial equations, and low exponent RSA vulnerabilities*, J. Cryptol. **10** (1997), 233–260.

[62] D. Coppersmith, M. Franklin, J. Patarin, and M. Reiter, *Low exponent RSA with related messages*, in **Advances in Cryptology**, EUROCRYPT '96, Springer-Verlag, Berlin, LNCS **1070** (1996), 1–9.

[63] D. Coppersmith, A. Odlyzko, and R. Schroeppel, *Discrete logarithms in $GF(p)$*, Algorithmica **1** (1986), 1–15.

[64] R. Cramer, I. Damgård, and S. Fehr, *On the cost of reconstructing a secret, or VSS with optimal reconstruction phase*, in **Advances in Cryptology**, CRYPTO 2001, Springer-Verlag, Berlin, LNCS **2139** (2001), 503–523.

[65] A.J.C. Cunningham and H.J. Woodall, **Factorization of $y^n \mp 1, y = 2, 3, 5, 6, 7, 10, 11, 12$ up to high powers (n)**, Hodgson, London (1925).

[66] J.A. Davis, D.B. Holdridge, and G.J. Simmons, *Status report on factoring* (at Sandia National Labs.), in **Advances in Cryptology**, EUROCRYPT '84, Springer-Verlag, Berlin, LNCS **209** (1985), 183–215.

[67] J.M. DeLaurentis, *A further weakness in the common modulus protocol for the RSA cryptoalgorithm*, Cryptologia **8** (1984), 253–259.

[68] B. Den Boer, *Diffie-Hellman is as strong as discrete log for certain primes*, in **Advances in Cryptology**, CRYPTO '88, Springer-Verlag, Berlin, LNCS **403** (1990), 530–539.

[69] D.E. Denning and G.M. Sacco, *Timestamps in key distribution protocols*, Commun. ACM **24** (1981), 533–536.

[70] Y. Desmedt, *Simmons' protocol is not free of subliminal channels*, in 9th Foundations Workshop, IEEE Computer Society, Kenmare, Ireland (1996), 170–175.

[71] Y. Desmedt and M. Yung, *Minimal cryptosystems and defining subliminal-freeness*, in Proceedings of the IEEE International Symposium on Information Theory (1994), 347.

[72] T. Dierks and C. Allen, *The TLS protocol, version 1.0*, Internet Request for Comments, 2246 (January 1999).

[73] W. Diffie, *The first ten years of public-key cryptography*, Proc. IEEE **76** (1988), 560–577.

[74] W. Diffie and M.E. Hellman, *New directions in cryptography*, IEEE Trans. Inf. Theor. **22** (1976), 644–654.

[75] W. Diffie, P.C. Van Oorschot, and M.J. Weiner, *Authentication and authenticated key exchanges*, Designs, Codes, Cryptogr. **2** (1992), 107–125.

[76] J.D. Dixon, *Asymptotically fast factorization of integers*, Math. Comp. **36** (1981), 255–260.

[77] B. Dodson and A.K. Lenstra, *NFS with four large primes: an explosive experiment*, in **Advances in Cryptology**, CRYPTO '95, Springer-Verlag, Berlin, LNCS **963** (1995), 372–385.

[78] W. van Eck, *Electromagnetic radiation from video display units: An eavesdropping risk?*, Comput. Security **4** (1985), 269–286.

[79] T. ElGamal, *A public key cryptosystem and a signature scheme based on discrete logarithms*, in **Advances in Cryptology**, CRYPTO '84, Springer-Verlag, Berlin, LNCS **196** (1985), 10–18.

[80] T. ElGamal, *A public key cryptosystem and signature scheme based on discrete logarithms*, IEEE Trans. Inf. Theor. **31** (1985), 469–472.

[81] M. Elkins, *MIME security with Pretty Good Privacy*, Internet Request for Comments 2015 (October 1996).

[82] J.H. Ellis, *The possibility of secure non-secret digital encryption*, GCHQ–CESG publication, January (1970), 7 pages.

[83] J.H. Ellis, *The history of non-secret encryption*, GCHQ–CESG publication (1987), 4 pages.

[84] P. Erdös, *On the converse of Fermat's theorem*, American Mathematical Monthly **56** (1949), 623–624.

[85] U. Feige, A. Fiat, and A. Shamir, *Zero-knowledge proofs of identity*, Proc. 19th Annu. ACM Symp. Theor. Comput. (1987), 210–217.

[86] U. Feige, A. Fiat, and A. Shamir, *Zero-knowledge proofs of identity*, J. Cryptol. **1** (1988), 77–94.

[87] N. Ferguson, *Single item off-line coins*, in **Advances in Cryptology**, Eurocrypt '93, Springer-Verlag, Berlin, LNCS **765** (1994), 318–328.

[88] A. Fiat and A. Shamir, *How to prove yourself: Practical solution to identification and signature problems*, in **Advances in Cryptology**, CRYPTO '86, Springer-Verlag, Berlin, LNCS **263** (1987), 186–194.

[89] FIPS 186, *Digital signature standard*, Federal Information Processing Standards Publication 186, U.S. Department of Commerce/N.I.S.T. National Technical Information Service, Springfield, Virginia (1994).

[90] M.K. Franklin and M.K. Reiter, *A linear protocol failure for RSA with exponent three*, Note for CRYPTO '95 rump session (1995).

[91] E. Fujisaki and T. Okamaoto, *Secure integration of asymmetric and symmetric encryption schemes*, in **Advances in Cryptology**, CRYPTO '99, Springer-Verlag, Berlin, LNCS **1666** (1999), 537–554.

[92] E. Fujisaki, T. Okamoto, D. Pointcheval, and J. Stern, *RSA-OAEP is secure under the RSA assumption*, in **Advances in Cryptology**, CRYPTO 2001, Springer-Verlag, Berlin, LNCS **2139** (2001), 260–274.

[93] M. Gardiner, *A new kind of cipher that would take millions of years to break*, Sci. Am. **237**, August (1977), 120–124.

[94] M.R. Garey and D.S. Johnson, **Computers and Intractability**, W. H. Freeman, New York, 22nd printing (2000).

[95] C. F. Gauss, **Disquisitiones Arithmeticae** (English edition), Springer-Verlag, Berlin (1985).

[96] R. Gennaro, T. Rabin, and H. Krawcyk, *RSA-Based undeniable signatures*, J. Cryptol. **13** (2000), 397–416.

[97] M. Girault, *Self-certified public keys*, in **Advances in Cryptology**, EUROCRYPT '91, Springer-Verlag, Berlin, LNCS **473** (1991), 490–497.

[98] O. Goldreich, **Modern Cryptography, Probabilistic Proofs and Pseudorandomness**, Springer, Berlin (1999).

[99] O. Goldreich, **Foundations of Cryptography — Basic Tools**, Cambridge University Press, New York (2001).

[100] S. Goldwasser, *The search for provably secure cryptosystems*, in **Cryptology and Computational Number Theory**, *Proceedings of the 42nd Symposium in Applied Mathematics* **42**, American Mathematical Society (1990), 89–113.

[101] S. Goldwasser and J. Kilian, *Almost all primes can be quickly certified,* Proc. 18th Annu. ACM Symp. Theor. Comput. (STOC), Berkeley (1986), 316–329.

[102] J. Gordon, *Strong primes are easy to find,* in **Advances in Cryptology**, EUROCRYPT '84, Springer-Verlag, Berlin, LNCS **209** (1985), 216–223.

[103] L.C. Guillou and J.-J. Quisquater, *A practical zero-knowledge protocol fitted to security microprocessor minimizing both transmission and memory,* in **Advances in Cryptology**, EUROCRYPT '88, Springer-Verlag, Berlin, LNCS **330**, (1988), 123–128.

[104] L.C. Guillou and J.-J. Quisquater, *A "paradoxical" identity-based signature scheme resulting from zero-knowledge,* in **Advances in Cryptology**, CRYPTO '88, Springer-Verlag, Berlin, LNCS **403** (1990), 216–231.

[105] L.C. Guillou, M. Ugon, and J.-J. Quisquater, *The smart card: A standardized security device dedicated to public cryptology,* in **Contemporary Cryptology: The Science of Information Integrity**, G.J. Simmons, ed., IEEE Press, Piscatoway, New Jersey (1992), 561–613.

[106] J. Guttman, *Correct cryptographic protocols provide authentication and confidentiality,* The EDGE–MITRE's Advanced Technology Newsletter, **5** (2001), 6–7, 10.

[107] J. Hastad, *Solving simultaneous modular equations of low degree,* Siam J. Comput. **17** (1988), 336–341.

[108] A. Herzberg, S. Jarecki, H. Krawczyk, and M. Yung, *Proactive sharing or: How to cope with perpetual language leakage,* in **Advances in Cryptology**, CRYPTO '95, Springer-Verlag, Berlin, LNCS **963** (1995), 339–352.

[109] R. Housley, *Cryptographic message syntax,* Internet Request for Comments 2630 (June 1999).

[110] R. Housley, W. Ford, W. Polk, and D. Solo, *Internet X.509 public key infrastructure certificate and CRL profile,* Internet Request for Comments 2459 (January 1999).

[111] ITU-T Recommendation X.500, *The directory — overview of concepts and models,* ITU, Geneva, Switzerland (1997).

[112] ITU-T Recommendation X.509, *Information technology — open systems interconnection — the directory: authentication framework* (June 1997).

[113] D.P. Jablon, *Strong password-only authenticated key exchange,* Computer Communications Review, ACM SIGCOMM **26** (1996), 5–26.

[114] D. Johnson, *The NP-completeness column: an ongoing guide,* J. Algorithms **5** (1984), 433–447.

[115] A.M. Johnson and P.S. Gemmell, *Authenticated key exchange provably secure against the man-in-the-middle attack*, J. Cryptol. **15** (2002), 139–148.

[116] A. Joux and R. Lercier *Discrete logarithms in GF(p)*, January 19, 2001, announced on the NMBRTHRY.Mailing.List.

[117] D. Kahn, **The Codebreakers**, Macmillan, New York (1967).

[118] S. Katzenbeisser, **Recent Advances in RSA Cryptography**, Kluwer Academic Publishers, Dordrecht, the Netherlands (2001).

[119] Lord Kelvin, *Nineteenth century clouds over the dynamical theory of heat and light*, Philos. Mag. **2** (1901), 1–40.

[120] P. Kocher, *Timing attacks on implementations of Diffie-Hellman, RSA, DSS, and other systems*, in **Advances in Cryptology**, CRYPTO '96, Springer-Verlag, Berlin, LNCS **1109** (1996), 104–113.

[121] A.G. Konheim, **Cryptography, a Primer**, John Wiley and Sons, New York (1981).

[122] E. Knudsen, *Elliptic scalar multiplication using point halving*, in **Advances in Cryptology**, ASIACRYPT '99 (Singapore), Springer-Verlag, Berlin, LNCS **1716** (1999), 135–149.

[123] D.E. Knuth, **The Art of Computer Programming**, Volume **2**, **Seminumerical Algorithms**, Third Edition, Addison-Wesley, Reading, Massachusetts (1998).

[124] D.E. Knuth, **The Art of Computer Programming**, Volume **3**, **Sorting and Searching**, Second Edition, Addison-Wesley, Reading, Massachusetts (1998).

[125] M. Kraitchik, **Théorie des Nombres, Tome II**, Gauthiers-Villars, Paris (1926).

[126] M. Kraitchik, **Mathematical Recreations**, Dover, New York (1953).

[127] E. Krananaskis, **Primality and Cryptography**, Wiley, New York (1986).

[128] H. Krawczyk, *The order of encryption and authentication for protecting communications (or: How secure is SSL?)*, in **Advances in Cryptology**, CRYPTO 2001, Springer-Verlag, Berlin, LNCS **2139** (2001) 310–331.

[129] K. Kurosawa, S. Obana, and W. Ogata, *t-identifiable (k, n)-threshold secret sharing schemes*, in **Advances in Cryptology**, CRYPTO '95, Springer-Verlag, Berlin, LNCS **963** (1995), 410–423.

[130] B. Lampson, *A note on the confinement problem*, Commun. ACM **16** (1973), 613–615.

[131] P. Lancaster and M. Tismenetsky, **The Theory of Matrices**, Second Edition, Academic Press, New York (1985).

[132] S. Landau, *Zero-knowledge and the department of defense*, Notices Amer. Math. Soc. **35** (1988), 5–12.

[133] D.H. Lehmer, **Selected Papers of D.H. Lehmer**, Volumes I–III, D. McCarthy, ed., The Charles Babbage Research Centre, St. Pierre, Canada (1981).

[134] D.N. Lehmer, *A theorem in the theory of numbers*, Bull. Amer. Math. Soc. **14** (1907), 501–502.

[135] E. Lehmer, *On the infinitude of Fibonacci pseudo-primes*, Fibonacci Q. **2** (1964), 229–230.

[136] H.W. Lenstra, Jr., *Factoring integers with elliptic curves*, Ann. Math. **126** (1987), 649–673.

[137] H.W. Lenstra, Jr., *On the Chor-Rivest knapsack cryptosystem*, J. Cryptol. **3** (1991), 149–155.

[138] A.K. Lenstra, H.W. Lenstra, Jr., M.S. Manasse, and J.M. Pollard, *The factorization of the ninth Fermat number*, Math. Comp. **61** (1993), 319–349.

[139] A.K. Lenstra and M.S. Manasse, *Factoring by electronic mail*, in **Advances in Cryptology**, EUROCRYPT '89, Springer-Verlag, Berlin, LNCS **434** (1990), 355–371.

[140] A.K. Lenstra and E.R. Verheul, *Selecting cryptographic key sizes*, J. Cryptol. **14** (2001), 255–293.

[141] R. Lidl and G.L. Mullen, *The world's most interesting class of integral polynomials*, J. Comb. Math. Comb. Comput. **37** (2001), 87–100.

[142] R. Lidl, G.L. Mullen, and G. Turnwald, **Dickson Polynomials**, Pitman Monographs and Surveys in Pure Appl. Math. **65**, Longman Scientific and Technical, Essex, England (1993).

[143] L. Lovász, **An Algorithmic Theory of Number, Graphs, and Convexity**, SIAM Publications, Budapest (1986).

[144] G. Lowe, *Breaking and fixing the Needham-Schroeder public key protocol using FDR*, in **Tools and Algorithms for the Construction and Analysis of Systems**, Springer-Verlag, New York, LNCS **1055** (1996), 147–166.

[145] J. Manger, *A chosen ciphertext attack on RSA optimal asymmetric encryption padding* (OAEP) *as standardized in PKCS # 1 v2.0*, in **Advances in Cryptology**, CRYPTO 2001, Springer-Verlag, Berlin, LNCS **2139** (2001), 230–238.

[146] J.L. Massey, *Contemporary cryptology: an introduction*, in **Contemporary Cryptology: The Science of Information Integrity**, G.J. Simmons, ed., IEEE Press, Piscatoway, New Jersey (1992), 1–39.

[147] M. Matsui, *Linear cryptanalysis method for the DES cipher*, in **Advances in Cryptology**, EUROCRYPT '93, Springer-Verlag, Berlin, LNCS **765** (1994), 386–397.

[148] M. Matsui, *The first experimental cryptanalysis of the Data Encryption Standard*, in **Advances in Cryptology**, CRYPTO '94, Springer-Verlag, Berlin, LNCS **839** (1994), 1–11.

[149] U.M. Maurer, *Fast generation of secure RSA-moduli with almost maximal diversity* in **Advances in Cryptology**, EUROCRYPT '89, Springer-Verlag, Berlin, LNCS **434** (1990), 636–647.

[150] U.M. Maurer, *Towards the equivalence of breaking the Diffie-Hellman protocol and computing discrete logarithms*, in **Advances in Cryptology**, CRYPTO '94, Springer-Verlag, Berlin, LNCS **839** (1994), 271–281.

[151] U.M Maurer, *Fast generation of prime numbers and secure public-key cryptographic parameters*, J. Cryptol. **8** (1995), 123–155.

[152] U.M. Maurer and S. Wolf, *Diffie-Hellman oracles* in **Advances in Cryptology**, CRYPTO '96, Springer-Verlag, Berlin, LNCS **1109** (1996), 268–282.

[153] U.M. Maurer and S. Wolf, *The relationship between breaking the Diffie-Hellman protocol and computing discrete logarithms*, SIAM J. Comput. **28** (1999), 1689–1721.

[154] U. Maurer and S. Wolf, *The Diffie-Hellman Protocol*, Designs Codes Cryptogr., Spec. Iss. Public Key Cryptogr. **19** (2000), 147–171.

[155] B. Mazur, *Rational points on modular curves*, in **Modular Functions of One Variable V**, Springer-Verlag, Berlin, LNM **601** (1977), 107–148.

[156] A.J. Menezes, P.C. van Oorschot, and S.A. Vanstone, **Handbook of Applied Cryptography**, CRC Press, Boca Raton, Florida (1997).

[157] M. Mignotte, *How to share a secret*, Cryptography, Workshop Proceedings, Springer-Verlag, Berlin, LNCS **149** (1983), 371–375.

[158] G.L. Miller, *Riemann's hypothesis and tests for primality*, J. Comput. Syst. Sci. **13** (1976), 300–317.

[159] Z. Mo and J.P. Jones, *A new primality test using Lucas sequences*, contained within the Ph.D. thesis by Mo completed at the University of Calgary (1995).

[160] R.A. Mollin, ed., **Number Theory and Applications**, NATO ASI **C265**, Kluwer Academic Publishers, Dordrecht, the Netherlands (1989).

[161] R.A. Mollin, ed., **Number Theory**, Proc. First Conf. of the Canadian Number Theory Association, Walter de Gruyter, Berlin (1990).

[162] R.A. Mollin, **Quadratics**, CRC Press, Boca Raton, Florida (1996).

[163] R.A. Mollin, **Fundamental Number Theory with Applications**, CRC Press, Boca Raton, Florida (1998).

[164] R.A. Mollin, **Algebraic Number Theory**, Chapman Hall/CRC Press, Boca Raton, Florida (1999).

[165] R.A. Mollin, **An Introduction to Cryptography**, Chapman Hall/CRC Press, Boca Raton, Florida (2000).

[166] R. Molva, D. Samfat, and G. Tsudik, *Authentication of mobile users*, IEEE Network **8** (1994), 26–34.

[167] P.L. Montgomery, *Speeding the Pollard and elliptic curve methods of factorization*, Math. Comp. **48** (1987), 243–264.

[168] M. Mouly, and M.-B. Pautet, **The GSM system for mobile communications**, Cell & Sys., Telecom Publishing, Olympia, Washington (1992).

[169] S. Müller, *A survey of IND-CCA secure public-key encryption schemes relative to factoring*, in **Public-Key Cryptography and Computational Number Theory**, K. Alster, J. Urbanowicz, and H.C. Williams, eds., De Gruyter, Berlin (2001), 181–196.

[170] M. Myers, R. Ankeny, A. Malpani, S. Galperin, and C. Adams, *X.509 Internet Public Key Infrastructure On-Line Certificate Status protocol — OCSP*, Internet Request for Comments 2560 (June 1999).

[171] S. Nasar, **A Beautiful Mind**, Touchstone, Simon and Schuster, New York (1998).

[172] National Institute of Standards and Technology, *FIPS 140-1, Security requirements for cryptographic modules* (January 11, 1994).

[173] R.M. Needham and M.D. Schroeder, *Using encryption for authentication in large networks of computers*, Commun. ACM **21** (1978), 993–999.

[174] R.M. Needham and M.D. Schroeder, *Authentication revisited*, Operating Syst. Rev. **21** (1987), 7.

[175] A.M. Odlyzko, *Discrete logarithms in finite fields and their cryptographic significance*, in **Advances in Cryptology**, EUROCRYPT '84, Springer-Verlag, Berlin, LNCS **209** (1984), 225–314.

[176] T. Okamoto, *Provably secure and practical identification schemes and corresponding signature schemes*, in **Advances in Cryptology**, CRYPTO '92, Springer-Verlag, Berlin, LNCS **740** (1993), 31–53.

[177] T. Okamoto and K. Ohta, *Universal electronic cash*, in **Advances in Cryptology**, CRYPTO '91, Springer-Verlag, Berlin, LNCS **576** (1992), 324–337.

[178] P. van Oorschot, *A comparison of practical public-key cryptosystems based on integer factorization and discrete logarithms*, in **Contemporary Cryptography: The Science of Information Integrity**, G. Simmons, ed., IEEE Press, Piscatoway, New Jersey (1992), 289–322.

[179] D. Otway and O. Rees, *Efficient and timely mutual authentication*, Operating Syst. Rev. **21** (1987), 8–10.

[180] N. Pippenger, *On the evaluation of powers and related problems – preliminary version*, in 17th Annual Symposium on Foundations of Computer Science, IEEE Computer Society, Long Beach, California (1976), 258–263.

[181] N. Pippenger, *The minimum number of edges in graphs with prescribed paths*, Math. Syst. Theor. **12** (1979), 325–346.

[182] N. Pippenger, *On the evaluation of powers and monomials*, SIAM J. Comput. **9** (1980), 230–250.

[183] PKCS1, *Public key cryptography standard no. 1 version 2.0*, RSA Labs.

[184] H.C. Pocklington, *The determination of the prime or composite nature of large numbers by Fermat's theorem*, Proc. Cambridge Philos. Soc. **18** (1914–1916), 29–30.

[185] S.C. Pohlig and M.E. Hellman, *An improved algorithm for computing logarithms in $GF(p)$ and its cryptographic significance*, IEEE Trans. Inf. Theor. **24** (1978), 106–111.

[186] D. Pointcheval, *Chosen-ciphertext security for any one-way cryptosystem*, Proc. PKC 2000, LNCS **1751**, Springer-Verlag, Berlin (2000), 129–146.

[187] J.M. Pollard, *Theorems on factorization and primality testing*, Proc. Cambridge Philos. Soc. **76** (1974), 521–528.

[188] C. Pomerance, *Analysis and comparison of some integer factorization algorithms*, in **Computational Methods in Number Theory**, H.W. Lenstra, Jr. and R. Tijdeman, eds., Math. Centre Tract **154**, Amsterdam, (1982), 89–139.

[189] C. Pomerance, *The number field sieve*, Proc. Symp, Appl. Math. **48** (1994), 465–480.

[190] C. Pomerance, J. Selfridge, and S.S. Wagstaff, Jr., *The pseudoprimes to* $2.5 \cdot 10^9$, Math. Comp. **35** (1980), 1003–1026.

[191] C. Pomerance, J.W. Smith, and R. Tuler, *A pipeline architecture for factoring large integers with the quadratic sieve algorithm*, SIAM J. Comput. **17** (1988), 387–403.

[192] B. Ramsdell, *S/MIME version 3 message specification*, Internet Request for Comments 2633 (June 1999).

[193] E. Rescorla, *Diffie-Hellman key agreement method*, Internet Request for Comments 2631 (June 1999).

[194] P. Ribenboim, *Selling primes*, Math. Mag. **68** (1995), 175–182.

[195] H. te Riele, W. Lionen, and D. Winter, *Factoring with the quadratic sieve on large vector computers*, J. Comput. Appl. Math. 27 (1989), 267–278.

[196] H. Reisel, **Prime Numbers and Computer Methods for Factorization**, Progress in Mathematics **126**, Second Edition, Birkhäuser, Boston (1994).

[197] R. Scheidler and H.C. Williams, *A public-key cryptosystem utilizing cyclotomic fields*, Designs Codes Cryptogr. **6** (1995), 117–131.

[198] B. Schneier, **Applied Cryptography**, Wiley, New York (1994).

[199] C.P. Schnorr, *Efficient signature by smart cards*, J. Cryptol. **4** (1991), 161–174.

[200] B. Schoenmakers, *An efficient electronic payment system withstanding parallel attacks*, Report **CS-R9522**, Centrum voor Wiskunde en Informatica, March 1995.

[201] P. Seelhoff, *Die Auflösung grosser Zahlen in ihre Factoren*, Z. Math. Phys. **31** (1886), 166–172. (French translation in Sphinx-Oedipe **7** (1912), 84–88.)

[202] A. Shamir, *How to share a secret*, Commun. ACM **22** (1979), 612–613.

[203] A. Shamir, *RSA for paranoids*, Cryptobytes **1** (Autumn 1995), 1–4.

[204] D. Shanks, **Solved and Unsolved Problems in Number Theory**, Chelsea, New York (1985).

[205] C.E. Shannon, *Communication theory of secrecy systems*, Bell Syst. Tech. J. **28** (1949), 656–715.

[206] Z. Shmuely, *Composite Diffie-Hellman public-key generating systems are hard to break*, Technical Report #356, TECHNION — Israel Institute of Technology, Computer Science Dept. (1985).

[207] V. Shoup, *OAEP reconsidered*, in **Advances in Cryptology**, CRYPTO 2001, Springer-Verlag, Berlin, LNCS **2139** (2001), 239–259.

[208] V. Shoup and R. Gennaro, *Securing threshold cryptosystems against chosen ciphertext attack*, J. Cryptol. **15** (2002), 75–96.

[209] G.J. Simmons, *Message authentication without secrecy: A secure communications problem uniquely solvable by asymmetric techniques*, in IEEE Electronics and Aerospace Systems Convention, EASCON '79 Record, Arlington, Virginia (1979), 661–662.

[210] G.J. Simmons, *Message authentication without secrecy*, in **Secure Communications and Asymmetric Cryptosystems**, G.J. Simmons, ed., AAAS Selected Symposia Series **69**, Westview Press, Boulder, Colorado (1982), 105–139.

[211] G.J. Simmons, *Verification of treaty compliance-revisited*, in Proceedings of the 1983 IEEE Symposium on Security and Privacy, IEEE Computer Society Press, Oakland, California (1983), 25–27.

[212] G.J. Simmons, *A "weak" privacy protocol using the RSA crypto algorithm*, Cryptologia **7** (1983), 180–182.

[213] G.J. Simmons, *The prisoner's problem and the subliminal channel*, in **Advances in Cryptology**, CRYPTO '83, Plenum Press, New York (1984), 51–67.

[214] G.J. Simmons, ed., **Contemporary Cryptology — The Science of Information Integrity**, IEEE Press, Piscatoway, New Jersey (1992)

[215] G.J. Simmons, *The subliminal channels in the U.S. digital signature algorithm (DSA)*, in Proceedings of the Third Symposium on: State and Progress of Research in Cryptography, Rome, Italy (1993), 35–54.

[216] G.J. Simmons, *An introduction to the mathematics of trust in security protocols*, in Proceedings: Computer Security Foundations Workshop VI, IEEE Computer Society Press, Franconia, New Hampshire (1993), 121–127.

[217] G.J. Simmons, *Cryptanalysis and protocol failures*, Commun. ACM **37** (1994), 56–65.

[218] G.J. Simmons, *Subliminal communication is easy using the DSA*, in **Advances in Cryptology**, EUROCRYPT '93, Springer-Verlag, Berlin, LNCS **765** (1994), 218–232.

[219] G.J. Simmons, *Subliminal channels: past and present*, Euro. Trans. Telecommun. **5** (1994), 459–473.

[220] G.J. Simmons, *Protocols that ensure fairness*, in Codes and Ciphers, P.G. Farrrell, ed., Royal Agricultural College, Cirencester, 1993 (1995), 13–15.

[221] G.J. Simmons, *The history of subliminal channels*, IEEE J. Selected Areas Commun. **16** (1998), 452–462.

[222] G.J. Simmons, *Results concerning bandwidth of subliminal channels*, IEEE J. Selected Areas Commun. **16** (1998), 463–473.

[223] M. Sipser, **Introduction to the Theory of Computation**, PWS Publishers (1997).

[224] R. Solovay and V. Strassen, *A fast Monte-Carlo test for primality*, SIAM J. Comput. **6** (1977), 84–85. *Erratum: A fast Monte-Carlo test for primality*, SIAM J. Comput. **7** (1978), 118.

[225] D.R. Stinson, **Cryptography**, CRC Press, Boca Raton, Florida (1995).

[226] C. Suetonius Tranquillus, **The Lives of the Twelve Caesars**, Corner House, Williamstown, Massachusetts (1978).

[227] E. Teske, *Speeding up Pollard's rho method for computing discrete logarithms*, in ANTS-III, Portland, Oregon (June 1998), J.P. Buhler, ed., LNCS **1423** Springer (1998), 541–554.

[228] E. Teske, *On random walks for Pollard's rho method*, Math. Comp. **70** (2001), 809–825.

[229] E.C. Tichmarsh, **The Theory of the Riemann Zeta Function**, Clarendon Press, Oxford (1951).

[230] Unknown author, *Final report on project C43*, Bell Telephone Laboratory, October (1944), p. 23.

[231] S.A. Vanstone and R.J. Zuccherato, *Short RSA keys and their generation*, J. Cryptol. **8** (1995), 101–114.

[232] S. Vaudenay, *Cryptanalysis of the Chor-Rivest cryptosystem*, J. Cryptol. **14** (2001), 87–100.

[233] M. Weiner, *Cryptanalysis of short RSA secret exponents*, IEEE Trans. Inf. Theor. **36** (1990), 553–558.

[234] W. Wen, T. Saito, and F. Mizoguchi, *Security of public key certificate based authentication protocols*, Proc. PKC '2000, LNCS **1751**, Springer-Verlag, Berlin (2000), 196–209.

[235] H.C. Williams, *Primality testing on a computer*, Ars Combinatoria **5** (1978), 127–185.

[236] H.C. Williams, *A modification of the RSA public-key encryption procedure*, IEEE Trans. Inf. Theor. **IT-26** (1980), 726–729.

[237] H. C. Williams, *A p + 1 method of factoring*, Math. Comp. **39** (1982), 225–234.

[238] H.C. Williams, *Some public-key crypto-functions as intractable as factorization*, Cryptologia **9** (1985), 223–237.

[239] H.C. Williams, **Édouard Lucas and Primality Testing**, Canadian Mathematical Society Series of Monographs and Advanced Texts, Vol. **22**, Wiley-Interscience, New York (1998).

[240] H.C. Williams, *Solving the Pell equation*, to appear in *Millennial Conference Proc.* **3**.

[241] H.C. Williams and B. Schmid, *Some remarks concerning the M.I.T. public-key cryptosystem*, BIT **19** (1979), 525–538.

[242] M.J. Williamson, *Non-secret encryption using a finite field*, GCHQ–CESG publication, January 21 (1974), 2 pages.

[243] M.J. Williamson, *Thoughts on cheaper non-secret encryption*, GCHQ–CESG publication, August 10 (1976), 3 pages.

[244] Working Group 4 International Telecommunication Union, Task Group 8/1, *Working document towards new recommendation: Security mechanisms and operating procedures for FPLMTS*, version 8, Doc. 8-1/150, Feb. 14 (1995).

[245] S.Y. Yan, **Number Theory for Computing**, Springer, Berlin (2000).

[246] G. Yuval, *How to swindle Rabin*, Cryptologia **3** (1979), 187–190.

[247] P. Zimmerman, **The Official PGP User's Guide**, MIT Press, Cambridge, Massachusetts, second printing (1995).

Index